Albert J. Bernays

The student's chemistry

Salzwasser

Albert J. Bernays

The student's chemistry

1. Auflage | ISBN: 978-3-84605-396-6

Erscheinungsort: Frankfurt, Deutschland

Erscheinungsjahr: 2020

Salzwasser Verlag GmbH

Reprint of the original, first published in 1869.

THE

STUDENT'S CHEMISTRY.

BEING THE

SEVENTH EDITION

OF

"HOUSEHOLD CHEMISTRY,"

OR THE

SCIENCE OF HOME LIFE.

BY

ALBERT J. BERNAYS, PH. DR., F.C.S.,

1869.

London: Lewis & Son, Printers, Swan Buildings, Moorgate Street.

PREFACE TO THE SEVENTH EDITION.

THE last edition of this work, which made its first appearance in 1851, has been so long out of print, that it has doubtless been thought the intention of the Author not to allow it to re-appear.

The delay in re-publication has only been caused by the changes that a careful consideration has shewn to be desirable. When, a twelvemonth since, the necessary revision was commenced, several months were wasted in endeavouring to retain the original plan of the book in its entirety.

The accumulation, however, of fresh notes, made it seem advisable that they should be *all* thrown into the text, in the hope that the student might thus be spared needless trouble, and the appearance of the book saved from an unpleasing eye-sore. For general reading, the less important matter has been marked by a system of brackets.

In the re-publication and re-casting of the contents of this book, the Author has been influenced by the conviction he has always entertained, that a student entering on the study of a Science, should not be presented at once with all the highest and most elaborated theories. His object is to give a comprehensive and clear account of the real grounds of Chemistry.

In no sense is this book intended to supersede larger works on the subject, but, by the intelligible form it has, it is hoped, assumed, the Author means it to incite the student, and by its matter to fit him, for the profitable study of such books.

The size of the book has been so increased by this plan that it has been found impossible to include the Chapters on Food, which are therefore reserved for a second Volume intended to embrace the theories of Chemistry, as well as Chemistry in its organic aspect. The title of "The Student's Chemistry" has been prefixed in order to indicate the character of the book; whilst the second titles have been preserved because all the Chapters on Inorganic Chemistry have been retained, and because the book is not a new one.

These Chapters are, however, now prefaced by others, giving the Chemistry of the subjects treated of more familiarly under the various headings of "the Atmosphere," "Combustion," "Gas," "the Chemistry of a Candle," "Water," "the Metals," &c.

Following the example of Professor Williamson, and in order to prepare the way for other and larger works on Chemistry, I have inserted a few suitable questions and answers, for which I am indebted to a former pupil, Mr. J. H. Parker.

Each Chapter is prefaced by a Table of Contents, on the principle of my "Notes for Students in Chemistry," published by Messrs. Churchill. This syllabus refers to the various paragraphs, and, it is believed, will be found useful both to teachers and pupils.

ST. THOMAS'S HOSPITAL, S.E.,
September, 1869.

CONTENTS.

CHAPTER I.

INTRODUCTION TO THE SUBJECT OF HEAT.

1. The Sun the chief source of Heat. 2. The Rays do not penetrate deeply. 3. Expenditure of a given Force results in the production of a given Heat. 4. Artificial Sources of Heat. 5. The mechanical equivalent of Heat. 6. Chemical Sources of Heat. 7. Thermo-electricity. 8. Heat is a Mode of Motion. 9. The temperature of a Body to be distinguished from its Heat. 10. Heat increases the volume of bodies. 11. Solids expand least. Unequal expansion of the Metals: Pendulums. 12. Gases expand most. Leslie's differential Thermometer. Co-efficient of expansion. 13. Liquids midway between Solids and Gases. 14. Thermometric Results without calculation, not quite accurate. 15. Self-registering Thermometers. 16. Pyrometers. 17. Maximum density of Water 4°. 18. Equilibrium of Temperature. 19. Heat, Molecular Motion. Conduction. Liquids and Gases bad Conductors. 19. Convection. The Gulf Stream. The Winds. Ventilation. 20. Radiation. Concave Reflectors. Dew. Daniell's Hygrometer. 21. Diathermancy. 22. Specific Heat. 23. Latent Heat. 24. Phenomenon of Boutigny.

1. The heat which we experience on the surface of the earth, and without which no living thing could exist, is derived almost entirely from the sun: the variations of temperature in the different seasons, at different periods of the day, and under the varying aspects of the sky, sufficiently attest the great influence of this source of heat. Its power is forcibly illustrated by a calculation of the late illustrious Faraday,—that, on a fine clear day in June, a single acre of ground, situated in the vicinity of London, will receive from the sun as much heat as would be obtained from the burning of six tons of coal.

2. The earth being but a bad conductor of heat, the sun's rays do not penetrate far below its surface. In our climate, the variations of temperature from external heat cease altogether at a depth of

about sixty-four feet. It is well known that cellars, but a few feet below the surface, retain a uniform temperature at all seasons; and, that a covering of some few feet depth of earth is sufficient to protect water-pipes, vegetables, &c., from the effects of the most severe frost.

3. From whatever source heat is artificially evolved, the amount produced is always in the exact proportion of the force employed. In other words, the expenditure of a given force will have for its result a given amount of heat.

4. Artificial sources of heat are of two kinds: physical and chemical. Two pieces of wood rubbed together rapidly, will become ignited, while the heat evolved in certain parts of machinery in motion is so great, as occasionally to produce serious inconvenience.

Count Rumford attempted to estimate the amount of heat generated by friction. He found, that when a brass cannon weighing 113lbs. was made to revolve under water at the rate of thirty-two times in a minute, while pressing against a borer of hardened steel with a force of about 10,000lbs., two gallons and a half of water at 60° F. (15°·5 C.) were made to boil in two and a half hours.

The power of compression to produce heat is well known. The sudden condensation of air (or of gases) in a syringe, like that in Fig. 1, occasioned by pressure on the piston, is sufficient to ignite the tinder fastened to the end of the piston, on its withdrawal from the cylinder.

Fig. 1.

And in the use of flint and steel, in the sparks struck by a horse's hoof from the road, the sparks from a knifegrinder's wheel, and from the rails when the break is applied,—we have familiar illustrations of heat produced by percussion;—the combined effects of friction and compression.

5. Work and heat are found to be inseparably connected. Dr. Mayer, of Heilbronn, and Mr. Joule, of Manchester, have both calculated the mechanical equivalent of heat. It was found, that "the quantity of heat which would raise one pound of water one degree Fahrenheit (1° F.) in temperature, is exactly equal to what would be generated if a pound weight, after having fallen through a height of 772 feet, had its moving force destroyed by collision with the earth. Conversely, the amount of heat necessary to raise a pound of water one degree in temperature, would, if all applied mechanically, be competent to raise a pound weight 772 feet high; or it would raise 772 lbs. one foot high. The term, 'foot-pound,' has been introduced to express in a convenient way the lifting of one pound to the height of a foot. Thus, the quantity of heat necessary to raise the temperature of a pound of water 1° F., being taken as a standard, 772 foot-pounds constitute what is called the mechanical equivalent of heat. If the degrees be Centigrade, 1390 foot-pounds is the equivalent."—*Heat, considered as a mode of Motion, by Professor Tyndall, F.R.S.*

6. Of chemical action, as productive of heat, we have daily evidence; every kind of combustion affords instances of it.

7. Electricity is also a notable source of heat, in the same manner as heat is a source of electricity. The subject of thermo-electricity (θερμὸς, thermos, heat) was discovered by Professor Seebeck, of Berlin. By the unequal heating of any two metals, a current of electricity is called forth: but as this current is found to be the greatest when the metals bismuth and antimony are employed, these two metals are used in the construction of the so-called thermo pile. On warming the place of junction between the two metals, a current is generated, the direction of which is from bismuth to antimony across the junction, or *vice versâ.* As an electric current will deflect a magnetic needle from its position, a galvanometer (see index) is employed, together with the thermo-pile, to detect alterations in the temperature.

8. Heat, itself, is motion, and nothing else. It is nothing material, for it neither adds to nor takes from the weights of bodies, although it so greatly alters their condition. It is mainly opposed to and loosens the force of cohesion; that force which binds together the

molecules (moles, a mass) of bodies. It unbinds the ice, and makes it liquid; it further separates the particles of water asunder, and converts them into steam. But a pound of ice will melt into a pound of water, and a pound of water will only produce a pound of steam.

Heat is but an accident of matter, namely,—a motion of its ultimate particles. Locke says: "Heat is a very brisk agitation of the insensible parts of the object, which produces in us that sensation from whence we denominate the object, hot: so, what in our sensation is heat, in the object is nothing but motion."

9. The temperature of a body should be distinguished from its heat. Temperature may be determined by the heating effects which a substance exerts upon others. It is indicated by the touch, and measured by the thermo-electric pile and galvanometer, or the thermometer, or the pyrometer (πῦρ pyr, fire, and μέτρον, metron, a measure).

The heat of a body may be said to be the actual amount of force or motion which it can give up to other bodies. The terms, hot and cold, are but relative, and express only the sensations experienced by ourselves; and these depend mainly on the character of the body.

10. As a rule, heat increases the bulk of bodies, or, in other words, the body occupies a greater space when hot than when cold. Thus: expansion becomes a measure of temperature, or of that modification of heat which is appreciable by the senses, and susceptible of measurement by the thermometer.

11. Solids expand least under the influence of heat, but their expansion is uniform, and generally they return, on cooling, to their original bulk. Figs. 2 and 3 afford illustrations of the expansion of solids. In Fig. 1, a bar of iron or brass, *a*, is made to fit, when cold, accurately to the gauge *c*, and to the opening *d*. When heated to redness, it will have increased so much in dimensions that it will no longer enter at either. In Fig. 3, a bar of metal, *a*, is fixed firmly by means of a screw, *b*, and the amount of its expansion, when heated, is measured by the indicator, *c*.

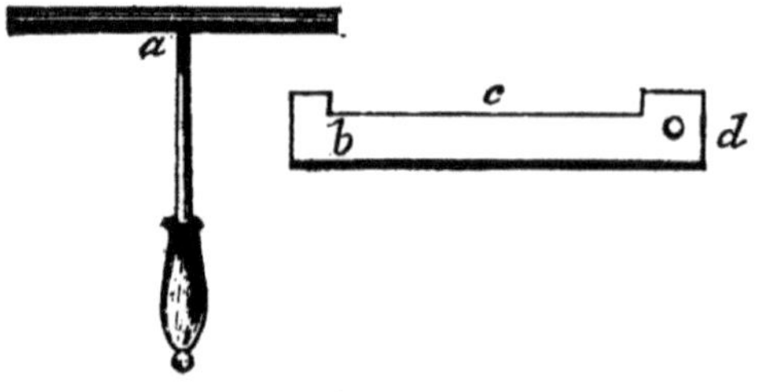

Fig. 2.

Different solids vary greatly in their degree of expansion under heat. Platinum and glass are nearly alike in this respect; hence is

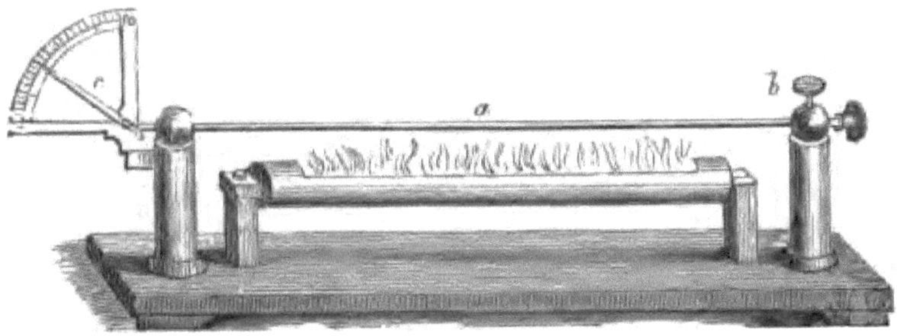

Fig. 3.

glass most suitable as a handle to platinum. The expansiveness of glass under heat is only half that of copper and one-third that of tin. Zinc and lead possess 3⅓ times the expansiveness of glass. The total or cubical expansion of a solid may be ascertained by multiplying its linear expansion by 3: thus, the linear expansion of zinc being found as 1 in 340, its cubical expansion will be 1 in 113. This difference of expansibility has received a very valuable practical application. The accurate measurement of time depends upon the uniform oscillation of a pendulum, which can only be insured by its remaining of the same length under all temperatures. But a pendulum of metal lengthens or contracts with changes of temperature, and, if composed of one metal, would cause a clock to lose in warm and gain in cold weather. This difficulty has been overcome by the contrivance of the gridiron, or compensating pendulum, Fig. 4. In this arrangement, the pendulum, with its bob, *a*, is formed of bars of steel and zinc: the shaded bars are supposed to be of steel, and the others of zinc. The expansion in one direction of the steel is compensated by that of the zinc in the opposite direction, and thus an invariable distance is maintained between the centre of gravity and the points of oscillation.

Fig. 4.

Fig. 5.

Fig. 5 represents another description of compensating pendulum, designed upon the same principle. The pendulum bob in this case consists of a glass vessel containing mercury, which, expanding upwards, compensates for the downward expansion of the steel rod, and preserves the uniform length of the pendulum in all changes of temperature.

Again, in watches, the circumference of the balance-wheel is made of two metals, having different rates of expansion, firmly soldered together, the more expansible being on the outside. The metallic thermometer of M. Breguet is on the same principle.

The effects of expansion and contraction require the careful consideration of the mechanician for the prevention of accident, or for turning the property to useful account. The variation in length between the summer and winter of a single tube of the Britannia bridge over the Menai Straits amounts to upwards of three inches, which, if not allowed for, would seriously affect the stability of the structure. In laying rails on a railroad, a small interval is left between the ends of the rails to admit of their free expansion.

The strength and tightness of steam-boilers is dependent upon the return to their original bulk of the rivets, which are applied red-hot, and, consequently, in an expanded state. So the wheelwright applies the tire to the wheel in a heated state, and, in contracting, it binds together the woodwork with considerable force.

Fig. 6.

12. Gaseous bodies expand most by heat. So great and so uniform is this expansion, that atmospheric air thermometers have long been made use of as an index of temperature. Between the temperature of freezing and boiling water, 1000 parts of common air will expand to 1366 parts. If the hand be placed on the bulb of the glass tube, Fig. 6, immersed in a bottle containing coloured water, the air at once expands under the heat of the hand, and rises in bubbles through the water; on the withdrawal of the hand the air returns to its original bulk, and the water rises in the tube, and fills the space previously occupied by the air that escaped in bubbles. On this principle depends the construction of the air-thermometer, an instrument which is of little or no value for general purposes,

partly on account of the small range of temperature which it indicates, but principally because of the great extent to which it is affected by atmospheric pressure.

Gases expand, between the freezing and boiling of water, $\frac{1}{491}$ for each degree upon our English scale of Fahrenheit, or $\frac{1}{273}$ for each degree of Centigrade. In other words, 491 parts of a gas at 32° become 492 parts at 33°, 493 parts at 34°, and so on. If a gas were heated 491° F., or 273° C., its volume would be doubled.

The fraction of its volume, by which a gas expands on being heated 1°, is called the co-efficient of expansion. This is 0·00203 on the English and 0·00366 on the Centigrade scale. Gases which are permanently elastic expand by almost precisely the same amount for each degree.

A pretty scientific illustration of the expansion of gases by heat is Leslie's differential thermometer, Fig. 7. It consists of two bulbs blown at the ends of a glass tube, bent twice at right angles, and containing a coloured liquid in the stem. It is not intended to indicate general changes of temperature, but only the differences between the air in the two bulbs. Thus, if the hand be brought near to the one bulb, the liquid rises at *a* and falls at *b*, owing to the expansion of the air over *b*.

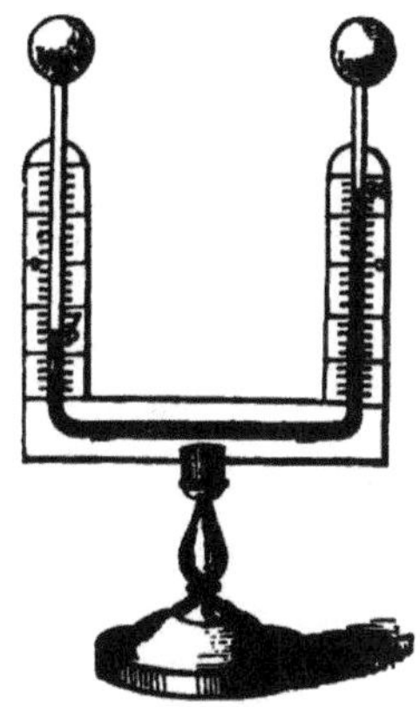

Fig. 7.

13. The expansibility of liquids under heat, stands midway between that of solids and gases; but that of the various liquids differs greatly. The lighter a liquid, the more expansible is it. Thus, spirit expands more than oil, oil than water, and water than mercury. In passing from the temperature of freezing to that of boiling water, pure spirit expands 1 volume in 9; oil 1 in 12; water 1 in 21·3; and mercury 1 in 55.

All ordinary thermometers depend for their use upon the expansion and contraction of the enclosed fluid upon the accession or loss of heat. The common mercurial thermometer is thus made. A tube is selected of perfectly uniform cylindrical bore, terminating in a bulb *a*, Fig. 8. Absolutely pure mercury, which has been boiled in order to deprive it both of moisture and of air, is introduced at *b* into the funnel-shaped opening of the tube. The air in the bulb and tube is driven out by the heat of a spirit-lamp, allowing the mercury to take its place as soon as the lamp is removed and the mercury becomes cool. The mercury is then again boiled, for the purpose of removing any air that may still remain in contact with it. When the tube is to be sealed, the mercury is made, by the application of heat, to expand to the top, which is then melted and closed by means of the flame of a blowpipe. If this is cleverly done, not a particle of air will remain in the thermometer, as is shown by the mercury flowing to the extreme end of the tube, with a click, when inverted.

Fig. 8.

All thermometers are graduated with reference to the freezing and boiling points of water, but different scales of subdivision of the intermediate range of temperature have been adopted in different countries, an arrangement much to be deplored, as rendering a comparison of scientific observations unnecessarily complicated. In Fahrenheit's thermometer, which is generally used in England, this range is divided into 180°, the freezing-point of water being marked 32°, and the boiling-point 212°. The point from which the reckoning starts is 0° or zero, corresponding with the greatest cold observed in Iceland.

In France, and among the scientific men of all nations, the thermometer of Celsius, known, from its mode of division, as the Centigrade, is in common use. In this instrument the freezing point of water is at 0° or zero, and the boiling-point at 100°.

In Germany, Reaumur's thermometer prevails, and in his scale, as in the Centigrade, the freezing-point of water is at 0° or zero, but the boiling-point is at 80°.

In all thermometers alike, all degrees below zero are recognised by the—prefixed. It will thus be seen, that 9° of Fahrenheit (9° F.) are equal to 5° of Centigrade (5° C.) or 4° of Reaumur (4° R.).

The conversion of degrees of F. into those of C. and R. is easily accomplished:—

F. to C. $\frac{5}{9}$ (F.$-$32°)=C. C. to F. $\frac{9}{5}$ C.+32°=F.
F. to R. $\frac{4}{9}$ (F.$-$32°)=R. R. to F. $\frac{9}{4}$ R.+32°=F.

As soon as the reader is familiarised with the Centigrade thermometer, the use of that of F. will be discontinued. Suppose it be required to know what would be the volume of a gas at 0° C., when it has been measured at 16° C. Since the expansion for each 0° C. is $\frac{1}{273}$, 273 volumes of a gas at 16° C. would become 273+16=289 volumes at 16° C. If, then, 100 volumes of a gas are taken at 16° C., the volume at 0° would be obtained by the following method of calculation:—

$$289 : 273 :: 100 : x; \therefore x=94{\cdot}463.$$

For high temperatures, and for such variations of heat as we have in England, the mercurial thermometer is best suited; but for the Arctic regions it is useless, inasmuch as mercury freezes at $-$39° F. ($-$39°·4 C.). In such countries a thermometer is used, in which spirit (which has never yet been frozen) is substituted for mercury.

14. As the glass tube expands as well as the liquid, the results given by a thermometer, of any construction, are not absolutely correct. But the expansion of the liquid is so much greater than that of the glass, that, for all practical purposes, the error may fairly be disregarded.

15. For recording the highest and lowest temperatures of the day and night, self-registering thermometers, of which there are many varieties, are in common use. The maximum thermometer generally employed is that of Rutherford. This instrument contains mercury, and includes in its stem a small steel or enamel index, which is pushed forward by the advancing mercury to the extreme point of its expansion, and is left behind when the mercury recedes, so as to indicate the highest temperature reached. The minimum thermometer contains spirit, and includes a similar index, buried in the liquid. As the spirit contracts by the cold, it carries with it the index as far as it recedes; but in expanding from that point, it passes the index without shifting it, leaving it to indicate the lowest temperature.

Negretti and Zambra's maximum thermometer is merely a thermometer placed horizontally with a contraction in the tube above the bulb, so as to allow the mercury to pass when it expands, but not to recede.

Though mercury does not boil till 662° (350° C.), its indications above 600° (315°·5 C.) are not to be depended upon, on account of the formation of vapour. A spirit-thermometer is only available for low temperatures.

16. For recording high temperatures, a Pyrometer is employed; the most accurate in use is that of Daniell, a description of which will be found hereafter.

17. Water forms a singular exception to the otherwise universal contraction of liquids under the influence of cold. It obeys the general law till it reaches the temperature of 39°·2 F. (4° C.), the point of maximum or greatest density; but, if cooled below that point, it begins to expand. The surface-water of a lake in a wintry atmosphere is lighter than, and floats on, the warmer liquid beneath. So, too, ice is lighter than water; and, but for this, the whole mass of lakes and other bodies of fresh water would, every cold winter, be lowered to the freezing-point, rendering the existence of animal and vegetable life in them impossible.

The presence of salt in water lowers the freezing-point. Thus, the waters of the ocean do not begin to freeze at 32° (0° C), but at 27°·4 (—2°·55 C.).

18. All bodies, after being heated, return with more or less rapidity to the temperature of surrounding objects; there being a constant tendency towards a common balance of temperature. This equilibrium may be effected in three ways, distinguished as Conduction, Convection, and Radiation.

19. When the sum of motion received is greater than that given out, a body becomes warmer; but chilling ensues, when the sum of motion given out is greater than that received. Heat itself is molecular motion. The molecules of bodies, when closely grouped as in solids, cannot, however, oscillate without communicating motion from one to the other. Conduction consists in each atom taking up the motion of its neighbours, and sending it on to others.

There is as great variety in the powers of Conduction in different substances, as in their expansibility; and it is mainly owing to this variety that bodies, equally heated, excite in us different sensations whether of heat or cold.

For the more readily a body conducts heat, the more readily does it also cool; or, in other words, impart its heat to other substances.

According to Wiedemann and Franz, the relative conductivities of metals, as compared with silver, is as follows:—

Silver	100	Brass	24	Platinum	8
Copper	74	Tin	15	German silver	6
Gold	53	Iron	12	Bismuth	2

The metals are by far the best Conductors of heat, all metals imparting or absorbing heat more readily than, for instance, does wood, but they differ greatly in this power amongst themselves. Silver is nearly four times as good a conductor as lead, and nearly twice as good as gold. Diamonds and other precious stones are much better conductors of heat than glass, and may be distinguished from imitations by feeling much colder than glass when brought into contact with the lips, which are very sensitive to a change of temperature. This difference of conducting power in different substances is turned to account in many ways. The mats on the dinner table, the kettle-holder, the ivory or bone inserted between the handle and the body of tea or coffee-pots, may serve as illustrations. At the siege of Gibraltar, red-hot cannon balls were carried on common wheelbarrows covered with sand.

Liquids and gases are very bad conductors of heat. Water may be boiled at its surface without communicating heat downwards. Porous non-conducting bodies, such as furs, wool, and eider-down, are in great part indebted to the air which they inclose for their value as warm clothing, Double doors and windows prevent, by the layer of interposed air, the escape of heat from, or the entrance of cold into, our apartments; while double doors to an ice-house prevent the access of heat to the ice on the hottest days of summer.

If the hand could be kept perfectly unmoved, it would remain uninjured even in hot water; but if it be put in motion, the constant accession of fresh particles of hot water would soon occasion all the symptoms of scalding. Arctic voyagers have been known to sustain without injury a temperature of −55° (−48°.33 C.), so long as the atmosphere was perfectly calm; but, in windy weather, the successive contact of fresh portions of air, even at zero, rapidly produced frost-bites and other serious injuries.

19. But, although neither air nor water (the types of gases and liquids) are much affected by Conduction, their physical property of fluidity enables them to be readily heated by Convection; meaning, the distribution of heat from one portion of a fluid to another, by a transfer of the heated particles themselves.

When a liquid is heated from below, it expands, becomes lighter and moves upwards; a fresh portion of the fluid descends, becomes heated in its turn and also rises. By this means a circulation of the particles is established, bringing about, by degrees, an uniform temperature in the whole mass of liquid. This principle is now generally applied in warming churches and other public buildings. The mode in which this movement of the heated parts of a fluid is effected, is well illustrated by a little apparatus, represented by Fig. 9. The flask having been filled with boiling water, and the cistern with cold, the hot and therefore lighter water, immediately begins to rise through the tube, which only dips into the flask, into the cistern, the cold water descending through the second tube to the bottom of the flask. The direction of the hot and cold water is represented in the Figure by two arrows.

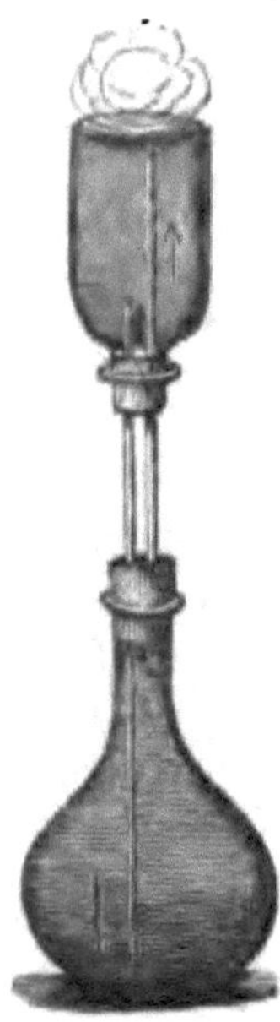
Fig. 9.

In the economy of nature the Convection of heat performs most important parts, in the ocean as well as in the atmosphere. This property operates on a gigantic scale, in the well-known instance of the Gulf Stream. In the tropics, the surface of the ocean becomes greatly heated. A large body of such heated water emerging from the Gulf of Mexico, passes northwards to the banks of Newfoundland, and thence eastwards to the coasts of Europe. Throughout its course, its water is many degrees warmer than that of the ocean, or either side of it. By its agency, heat is transported to the north of Europe direct from the tropics, and tends in a very great degree to mitigate the severity of its climate. To this cause is partly due the clemency of winter in England, as compared with its severity on the other side of the Atlantic. In Canada, the whole of which is in a lower latitude than England, there is perpetual frost throughout the winter.

To an ordinary observer, the Winds appear so capricious and irregular, that their fickleness has passed into a proverb, and the weather-cock, which indicates their direction, is taken as the type of an inconstant man. But this capriciousness is only apparent. Under their irregularity is a never-failing order. They are as obedient to laws as the movements of the earth itself; and of late years these laws have been traced with such success, that the changes of the wind, with their accompanying changes of weather, can, in many instances, be foretold with marvellous certainty.

Wind is simply a movement of air from one part of the earth's surface to another, caused by differences in its density, and, therefore, in its weight in different parts. There are two causes of such difference—heat and moisture; and on the influence of one or the other of these every wind must depend. Heat, however, is far more influential than moisture. The mode in which heat, by expanding the air, and, causing it to rise, creates a draught of colder and heavier air from all sides, will soon be more fully understood and appreciated. There is, indeed, no difference whatever in principle between the draught towards a fire in a room and the wind. The draught is a wind. And as in the room, the cooler air which supplies the fire must be replaced by other air, so it is with the winds. The air which has come from the cold parts of the earth to the hot must return, otherwise there must be an accumulation of air over the hot parts, and a want of it on the cold side: the balance of the atmosphere would be destroyed. And it does return; not along the surface, but in the upper regions, whither it has mounted to cool itself. The further rise of warm air pushes it aside, and forces it into the cooler regions.

So, then, every surface-flow of air must be compensated by a return-flow, or, as it is called, a counter-current, in the upper regions. This principle may be easily seen at work in a warm room.

The influence of vapour in lightening the air has yet to be referred to; but a remembrance of these three facts—the expansion of air by heat, its further expansion by vapour, and the compensation of every current by a counter-current—will explain every wind that blows. Thus the upward and downward movements of the air, though we have at present no means of measuring them, are the causes of all its lateral movements, which alone we can estimate.

The great centre of heat on the earth, or, as it may be called, the great chimney of the earth, is, of course, the Equator. Accordingly, there is always a draught towards the Equator. And if there were no causes in operation to interfere with the regularity of this draught, there would be a simple flow of cold air from the poles to the equator along the surface of the earth, and a return of hot air from the equator to the poles in the upper regions. As it is, this flow is found with great regularity only in the neighbourhood of the tropics. Now, as, in consequence of the shape of the earth, the parts about the equator have, in their daily rotation, to form a larger circle than those in the higher latitudes and, therefore, to travel faster, it follows that air coming from those higher latitudes and having their rate of travelling, will, as it approaches the tropics, lag behind. The earth here moves faster than the air; it slips, as it were, from under the air, which, therefore, appears to be blowing in the opposite direction to the earth, or from east to west. The wind, therefore, in the tropics, is made up of these two movements—from the Poles to the Equator, and from East to West; north-east in the northern hemisphere, south-east in the southern. These winds are called Trade-winds. In some parts of the tropics they blow for 329 days out of the 365; they extend on either side of the Equator for about 30 to 35 degrees. It is clear that they must meet each other near the equator, to which, as to the hottest place, they are both rushing. When they meet, they neutralize each other and produce a calm-belt which completely surrounds the globe. This belt, known as the Equatorial Calm-belt, or Equatorial Doldrums, is about 400 miles in width. It is dreaded by sailors as a region of perpetual calm; doubly trying in consequence of the insufferable heat and moisture of the atmosphere. Here the hot air ascends, and, spreading out, passes as an upper-current towards the poles. The reverse reason to that which has caused the trade-winds to blow from east to west, causes this return-current to blow from west to east; its course then is exactly opposite to that of the surface-current. At the latitude 30° to 35° this current meets another from the poles, which, for some reason not sufficiently explained, appears, although cold air, to travel from the poles as an upper-current instead of along the surface. These two currents, pressing against each other, cause a great accumulation of air, and a calm. The belt of calm-air so caused (from 15° to 20° in width) is known as the Calm of Cancer in the northern hemisphere, and the Calm of Capricorn in the southern. By seamen the former is vulgarly known as the "Horse" latitudes, because, in the old days of long voyages, they were often detained here so long that the forage of the horses, which they had as freight, was exhausted, and they were obliged to throw the animals overboard. The great accumulation of air is shown here by the fact, that the barometer has a mean height greater than that in any other latitude. From the bottom of this bank of air two surface currents flow in opposite directions: one, the trade-wind already traced; the other, towards the poles. This latter, continuing its former direction, is known as the counter-trades, and extends to about latitude 60°. The direction of these winds, however, is interfered with, at least in our hemisphere, by so many minor causes, that it is very far from regular; nevertheless, the prevailing wind in this region is what we may call the normal one, that is, south-west. For example, we in England have twice as much west wind as easterly. Above the latitude of 60°, these currents all round the earth meeting together, pile up about the poles, whence they flow out on all sides, as a now cold upper-current, to return towards the equator.

Such is an outline of the circulation of the winds on the earth. But its regularity is liable to numerous interruptions from local causes. The most important of these is the unequal heating of land and sea. In all warm climates the land becomes so much more heated during the day than the sea that it acts as a chimney, and draws towards itself the cooler sea air. Thus is produced the Sea-breeze, which is common enough even in England during the summer; but in the tropics is so regular, that it goes far to mitigate the discomforts of the climate, and is so powerful that it not only overcomes the force of, but blows in many cases counter to, the trade-winds, and it is turned to regular account in navigation. At Valparaiso it is actually furious, so that pebbles are whirled about the streets; and yet, as the sun goes down, it drops suddenly, and there is such a sweet lull that ladies walk out in ball costume to enjoy the evening. But presently, when the land has had time to radiate its heat, it becomes cooler than the sea. Then the current sets in the opposite direction, and the Land-breeze lasts till just before sunrise. In accordance with their origin, these land and sea-breezes are interrupted more or less during cloudy and rainy seasons.

This unequal heating of land and sea, acting on an extended scale, actually affects the whole of the trade-winds; for the centre of the region in which they meet is not, as might be supposed, the equator, but 5° north of it. The heating of the great excess of land in the northern hemisphere over the southern is actually able to hold back, as it were, the trade-winds to this extent. Now, in some parts of the world, a great extent of dry land will heat the air to such a degree as to reverse its motion during the whole of the hot season, both by night and day. This is the case with India, and the large tract of desert-land north of it. Although India is in the region of the north-east trade-wind, the wind blows regularly over that country, throughout the summer, from the south-west. Winds so caused are called Monsoons.

The descent of a large body of cold air from the upper regions accounts frequently for the accession of sudden cold. In Texas, such attacks of cold are frequently so sudden, that the inhabitants have scarcely time to protect themselves by change of clothing; and animals, and even men, have been known to be frozen to death by them.

Fig. 10.

The motion in gases brought about by convection admits of constant application in the phenomena of every-day life. All ventilation of buildings depends on the expansion of air by heat; as, too, does the burning of the fire in a grate. The air over the fire (Fig. 10), becoming specifically lighter as it expands from the heat ascends, and its place is taken by cold air drawn from the room. Thus, by the successive expansion and ascent of fresh portions of air, a perpetual upper current, as indicated by the arrow, is maintained in the chimney, by which the fire is supplied with the fresh air it requires for the combustion of the fuel.

This explains why an open fire-place assists in ventilating a room, and why the best bellows for a flagging fire is a piece of paper sufficiently large to close the space between the fire and the chimney, compelling all the air to pass from below directly through the fire. In cases when it is necessary, such as in rooms where gas is burnt, this current may be turned to further account in ventilation by the introduction of a properly constructed valve at *c* (such as an Arnott's ventilator). The air of the room, as it becomes vitiated by respiration and combustion, rises to the ceiling, and is drawn by the current through the ventilator up the chimney.

These currents may be readily shown by a very simple experiment. If a chimney-glass be placed upon the table over a small lighted taper, the flame will be soon extinguished; but if the chimney-glass be divided vertically by a card, as in Fig. 11, the taper will continue burning. The air heated by the flame will ascend by one division, while fresh air to supply its place will descend by the other.

Fig. 11.

Fig. 12.

Such an arrangement is of common application in the shafts of coal mines, and it may be well understood from a consideration of Fig. 12. The cold air descends by *a*, and the vitiated air escapes at *b*. If, instead of placing the chimney-glass without the division on the table, it is supported on two pieces of wood at the bottom, as in Fig. 13, the light will continue to burn, because the fresh air required to sustain the current can be supplied from below.

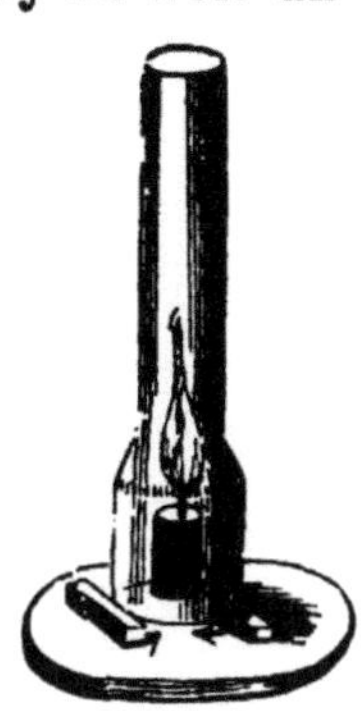

Fig. 13.

The admission of air by windows should be regulated, as far as possible, so as to avoid partial currents, or "draughts." Therefore, especially in large rooms, the more freely they are open, the better, provided, of course, that the external temperature permits it. Every large room should be so provided with windows as to allow of a thorough draught.

A very easy and effective mode of ventilating small rooms, without creating a draught, has been suggested by the late Mr. P. Bird. He has a board cut of the width to which it is desired to open the window and just the length of the sash. When the window is open, this board fills the vacant space beneath it, while the air diffuses itself upwards into the room through the chink left between the two sashes.

20. There is a third way in which heat travels, for we know that heated bodies cool even *in vacuo*: this is by so-called Radiation. The particles of a heated body are in a state of motion; that motion is communicated to the infinitely-elastic ether, which is supposed to fill all space, and to be transmitted through it with a velocity equal to that of light (192,000 miles a second). When a heated ball cools, the atoms oscillate in a resisting medium, which transmits motion on all sides: just in proportion as it loses motion, does it cool down.

A body cannot be heated, unless it accepts motion. When the sum of motion received is greater than it loses, the body becomes warm; otherwise it becomes cool.

Radiant heat and light obey the same law. The angle of incidence is equal to that of reflection. The heat-rays diminish in intensity exactly as light diminishes. The intensity of light diminishes as the square of the distance increases. If the same quantity of light is diffused over four times the area, it must be diminished by one-fourth of its intensity. Polished objects reflect the greatest part of the rays falling upon them, so that their temperature does not rise. The best reflectors are the metals, and amongst them, silver stands first. The reflecting power lies entirely in surfaces, as a film of gold-leaf $\frac{1}{250000}$ of an inch in thickness, will as effectually prevent paper from scorching under the rays from a fire as a mass of solid gold.

By means of concave reflectors, the rays of heat may be concentrated and brought to a focus, just like rays of light, so as to ignite combustible substances placed at the point of concentration. The rays in this case take the course shown in Fig. 14.

If two concave mirrors be placed on stands at a distance of about

12 feet from each other, a red-hot cannon-ball placed in the focus *a* of one, will scorch paper or ignite any readily-combustible matter, such as gun-cotton or gunpowder, placed at the focus *b* of the other.

Fig. 14.

When a ray of light is made to pass through a double convex lens and through a glass prism, the white beam is disentangled, and reduced to the colours, red, orange, yellow, green, blue, indigo, and violet; but there is radiant action beyond the red and beyond the violet. The obscure rays beyond the violet—the actinic rays—produce chemical action; those which fall beyond the red are thermal, or heat rays. Dark-coloured and rough bodies are the most absorbent of luminous heat; but in regard to the obscure rays of heat, radiation is nearly as copious from white as from black surfaces. A body that heats rapidly will cool with equal rapidity. When the loss of heat by radiation ceases to be made good to the plant or ground by sunshine, the temperature of the portion of air in contact with it begins to fall, and soon passes so much below that necessary for sustaining in the air the moisture always present, that a portion of it is deposited on the plants in drops of Dew.

It has been already hinted that there is so intimate a relation between the heat and moisture of the atmosphere, that they cannot be individually understood, except in relation to each other. Now, the atmosphere always contains a certain amount of water, in the shape of invisible vapour; as water evaporates at all temperatures, wherever dry air passes over water, whether the sea, a lake, or wet soil, it will take up some of its moisture.

Now, the quantity of vapour which the air can hold in solution, depends entirely on its temperature; the higher the temperature, the more vapour it contains. A very homely illustration of these two facts—the existence of vapour in the air, and its variation with the temperature—is daily seen. If a decanter, filled with cold water, is brought into a warm room, its surface is immediately covered with dew. The air in contact with the bottle, being cooled down by the water, is no longer able to contain the vapour present in it, which, therefore, condenses on the glass. A still more familiar illustration of the kind, is the condensation of moisture in the air, exhaled from the lungs. In summer, when the temperature of the external air and that from the lungs is more nearly equal, the moisture from the breath is scarcely visible; but, in winter, when the external air is much colder than the breath, the moisture, condensed immediately it leaves the mouth and nostrils, is seen as a small cloud.

This law is of immense importance in the economy of nature. It is in daily operation in all parts of the world. To it is due the transference of moisture, from one place where it abounds, to another where it may be wanted. For, whenever a body of air, saturated with moisture, is cooled, it must set free a portion of that moisture, in some form or other,—either as rain, or snow, hail, mist, or dew. Now, as this process of separation is for ever going on in some part of the globe, to such an extent that the mean annual rain-fall, for the whole globe, is estimated at about 33 inches, whereas all the moisture contained in the atmosphere at any one time would cover the earth's surface to a depth of not more than 5 inches, it is manifest that this moisture must be replenished by constant evaporation. Restored to the ground, in one or other of the above forms, it ministers to vegetation; while all that is not so employed either evaporates again, or sinks into the soil, to re-appear in springs, which pour their contents into rivers, by means of which the water ultimately finds it way back again to the sea.

The simplest form of deposit is Dew. In accordance with the

laws of absorption and radiation of heat, the deposit of dew is most copious on dark-coloured substances, such as the darker soils and plants. These, so long as the sun shines, absorb more heat than they radiate and consequently become warmer; but, as soon as the sun sets, radiation from them continues without any sensible addition of heat, they cool rapidly and, reducing the temperature of the air in immediate contact with them, cause it to deposit on their surface a portion of its moisture in drops of dew.

Dew is deposited most abundantly on cloudless nights; for, when the sky is obscured by clouds, these will radiate back to the earth a portion of the heat received from it, and prevent any considerable lowering of the temperature of the air near the surface of the ground. Nocturnal radiation frequently becomes most destructive to vegetation in spring and autumn, by lowering the temperature of the air below the freezing point of water, when the dew is deposited in the form of rime or Hoar-frost. Before it was known that bodies on the earth's surface became cooler than the surrounding air, it was impossible to understand how the flimsy coverings employed by gardeners could act so effectually in protecting vegetables and fruit-trees from the injury of frost. The natives of Peru having observed that, in that country, it freezes only when the sky is clear, contrive to create artificial clouds whenever, from the appearance of the sky, they apprehend a frost. This they do by setting fire to heaps of damp straw, which raise clouds of smoke over their crops, and, as Boussingault states, afford them complete protection.

By means of Daniell's hygrometer (ὑγρόν, hygron, moisture, and μέτρον, metron, a measure), the "dew-point," or the temperature at which the atmosphere begins to deposit moisture is best ascertained. It consists of a kind of cryophorus (κρυός, kryos, frost, and φορός, phoros, bearer), which is constructed on the principle of freezing water by its own evaporation, and very much resembles Leslie's differential thermometer (p. 7, Fig. 7), except that the bulbs are turned downwards, and one stem is much foreshortened. The hygrometer, however, contains ether instead of water; the one bulb is blackened, and the other covered with white muslin. All the ether is collected in the blackened bulb, and into it dips a sensitive thermometer. By first observing the temperature of the outer air, then cooling down the muslin-covered bulb with ether until moisture deposits on the blackened bulb, and ascertaining the difference between the readings of the thermometer without and within, we can tell at what temperature dew would be deposited, supposing the air to be cooled down. The atmosphere can take up moisture in proportion to the difference between the prevailing temperature and the dew-point.

In the application of heat to the purposes of daily life, these properties are further turned to account. The Meat-screen and the Dutch oven reflect the heat upon the articles of food placed between them and the fire; and our tea, coffee, or hot water, are preserved in vessels of polished silver or other bright metal. The soot-covered kettle is admirably adapted for heating water rapidly in the kitchen, but loses its heat as quickly by radiation; therefore, the parlour-kettle or urn, which is intended to retain heat, is to be kept bright. The pipes of a hot-water apparatus are kept bright in passages not intended to be heated, but are left rough and black in the chambers they are designed to warm. So rapidly will radiation reduce the temperature, that water will remain hot for a longer time in a highly-polished vessel, than in a similar one coated with lamp-black.

21. Some bodies are transparent. They are such, because the ether and the atoms of these bodies are so related to each other, that the waves which excite light can pass through them, without transferring their motion to the atoms. In coloured bodies, certain waves are broken or absorbed; but those which give the body its colour, pass without loss. Transparency to light does not necessarily imply transparency to heat. A body highly transparent to the luminous undulations, may be highly opaque to the non-luminous. Among solid bodies, rock-salt alone is equally transparent to heat from the different sources with which Melloni experimented.

Those bodies which allow heat freely to traverse them, are called diathermanous (from the Greek διὰ, dia, through, and θερμὸς, thermos, heat); those which do not, are called a-diathermanous. Ice is about the most a-diathermanous body among solids; distilled water and albumen (white of egg) amongst liquids; and perfectly dry air freed from carbonic anhydride amongst gases.

Rock-salt transmits all rays, luminous and obscure. Of the heat radiated from the flame of an oil-lamp, 90 per cent. is due to obscure rays; while of the heat radiated from the alcohol flame, 99 per cent. is due to obscure radiations. (Tyndall.)

22. All bodies are affected by heat, but equal weights of different kinds of matter are of course very unequally affected and require different degrees of heat to raise them through a given range of temperature.

The quantity of heat necessary to raise the temperature of a pound weight of a body by 1° F., as compared with that required to raise an equal weight of water by 1°, is called its Specific Heat; being, as the word denotes, the amount of heat specific or peculiar to the body in question. The capacity of bodies for heat depends chiefly upon their density; when that is increased, their specific heat is diminished. Sudden compression causes an evolution of heat, and sudden expansion the phenomenon of cold.

Of the one fact, Fig. 1, p. 2, affords a good illustration. The sudden compression of the air in the syringe ignites the tinder when the piston is withdrawn. In this case, the heat required to keep up the volume of the gas in the normal state, is no longer required.

But when the volume of a compressed gas, as in the case of the carbonic anhydride gas contained in soda water, is enlarged, heat is required to expand the particles; consequently cold is produced. The "cloud," visible when a bottle of soda-water is opened, is due to this fact. The escaping and expanding gas draws upon the vapour contained in the air, in an invisible form, for the requisite heat; the vapour is liquefied, and appears as a cloud. Of necessity, those bodies which have the greatest specific heat are the slowest to undergo changes of temperature. The low specific heat of mercury renders it very sensitive to changes of temperature,—a property which specially recommends it for use in the thermometer. Consequently, the mercury expands rapidly when the temperature rises, and falls with corresponding rapidity when a change takes place.

Water has the greatest capacity for heat, and therefore affords the most convenient standard of comparison with other bodies; it possesses nearly double the specific heat of ice. In the case of water, this property is of immense importance. Were it like mercury in respect of specific heat, the ocean would heat so rapidly in summer and cool so quickly in winter, as to render impossible the existence in it of animal life.

There are three methods by which the specific heat of bodies are obtained. Firstly: by heating given weights to a fixed temperature, and ascertaining the rate of cooling, Secondly; by heating given weights, and ascertaining how much ice they will melt. And, thirdly; by heating given weights of bodies, and discovering by how much they raise a given weight of water at a fixed temperature.

A pound of water at 40° F., mixed with another pound at 156° F., gives two pounds of water at 98°—the exact mean of the two temperatures; but a pound of mercury at 40°, and a pound of water at 156°, gives a mixture having a temperature of 152°·3. The mercury gains 112°·3, while the water loses but 3°·7. As the gain of the mercury is to the loss of the water, so is the specific heat of the latter to that of the former: 112·3 : 3·7 : : 1=0·0329. The specific heat, therefore, of water to that of mercury is as 1 to 0·0329, or about as 30 to 1.

The following table of important Specific Heats is by Regnault:—

Water	1·0000	Phosphorus (red)	0·1700	Arsenicum	0·0814
Lithium	0·9408	Potassium	0·1696	Silver	0·0570
Ice	0·5050	Diamond	0·1469	Cadmium	0·0567
Sodium	0·2934	Manganese	0·1217	Tin	0·0562
Magnesium	0·2499	Iron	0·1138	Iodine	0·0541
Carbon	0·2414	Bromine	0·1129	Antimony	0·0508
Boron	0·2352	Nickel	0·1086	Mercury	0·0333
Aluminum	0·2143	Cobalt	0·1067	Gold and Platinum	0·0324
Sulphur	0·2026	Zinc	0·0955	Bismuth	0·0308
Phosphorus	0·1887	Copper	0·0952		
Silicon	0·1774	Selenium	0·0827		

An illustration of the importance of the bearing of specific heat upon the welfare of man may be given in the following statement. Taking equal weights of water and of air, their specific heats are in the ratio of 1 to 0·237. A pound of water in cooling down 1° would heat 4·2 lbs. of air one degree. But as water is 770 times as heavy as air, the loss of 1° F. on the part of a cubic foot of water would impart the same temperature, 1° F., to 3234 cubic feet of air!

23. In considering the subject of Specific Heat, it must have been noticed that the bodies undergo no change of condition; they remain respectively solid, liquid, and gaseous. But a little inquiry into the circumstances attending the conversion of water from one of its forms into another, say from the state of ice into that of water, will enable

us to understand the peculiar relations of heat which accompany Changes in the State of Cohesion.

When ice is melting, the temperature of the water from it cannot be made to rise above 32°, so long as any ice remains. All the heat that may be employed is required for the melting of the ice; *i.e.*, for its conversion from the solid to the liquid state. And how much is requisite for this purpose may be readily ascertained.

If a pound of ice at 32° be melted by a pound of water at 212°, two pounds of water are obtained of the temperature of 51°. The water has lost 161°, the ice has gained but 19°. So that, so far as any indication of the thermometer goes, 142° of heat may be said to have been lost in the process (79°·4 C., if 143° be accepted).

And yet these 142° of heat are not lost, for they are again given out by the water whenever it freezes.

There is a similar apparent loss of heat in the conversion of boiling water into steam, an absorption amounting to 966°·6. This is ascertained by an experiment analogous to the above, viz., by passing a given weight of steam into a given weight of water, and observing the resulting temperature.

The heat absorbed is said to be "Latent," and the phenomenon is known by the name of "Latent Heat." The heat absorbed is employed in overcoming cohesion, in pushing the particles asunder, in moving them into new positions.

The latent heat of water at the ice-point is 142° (79° C.); that of steam at 212° is 966°·6 (537° C. at 100°), which, being interpreted, means, that a pound of ice at 32°, in thawing requires 142°. And a pound of boiling water 966°·6 to convert it into steam of the same temperature.

As a consequence of this phenomenon, no liquid can evaporate without occasioning cold, nor can any solid body be dissolved in water, provided it acquires no new properties (simple solution), without also lowering the temperature. And, indeed, the more rapidly solution is accomplished, the greater is the cold produced.

In the ordinary operation of distilling, which is shown in Fig. 15, fresh supplies of cold water are constantly passed into the refrigerator, as the worm and tub are commonly called. The vapour from the boiler passes into the worm at *b*, and is surrounded by a great body of cold water. The heat being greatest in the upper coils, heats the upper layers of water first, so that if the supply of cold water is allowed to enter at the bottom of the tub, the heated portions overflow at the top.

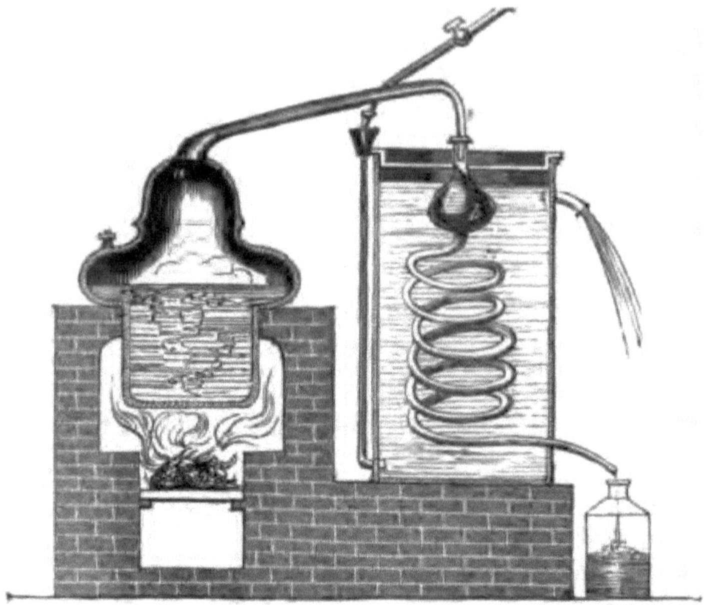

Fig. 15.

But the great benefit of the law of latent heat, that in all cases of the conversion of a solid into a liquid, or a liquid into a gas, a quantity of heat not indicated by the thermometer is absorbed, is best seen in the ordinary operation of boiling water. For when water is converted into vapour, it occupies a space 1689 times greater than before: the heat is consumed in expansion. Now, if it were all converted into steam at the moment of boiling, no vessel could resist its pressure, nor could any opening allow of its escape with sufficient rapidity.

24. Connected with the production of vapour is the so-called phenomenon of Boutigny or Leidenfrost. When water is dropped into a red-hot basin, the drops roll about in a spheroidal form, without actually touching the bottom of the containing vessel. The drops are balanced upon their own vapour, and the temperature of the spheroid is less than that of the boiling liquid. So likewise if the

hand be moist, it may be dipped into molten lead, the vapour preventing actual contact with the metal.

There are many points connected with heat so completely bound up with the subject of water, that all further information on this point is reserved. Such knowledge of the leading properties of heat, as has been sought to be conveyed in the preceding pages, will form a fitting introduction to the physical properties of the atmosphere, to which the following chapter is devoted.

For more complete information the reader is directed to the work of Professor Tyndall, the standard authority on the subject at the present time.

Examples for Calculation.

a. 135° C., how much on the scale of Fahrenheit? $135° = \frac{5}{9}$ [F.—32°].

$$\therefore\ 243° = F. - 32°;\quad \therefore\ F. = 275°.$$

b. —4° F., how much on the scale of Centigrade?

$$C. = \frac{5}{9}\,[-4° - 32°] = 5 \times -4 = -20°\,C.$$

c. The latent heat of water is 143° F., what is this upon the scale of C.?

$$C. = \frac{5}{9} \times 143;\quad \therefore\ C. = 79{\cdot}4.$$

d. 10 lbs. of water at 100° C., are mixed with M lb. of ice at 0° C.; the temperature of the mixture is found to be 50° C. Find the value of M in lbs.

$M \times 79{\cdot}4 + M \times 50$ = amount absorbed by the ice. $10\,(100° - 50°)$ loss of heat of water.

$$\therefore\ M \times 129{\cdot}4 = 500;\quad \therefore\ M = 3{\cdot}86\ \text{lbs.}$$

CHAPTER II.

ON THE PHYSICAL CONDITION OF THE ATMOSPHERE AS AFFECTED BY PRESSURE, MOISTURE AND HEAT.

25. Colour of the Atmosphere. 26. Weight of the Air. 27. Barometers. 28. Variations in weight. Law of Marriotte. 29. Weight of the Air calculated at a fixed temperature. French Weights and Measures. 30. Air-pump. Magdeburgh Hemispheres. 31. Limit of the Atmosphere at 43,334 yards, or 39,624 metres. 32. Moisture in the Air as mist and fog. 33. Clouds. 34. Height of Clouds. 35. Cirrus, Cumulus and Stratus. 36. Registration of atmospheric phenomena. 37. Occurrence of Clouds. 38. Effects of Rain upon the Atmosphere. 39. Snow. 40. Hail. 41. Resistance to the fall of Hail and Rain by the Atmosphere. 42. The relation of Winds to Moisture and Rain. 43. The transfer of Moisture accompanied by a transfer of Heat.

25. From what has been already said, it will be evident that heat plays a most important part in the economy of nature. The form which bodies assume depends, in great part, at least, on the amount of heat which they contain ; the gaseous form requiring the most.

Of matter in this form we have the completest type in the atmosphere which surrounds us. "The atmosphere rises above us with its cathedral dome, arching toward the heaven, of which it is the most familiar synonyme and symbol. It floats around us like that grand object which the Apostle John saw in his vision—'a sea of glass like unto crystal.' So massive is it, that when it begins to stir, it tosses about great ships like playthings, and sweeps cities and forests like snow-flakes to destruction before it. And yet it is so mobile that we have lived years in it, before we can be persuaded it exists at all ; and the great bulk of mankind never realize the truth, that they are bathed in an ocean of air. Its weight is so enormous that iron shivers before it like glass, yet a soap-ball sails through it with impunity, and the tiniest insect waves it with its wings. It ministers lavishly to all the senses. We touch it not, but it touches us; its warm south wind brings back colour to the pale face of the invalid; its cool west winds refresh the fevered brow, and make the blood mantle in our cheeks; even its north blasts brace into new vigour the hardened children of our rugged clime. The eye is indebted to it for all the magnificence of sunrise, the full brightness of mid-day, the chastened radiance of the gloaming, and the clouds that cradle near the setting sun. But for it, the rainbow would want its triumphal arch, and the winds would not send their fleecy messengers on errands round the heavens; the cold ether would not shed its snow feathers on the earth, nor would drops of dew gather on the flowers. The kindly rain would never fall, hail-storm nor fog diversify the face of the sky. Our naked globe would turn its tanned, unshadowed forehead to the sun, and one dreary monotonous blaze of light and heat dazzle and burn up all things.

"Were there no atmosphere, the evening sun would in a moment set, and, without warning, plunge the earth in darkness. But the air keeps in her hand a sheaf of his rays, and lets them slip but slowly through her fingers; so that the shadows of evening gather by degrees, and the flowers have time to bow their heads; and each creature space to find a place of rest, and nestle to repose. In the morning, the garish sun would at one bound burst from the bosom of night, and blaze above the horizon; but the air watches for his coming, and sends at first but one little ray to announce his approach, and then another, and, by-and-by, a handful; and so gently draws aside the curtain of night, and slowly lets the light fall on the face of the sleeping earth, till our eyelids open, and, like man, she goeth forth again to her labour until the evening."—(*From the Quarterly.*)

The air being perfectly transparent permits the rays of light to pass through it, and is, therefore, generally invisible. Nevertheless, the sky, when free from clouds, always appears more or less blue. The blue colour is due to the refraction of the rays of light, by the particles of vapour diffused through the air. Its intensity, therefore, is in proportion to the amount of vapour. Accordingly, the bluest skies are seen in the summer, and in the warmer regions of the globe. In fact, all the varied colours of the atmosphere are due to a similar cause. The clouds reflect to us, not isolated rays only, but the undecomposed sunbeam, consequently they appear to our vision like banks of snow. The glories of sunrise or sunset depend entirely upon the quantity of watery vapour which is mixed with the air, and its state of condensation. It has been observed that steam, at night, issuing into the atmosphere under a pressure of twenty or thirty pounds to the square inch, reflects and transmits orange-red light. We are, therefore, led to conclude that this is also the property of such a condition of mixed vapour and air as prevails when the rising or the setting sun is shedding over the horizon the glories of its coloured rays.

26. In common with every form of matter, the atmosphere is possessed of weight. This property in air was first demonstrated by the celebrated Italian, Torricelli, whose attention was called to it by the following circumstance. In sinking a well at Florence in 1643, it was found that after reaching a certain depth, the common suction pump ceased to bring water to the surface. It had been previously known, that when a tube was dipped in water and the air above it withdrawn, the water would rise in it to a very considerable height; but the only explanation of this phenomenon that was offered, was the vague assertion that "nature abhorred a vacuum," and that the water rose in the tube to fill the vacuum caused by the exhaustion of the air.

The attention of Torricelli having been called to the fact that the water in the suction pipe of the pump would not rise beyond a height of 34 feet, it occurred to him that its rising at all was to be attributed to the pressure of the atmosphere external to the pipe, forcing the water to occupy the vacuum caused by the exhaustion within it. Torricelli presumed that, as the volume of water was sustained by the pressure of the air, the two must exactly balance each other; and, therefore, that a volume of water 34 feet high must be of equal weight with a similar column of the whole height of the atmosphere Supposing this to be the case, then a denser fluid substituted for the water, ought to rise in the exhausted tube less than the water, in exact proportion to the greater density of the fluid employed. To test this theory, Torricelli filled a glass tube. Fig. 16 *d*, three feet long, and closed at the lower end, with mercury, and inverted it at *c*, in a basin of the same liquid. The mercury immediately sank 6 inches in the tube, in fact to *a*, leaving a column, *b*, 30 inches in height. Now, as the weight of mercury is 13·5 or 13½ times greater than that of water, the truth of his conjecture was fully proved, viz., that the weight of a column of the atmosphere was equal to a similar column of water 33¾ feet, or of mercury 30 inches, high.

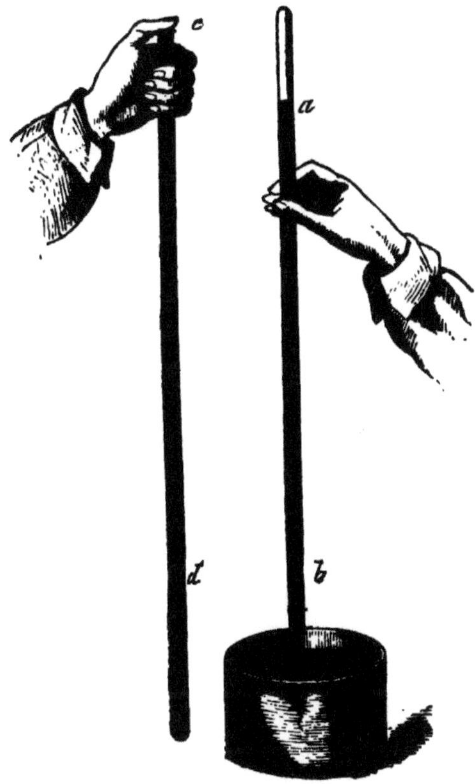

Fig. 16.

In 1648, Pascal repeated Torricelli's experiment, and varied it by the substitution of fluids of different specific gravities. But as many still affected to doubt whether the rise of the fluid was really due to atmospheric pressure, he determined to bring the question to a definite issue. Having filled two tubes with mercury, at the base of the Puy de Dome, a mountain in Auvergne, he inverted them as in the previous experiment, and saw that the mercury stood at the same height in both. One tube he left at the foot of the mountain, and carried the other to its summit. As he ascended the mountain, the mercury fell steadily in the tube, till, on reaching the summit, it stood at $3\frac{3}{10}$ inches lower than it did at the base. On descending, the mercury steadily rose till he reached the point whence he started, when the column had returned to the exact height at which it originally stood. He was forced to the conclusion, that as, in ascending the mountain, the atmosphere above him was diminished in proportion to the height ascended, its power of supporting the mercury was diminished in like proportion, and therefore the column in the tube must always represent the exact weight of the superincumbent atmosphere.

27. Thus was invented that most valuable instrument, the Barometer, from the Greek βάρος, baros, weight, and μέτρον, metron, a measure. As the range of the rise and fall of the mercury at the level of the sea never exceeds or even reaches 3 inches, the division of the scale is made only on that part of the column between 27 and 31 inches.

The vacant space above the quicksilver in the closed end of the tube is called the Torricellian vacuum, from its having been observed by Torricelli to be free from air.

The mean height of the barometer is about 30 inches at the level of the seâ. As the height of the column of mercury is somewhat affected by its different expansion at different temperatures, it is necessary, when accuracy is required, to calculate and make allowance for the expansion; and, for this purpose, a thermometer is always attached to a good barometer.

Fig. 17.

On account of the cohesion of the mercury, there is a depression in the tube which renders a correction necessary in reading off the height of the column, unless the tube be above a half inch in bore, when it may be disregarded. The reading is from the surface of the mercury. Before taking an observation, the instrument should be gently tapped to prevent the adhesion of the mercury to the tube; and the eye of the observer should be on a level with the top of the column. Very little attention should be paid to the words Fair, Set Fair, Stormy, &c., which are usually engraved on the scale, as it is not the absolute height of the column so much, as its rise or fall, which indicates the character of the weather. Extreme elevations or depressions rarely occur suddenly, especially in this climate; the mercury being usually, for several days together, in a rising or falling state.

The expansion of mercury is $\frac{1}{9990}$ of its bulk for each degree of Fahrenheit between 32 and 212. In the best barometers there is also an apparatus for correcting another source of error, viz., the different height at which the mercury stands in the cistern, varying with its rise in the tube.

There are many modifications of the Barometer. In the common wheel barometer, Fig. 17 (which is more ornamental as an article of furniture than useful), the variations are shown by the index, *a*, which traverses the dial plate. A weight at *c* (attached to a thread passing over a pulley, to which the index is attached and nearly counterpoised at *b*), is made to rest on the surface of the mercury in the open syphon end of the tube. It consequently rises and falls as the column becomes longer or shorter, the motion of the thread turning the pulley and moving the index through the arc of a graduated circle.

The Aneroid (from *a*, *α*, privative and *νηρός*, neros, liquid) Barometer depends for use on the effect of atmospheric pressure upon a metallic box, from which the air has been exhausted. The changes in the weight of the air are expressed by an index traversing a dial.

For use at sea a marine barometer is made, having a portion of the tube narrowed, to prevent the oscillation or pumping up and down of the mercury with the movements of the ship. This contraction of the tube so far interferes with the working of the instrument, as to make its indications slow, and, therefore, less delicate. This defect is remedied by the use of the Sympiesometer, or air-barometer. In this, the varying weight of the air is measured by the compression of a gaseous column, which occupies more or less space according to the pressure, a correction being made for the effects of temperature at the time of observation. It is not nearly so correct as the mercurial barometer, but is very readily affected, and so serves to direct attention to the other.

28. For the most part, variations in the weight of the atmosphere, as indicated by the fluctuations in the barometric column, depend upon differences in its moisture and heat. These differences are so bound up with the subject of the causes of wind and rain, that they will be considered together with them. But the practical application of the barometer to the measurement of the height of mountains introduces us to an important law affecting air in common with all gases.

In virtue of its perfect elasticity, by which its particles tend to repel each other, it is enabled to adapt itself exactly, under any degree of pressure, to the space which it is required to fill. If, therefore, the pressure is altered, the bulk of the air is altered too.

In the language of science, the volume occupied by a gas is inversely proportional to the pressure to which it is subjected. If the pressure be doubled, the volume of a gas is halved; if the pressure be halved, the volume of a gas is doubled. This great fact in Physics is known as the law of Marriotte, and is well illustrated by the following experiment. Fig. 18 represents a bent glass tube, about 7 feet long, furnished at *b* with a tap opening up to the air. The tap being open, some mercury is poured into the long limb of the tube, and sufficient in quantity to cut off the communication between the long and short limb, denoted in the figure by the commencement of the scale. The mercury being subject to equal atmospheric pressure will thus stand at the same level in both tubes. If the tap be then closed and mercury be poured into the other tube to an additional height of 30 inches (equivalent to the weight of another column of the atmosphere), the bulk of the air at *a* will be diminished one-half. If a further 30 inches of mercury be added, there will be a pressure on the air in the closed tube equal to three atmospheres and its bulk will be reduced to one-third of its original volume.

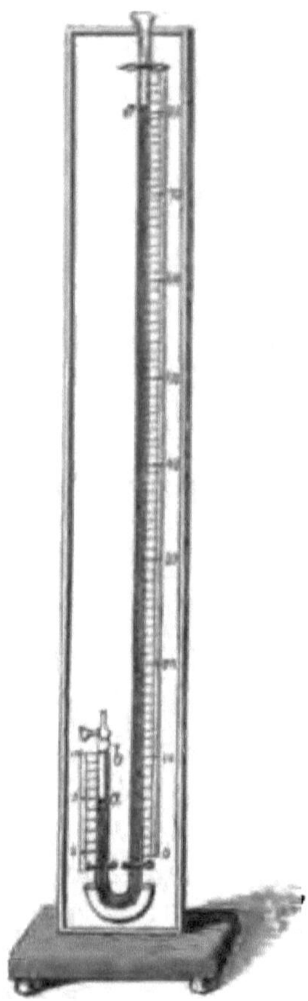

Fig. 18.

A necessary consequence of this law of elasticity is the increasing rarefaction of the air as we ascend mountain heights. If we suppose the atmosphere to be divided into strata of equal weights, the upper stratum will press downwards with its own weight only; the second, with its own weight increased by the weight of the first; the third stratum will have to support its own weight added to that of the two upper strata, and so on. Thus, the density of the atmosphere must decrease in equal ratios in ascending through equal heights, and is, therefore, capable of measurement by a corresponding fall of the barometric column.

In the following table, the height in miles above the sea-level is taken in terms of an arithmetical progression; the bulk of an equal weight of air increasing, whilst its density decreases, in terms of a geometrical progression:—

Miles above the sea level.	Bulk of equal weights of air.	Density.	Height of Barometer in inches.
0·0	1	1	30·00
3·4	2	$\frac{1}{2}$	15·00
6·8	4	$\frac{1}{4}$	7·50
10·2	8	$\frac{1}{8}$	3·75
13·6	16	$\frac{1}{16}$	1·87
17·0	32	$\frac{1}{32}$	0·93

29. As the bulk of air is so much affected by pressure and heat, its weight is always calculated at a fixed temperature and pressure. At a temperature of 60° F., and a barometer pressure of 30″, one hundred cubic inches of dry air weigh 31·0117 grains. A simple method of calculating the absolute weight of a given volume of a gas, exists in the knowledge of the fact, that the combining weights of all the Elements (P. 50.) existing as gases are identical with their specific gravities.

In France, all volumes of gases are compared at the standard temperature 0° C and under the standard pressure of 760 millimetres of mercury. Now 760 millimetres are equal to 29″·92 Bar.; 100 cubic inches of dry air, measured at 0° C. and 760 millimetres Bar. weigh 32·586 grains.

A few words on French Measures and Weights may with advantage be introduced in this place. One Metre is the unit of Linear Measures: One Cubic Metre, is the unit of Cubic Measures: One Litre or Decimetre Cube, is the unit of Measures of Capacity, and one Gramme is the unit of Measures of Weight.

One Metre is = 39·37 inches. One Litre is = 61·027 cubic inches, or 1·76 Imperial Pint. One Gramme =15·432 grains.

One Cubic Metre of dry air at 0° C. and 760 mm. Bar., weighs 1·2932 Kilogramme.

The following table speaks for itself and will be found useful for reference.

TABLE
OF
FRENCH MEASURES
OF
LENGTH, SURFACE, CAPACITY AND WEIGHT.

METRICAL MEASURES OF LENGTH.

	In English Inches.	In English Feet = 12 Inches.	In English Yards = 3 Feet.
Millimetre	0·03937	0·0032809	0·0010936
Centimetre	0·39371	0·0328090	0·0109363
Decimetre	3·93708	0·3280899	0·1093633
Metre	39·37079	3·2808992	1·0936331
Decametre	393·70790	32·8089920	10·9363310
Hectometre	3937·07900	328·0899200	109·3633100
Kilometre	39370·79000	3280·8992000	1093·6331000
Myriometre	393707·90000	32808·9920000	10936·3310000

MEASURES OF SURFACE.

	In English Square Feet.	In English Sq. Yards = 9 Square Feet.
Centiare or sq. metre	10·7642993	1·1960333
Are or 100 sq. metres	1076·4299342	119·6033260
Hectare or 10,000 sq. metres	107642·9934183	11960·3326020

MEASURES OF CAPACITY.

	In Cubic Inches.	In Pints = 34·65923 Cubic Inches.
Millilitre, or cubic centimetre	0·061027	0·001761
Centilitre, or 10 cubic centimetres	0·610271	0·017608
Decilitre, or 100 cubic centimetres	6·102705	0·176077
Litre, or cubic decimetre	61·027052	1·760773
Decalitre, or centistere	610·270515	17·607734
Hectolitre, or decistere	6102·705152	176·077341
Kilolitre, or stere, or cubic metre	61027·051519	1760·773414
Myriolitre, or decastere	610270·515194	17607·734140

MEASURES OF WEIGHT.

	In English Grains.	In Avoirdupois Lbs. = 7,000 Grains.
Milligramme	0·015432	0·0000022
Centigramme	0·154323	0·0000220
Decigramme	1·543235	0·0002205
Gramme	15·432349	0·0022046
Decagramme	154·323488	0·0220462
Hectogramme	1543·234880	0·2204621
Kilogramme	15432·348800	2·2046213
Myriogramme	154323·488000	22·0462126

As the weight of a column of mercury, 30 inches (30″) in height and covering one square inch, amounts to 15 lbs., and as the air presses upon the earth's surface with that force, it follows, that objects of all kinds, at the sea-level, have to support an average pressure of 15 lbs on each square inch of surface. As there are about 2000 square inches surface of a man's body, the pressure of the air upon it must amount to about 13⅓ tons; yet, owing to the perfect equality of its pressure, and to the fact that every cavity within him is expanded by air of the same elastic force, he is enabled to sustain this enormous load, not only without inconvenience, but generally without even a consciousness of its existence.

30. From the perfect mobility of the particles of air, the force of its pressure is evenly communicated throughout and exerted equally in all directions. If, therefore, the pressure on one side of any body,

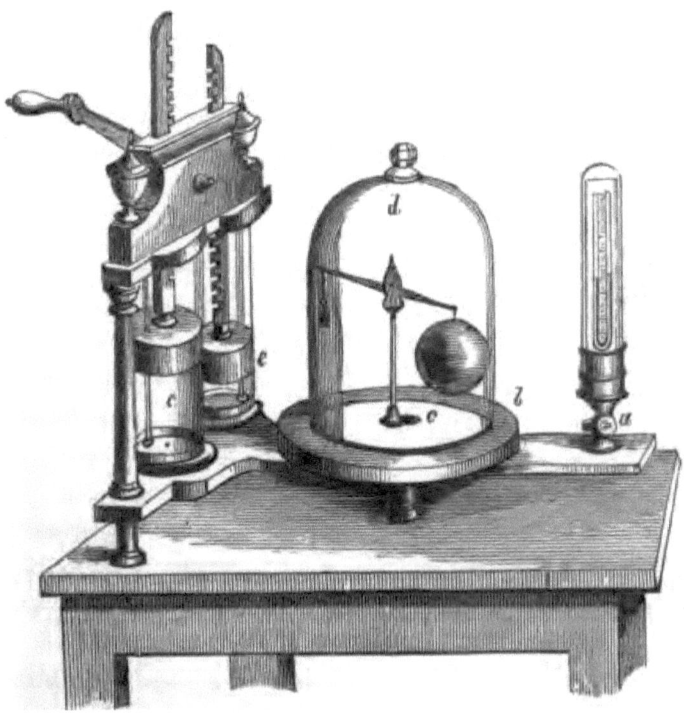

Fig. 19.

as a glass receiver, be diminished or destroyed, as may be readily done for purposes of experiment by the air pump, the pressure of the air on the opposite side immediately makes itself manifest, inasmuch as the first stroke of the pump fixes the receiver to the plate on which it rests.

Fig. 19 represents the common double-barrel air-pump of French manufacture, with its barrels made of glass. In it, two cylinders are combined and made to act alternately by means of a rack and pinion. In the illustration, a very interesting experiment is supposed to be in the act of performance. If a large ball of cork exactly counterbalances a small metal weight in the air, it will nevertheless outweigh it under the receiver of the air-pump, in proportion as the air is exhausted; inasmuch as the loss of weight of a body in air is proportional to the amount of air displaced.

Fig 20 represents the air-pump in vertical section, and will enable the reader better to understand its construction. On working the piston *e*, the elasticity of the air within the receiver *d*, Fig. 19, raises the valve *b*, and the expanded air enters the vacuum formed between *b* and *l* in the lower portion of the cylinder by lifting the piston; while the downward stroke of the piston *d*, by the increased pressure of the air contained between *k* and *a*, closes the valve *f*, overcomes the elasticity of the outer air and thus allows the escape of the compressed air through the valve *k* in the piston. This action is continued with the two barrels alternately, until the elasticity of the air within becomes so reduced as to be incapable of opening the valve.

Fig. 20.

The air-gauge *a*, Fig. 19, affords a means of estimating the amount of vacuum produced.

The effects of diminished pressure are also well shown in the so-called Magdeburgh hemispheres. Fig. 21 represents two hollow brass hemispheres, with the sectional parts made to fit on one another perfectly and with a perforation at the bottom of the lower one (shown in the illustration by removal of a portion of the hemisphere) fitted to allow it to be screwed to the plate of the air-pump. If the two sections are brought into close contact and the lower

Fig. 21.

hemisphere secured to the air-pump, a few strokes will suffice to remove as much of the air from within, and consequently diminish its pressure, that the outward pressure will keep the hemispheres firmly together. A few more strokes of the pump and it will be impossible to separate them without the exertion of a considerable amount of force; in fact, of just so much as is equal to the difference of pressure between the air without and within the hemispheres.

In like manner, in ascending high mountains, or in rising in a balloon, we for a time destroy the equilibrium between the air within and without our bodies. As we ascend in the air, and the atmosphere outside presses on us with less and less force, that within us takes time to adjust itself to the diminished pressure, and we become sensible of great discomfort from its distension, which may even cause rupture of the delicate blood-vessels of the lips, eyes, and nose.

So, too, persons in delicate health are wonderfully sensitive to rapid fluctuations of the barometer; that is, to an increase or diminution of the pressure on their bodies.

31. From what has been said on the subject of the elasticity of the air, it might be inferred that, although by far the greater weight of the atmosphere would be found within a few miles of the earth, it must yet, from the fact of its elasticity, extend into boundless space. But, as all astronomical observations demonstrate that no atmosphere resembling our own surrounds the larger planets, we are led to infer that there is a distinct limit to it, when it is prevented by its own weight from expanding yet more into the vacuum beyond it. Such a limit is supposed to be attained at a distance of 43,334 yards or 39,624 metres from the earth's surface.

32. It has been already hinted that there is so intimate a relation between the heat and moisture of the atmosphere and the winds and rains, that they cannot be individually understood, except in relatin to each other.

Under certain circumstances, the condensed moisture instead of immediately falling as dew, hangs about the air as Mist or Fog. There is much difference of opinion as to the exact nature of mist. The most plausible explanation of it is, that the water is in the form of little bubbles, which are kept floating in the air by the lightness of the watery vapour which fills them. It may be, that in the black fogs of London, which are so unpleasant and even injurious to respiration, the innumerable soot-particles act as so many radiating centres, favouring the deposit of these water-bubbles on them.

33. But whatever be the nature of mist, it is certain that Clouds are nothing more than masses of mist in the higher regions of the air. Now, there are three causes for the more frequent formation of clouds in these higher regions, than near the ground. In the first place, the gradual rarefaction of the air as it ascends, at last cools it down, by decreasing its capacity for heat, to its "dew-point"; that is, to the temperature at which it is obliged to part with some of its moisture. Secondly, it is chiefly in the upper regions that different currents of air meet and mingle, If two currents of different temperatures, both charged with moisture, meet, although their temperature becomes uniform, the mixed air at this mean temperature can no longer contain all the moisture, and some of it must separate.

Or, a large body of air may be cooled down in passing from a warm climate to a cold one, and the result is the same. Thirdly, mountain tops, by their cooling influence, condense the vapour in the air that sweeps over them. In common language, they attract the clouds; in fact, they do not attract, but cause them. Hence, mountainous regions are those in which most rain falls.

34. The height at which the clouds are found is very various. Thin, light clouds soar above the loftiest mountains, while thick and heavy clouds approach very near the low-lands. Their average height may be about two miles and a half, and they are not supposed to rise more than double that height above the earth. Their size is very varied; some extending over an area of 20 square miles, with a thickness of a thousand feet, while others are of very insignificant dimensions.

The effect of clouds on atmospheric pressure, *i.e.* on the indications of the barometer, depends on the circumstances accompanying their formation. The weight of an atmosphere of vapour is but $\frac{6}{10}$ of that of an equal bulk of dry air. It is clear therefore that, cæteris paribus, the more vapour the air contains, the lighter it must be. A striking confirmation of this statement is found in the equatorial calm belt (p. 14, par. 19), a region of eternal cloud and rain, where the mean height of the barometer for the year is only 29·″915. On the other hand, the mean barometer height in the higher trade wind regions, where the air is comparatively dry, is 29·″995. Suppose, then, that a body of dry air is invaded by moist, with the formation of cloud, a fall of the barometer will be the result. In the opposite case of a body of cold dry air coming to the moist, a different effect will be produced on the barometer. But in either case the change must extend over a considerable district to cause an appreciable change of atmospheric pressure. Mere slight local causes, such as give rise to the lighter forms of cloud, can have no such influence. And, of course, a long continuance of cloud, unaccompanied by changes in the moisture of the air, is quite compatible with a steady barometer.

35. There are three great and characteristic forms of cloud—the Cirrus, Cumulus, and Stratus.

The lightest of clouds, that, therefore, which ascends to the greatest heights, and is seen at almost all seasons of the year, is the Cirrus. It is denoted in Fig 22 by four crows. From its height,

Fig. 22.

it is often lighted up by the rays of the rising or setting sun, while the lower clouds are in darkness. It lies in patches on the deep blue sky, while its motion is generally slow, and its appearance fibrous, delicate, and thread-like.

The Cumulus is the cloud of spring and summer, and is represented in the picture by three flying crows. Formed in the daytime, its density keeps from the earth the sultry rays of the noonday sun. Towards evening it gradually disappears, owing to the increased temperature of the superior strata of air. Its form, when viewed sideways, increases from above in dense convex heaps; in hot weather it often appears stationary, with a flattened base, its rock-like summits showing with a silvery light. When the cumulus increases after sunset and shines with a ruddy hue, it indicates the coming of a thunder-storm.

The Stratus is the horizontal sheet of cloud nearest to the ground, and represented pictorially by two crows. It occurs chiefly at night, or in the cool of a calm evening. It frequently appears on fine autumnal mornings and evenings, sometimes resting on the ground, sometimes hovering a few feet above it. The Stratus deposits moisture which is often cooled into hoar-frost, rendering the face of nature exquisitely beautiful: it is also the cause of the mists which are scattered by the rising sun. It is most frequently seen from September till January.

There are several well-defined modifications of the three forms of cloud already described, which can be recognized in cloud-land.

The Cirro-stratus, like the Cirrus, is to be found in the higher cloud-region, and is to be seen at all seasons. Sometimes it appears dark and threatening; at others, attenuated. The Cirro-stratus accompanies a fall in the barometer and is the precursor of rain and wind.

The Cirro-cumulus is often formed from the cirrus. It is the warm-weather cloud and attends a rising barometer. When seen alone in the higher regions of the atmosphere, it assumes a dappled appearance and is commonly known as a mackerel-back sky. More frequently it bears the form of small, round patches of cloud, leaving between them openings of clear blue sky. The glorious crimson contrasting with the delicate blue of a fine sunset, and the rich golden hues pervading the bright blue of a sunrise in summer, are due to the Cirrus and Cirro-cumulus.

The Cumulo-stratus is the most dense form of cloud, and causes, in passing, a sensible reduction of temperature. It is made up of the Cumulus, the Cirro-cumulus, and the Cirro-stratus. It forms the thunder-cloud and indicates changes of wind, hail, and storm.

The Nimbus, represented in the illustration by a single crow, is a modification of the Cumulo-stratus cloud seen in profile during a shower; and is never seen with high barometer. Its attendant is the glorious rainbow. Scud, is, like Stratus, of the clouds nearest the earth. It is common during the winter months and is witnessed more frequently after rain than at other times.

The clouds may be divided into two groups, the first comprising Cirrus, Cirro-stratus, Cirro-cumulus, and Scud, which descend, producing rain and wind; the second, including Stratus and Cumulus, which ascend and are the attendants of fine weather.

36. In the registration of atmospheric phenomena, it is usual to record the amount of cloud visible at the time of observation. A

clear sky is marked 0, and a cloudy sky 10, the degrees of cloudiness being indicated by the intermediate figures.*

37. The sky is most free from cloud at night, and most covered at noon. In our climate three-fourths of the sky are generally covered during November, December, January, and February; and in the finest months of the year three-fifths of the sky are clouded over.

38. The variations in the Rain-fall remain to be noticed, but a few words must be said as to the effects of rain on the atmosphere. Rain is a very pure form of water and, as such, has of course great powers of solution. Hence, in falling through the air, it literally washes it and frees it from those local gaseous and other impurities which would, in time, render it corrupt and unwholesome. These substances, which would be most injurious if left in the air, it dissolves and conveys through the soil into the circulation of plants, which require them for their nourishment.

39. When the temperature of the upper regions of the air is below the freezing point, the moisture freezes and descends as Snow. Whether the snow-flakes are formed at once, by the congelation of the moisture in the cloud, or whether the frozen particles unite with a myriad of others in their fall, has not been fully ascertained. However this may be, they are formed of crystals of pure water. The exquisitely beautiful forms assumed by the flakes of snow must be closely examined to be at all appreciated; some of them are here delineated in Figs 23 and 24.

The effects of the physical changes of the vapour of the air, from water to snow, are of great importance. From what has been stated at p. 25, par. 23, it will be seen that the formation of snow must be attended by the liberation of heat; and experience tells us that a

* Much of the information on the subject of the clouds, has been culled from a little work on the Barometer, by Mr. Belville, the use of which was kindly granted by Messrs. Elliott, of the Strand.

Several of the illustrations are taken from the French Edition of Professor Ganot's "Elements de Physique," and were prepared for and used in the Author's "Science of Home Life, or Household Chemistry" before Dr. Atkinson's Translation made its appearance.

heavy fall of snow is always succeeded by a rise in the general temperature of the air. So also a good rainfall is commonly associated for a time with a rise in temperature.

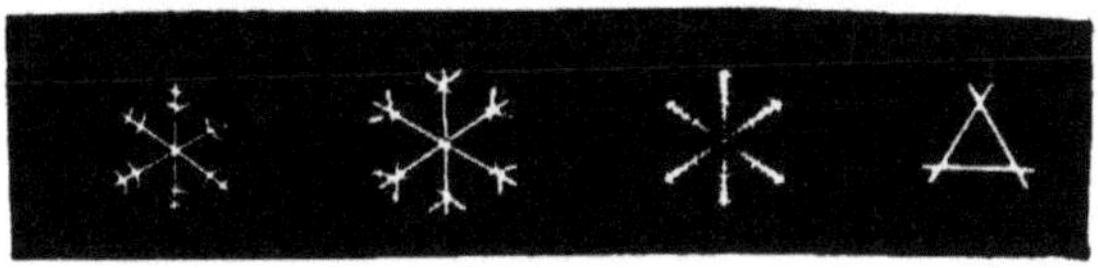

Fig. 23.

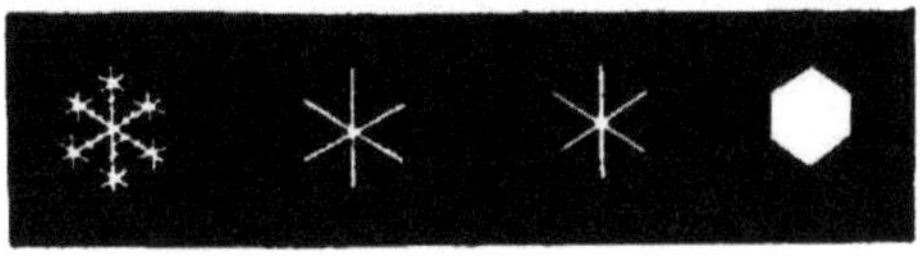

Fig. 24.

A fall of snow, also, has peculiar effects on vegetation. From its porous character, the snow, in its fall from the sky, entangles much air, thus, by its non-conducting property, protecting the tender shoots from the frost: neither are they injured by its subsequent thaw, which is sufficiently gradual to prepare them for the change. It is a curious circumstance that the snow contains a larger proportion of oxygen than is found in the atmosphere, which, in melting, it carries into the ground, fertilizing the soil by aiding the solvent action of the water upon its most important mineral constituents. A fall of snow one inch in depth affords in melting $\frac{1}{10}$ of an inch of water.

40. Of Hail, another form in which water is condensed from the atmosphere, little is known beyond the fact that electricity is somehow concerned in its production. It seems to be formed by a moist ascending current of air, greatly cooled by rarefaction, and having an upward velocity sufficiently strong to counteract the downward tendency of the hailstones, until they have attained a considerable size.

41. The fall of hailstones and of rain illustrates, in the most striking manner, one of the physical characteristics of the air: by its resistance to their passage, it retards their descent and mitigates the force of their fall. How quietly and softly does the rain descend

upon the tender foliage of the trees and the delicate petals of flowers, refreshing, without injuring, their fragile textures! In separating from the clouds, the drops fall faster and faster, till the increasing resistance of the air becomes nearly equal to their weight, when they continue to fall at an even rate. The destructive effects of hailstones are often sufficiently striking; but what incalculable devastation would ensue were the atmosphere to offer no resistance to their descent!

42. The relation of winds to Moisture and Rain remains to be considered. The trade-winds in the higher latitudes, where they first reach the earth's surface, consisting of cold air from the polar and temperate regions, having parted with their moisture in those regions, are dry. As they sweep over the ocean they suck up moisture, and gradually become saturated with it. Now, as this air is passing from a colder to a warmer climate, its capacity for moisture increases, and, therefore, it deposits none. Accordingly, there is seldom rain in the trade-wind regions. But when the two streams meet, the saturated air, ascending, begins to cool, and immediately deposits some of its moisture. Hence the equatorial calm is a region of perpetual rain. Now, this region is not stationary. As it must always be in the hottest region of the earth (or a little north of it), it travels up and down, according to the position of the sun at different times of the year, over nearly 1000 miles in latitude. Wherever it goes, it creates a rainy season. Accordingly, the rainy season differs at different places, and some places over which this calm-belt passes twice in the year, have two rainy seasons.

The air which ascends from this calm-belt contains the moisture which is to supply the whole world with rain. As this air passes towards the poles, from hotter to cooler regions, it gradually loses its moisture. Accordingly, the rain-winds of every country are those which come from the equator. Everyone knows that the rain-winds of England, which lies in the counter-trade region, are the Southwest. It is the same in America. But different places and countries differ widely in their rainfall. In other words, the passage of the air from warm climates to cooler, is not the only cause of the deposit of its moisture.

It has been already explained, that the cooling of the air by mountain tops is a very common cause of such deposit. So common, indeed, and so powerful a cause is this, that the most rainy regions in the world are on the slopes of those mountains, which are first struck by the water-laden air of the trade winds or monsoons. Thus the whole of Brazil is a very rainy district, in consequence of the influence of the Andes. But the Andes are so far from the sea and the rain is therefore distributed over so large an area, that the influence of the mountains is less manifest than in some other regions. On the western slopes of the Ghaut Mountains, in India, there is a rainfall during the monsoons equal to 86 feet in the year. But even this is surpassed by Patagonia, where the South-west counter-trade winds are intercepted, very near the sea, by the Andes. Here Captain King found the astonishing fall of 151 inches in 41 days, or at the rate of 114 feet in the year. The effect of land, in causing precipitation, is seen even in England. On the West coast of England, the annual rainfall is about 34½ inches: on the East coast where it is reached by the rain-wind not till after it has crossed the island, only about 24 inches.

The rainy side of a mountain, then, is its windy side. And if the mountain is sufficiently lofty, and, therefore, sufficiently cold at the summit to condense all, or nearly all, of the vapour in the air, it is manifest that the country beyond can get no rain at all. This is the case with the Andes; and Peru, which is on the lee-side of that range of mountains, is, therefore, a rainless country. If the mountain is not so high, the air crossing it leaves behind only a portion of its moisture. The next range of mountains will take more, and so on, till the air is gradually deprived of the greater part of it. This is the case in India; and north of India is another large rainless district. A third rainless district is found in the deserts of Africa. This depends not so much on the dryness of the air, as the absence of all causes of condensation in the district itself.

Now, as, on the one hand, the principal causes of the cooling of the air are found over land, and the northern hemisphere contains by far the greater quantity of land; and, on the other hand, the southern hemisphere, containing more sea, gives rise to the greater quantity of vapour—in other words, the northern hemisphere being the greater region of precipitation, the southern of evaporation—it is clear that the vapour must pass in some way from the former to the latter. Hence, it is supposed, that in the calm-belts, there is a crossing of the winds in the upper regions from the one hemisphere to the other. This conjecture is supported by a highly-interesting fact. The red fog which is met with in the Mediterranean and about Madeira, is found to be caused by a red animalcule, whose habitat is South America. It must, therefore, have been blown from one place to the other.

43. The transfer of moisture from hot climates to cold, is accompanied by a similar transfer of heat. The vapour which rises in the tropics renders latent an immense amount of heat: this heat is given out in the colder regions of the earth wherever the vapour is condensed as rain. In this way the winds tend to equalize the temperature of the earth, by carrying heat from the places where it is in excess, to those where it is wanted.

Problems Solved.

e. A cannon-ball weighing 1 lb. is placed in a vessel containing 1 lb. of water at 0° C.; without any evaporation, the water rises to 100° C. Find the temperature of the cannon-ball.

Let T be the temperature. $\therefore$ T—100 = loss of temperature of the cannon-ball. Also: 100 = the gain of temperature on the part of the water.

$$\therefore (T-100^\circ) \times 1 \text{ lb.} \times 0{\cdot}1138 = 100 \times 1 \text{ lb.} \times 1.$$

(0·1138 and 1, being respectively the specific heats of iron and water).

$$\therefore T = 978^\circ{\cdot}73.$$

CHAPTER III.

INTRODUCTION TO MODERN CHEMISTRY.

44. Elements. 45. Sixty-four elements. Metallic and non-metallic elements. Some very rare. All very unevenly distributed. 46. Non-metallic elements, 13 in number. 47. The elements represented by the initial letters of their Latin names. 48. Atoms. Hydrogen the standard of unity. 11·19 litres the absolute volume occupied by 1 gramme of Hydrogen. 49. Equal volumes of elementary gases contain the same number of atoms. 50. Atoms and Molecules. 51. Compounds in the state of vapour or gas occupy the space of two volumes of Hydrogen. 52. Radicals. 53. Compounds occupy the space of two volumes. 54. Monads, Dyads, Triads, Tetrads and Hexads. Also Table of the Elements.

44. MANIFOLD as are the substances which we meet with on every hand, there are but few bodies which cannot be resolved by heat or otherwise into simpler forms. Those bodies which resist all attempts at decomposition are called "elements" or simple bodies.

45. The elements are sixty-four in number; they may be divided for convenience into "Non-metallic elements" and into "Metals." The non-metallic elements are more or less imperfect conductors of heat and electricity and their chief tendency is to unite with the element Hydrogen and to form acid compounds.

Several of the elements are very little known, such as Indium, Erbium and Terbium: others are of no present importance, as Caesium, Rubidium, Ruthenium, Cerium, Didymium, Lanthanum and Thorinum.

The elements are very unevenly distributed throughout the earth and the air. The most abundant in the earth are Oxygen, Silicon, Aluminum, Iron, Calcium, Magnesium, Sodium and Potassium.

46. The non-metallic elements are 13 in number: five are gaseous —viz., Hydrogen, Oxygen, Nitrogen, Chlorine and Fluorine; one is a liquid, Bromine; the other seven are solids—viz., Carbon, Silicon, Sulphur, Selenium, Phosphorus, Iodine, and Boron.

47. Chemists are in the habit of representing the elements by symbols: thus, the initial letters of the Latin names represent the smallest proportion in which one element will unite with another. When two elements commence with the same initial letters, the one is distinguished from the other by an affix. Thus, S stands for Sulphur; Se for Selenium; Si for Silicon.

48. Each element is supposed to consist of very minute particles which cannot be further sub-divided; these particles are called *atoms*, from the Greek ἀ privative, and τεμνω, *temno*, I cut. The atoms of the various elements possess distinct weights; thus the atom of Oxygen is found to be sixteen times heavier than the atom of Hydrogen. Now, as Hydrogen is accepted as unity, the atomic weight of Hydrogen being 1, that of Oxygen is 16, of Nitrogen 14, of Chlorine 35·5, of Fluorine 19. One gramme of Hydrogen, at the normal temperature and pressure, fills a volume of 11·19 litres. The same space is occupied by 16 grammes of Oxygen, 14 grammes of Nitrogen, 35·5 grammes of Chlorine and 19 grammes of Fluorine.

49. Equal volumes of most of the elementary gases, when compared under the same conditions of temperature and pressure, are supposed to contain the same number of atoms.

50. The smallest proportion of an element found in combination is called its "Atom," whilst the smallest proportion of it in a free state is called its "Molecule." The atom of Hydrogen being $H=1$, its molecule must be HH or $H_2=2$. So also with Oxygen, its molecule must be OO or $O_2=32$.

51. Every compound consists of the smallest cluster of molecules; thus, water or hydrogen oxide, being a compound of two atoms of Hydrogen with one atom of Oxygen, is represented by the symbol $H_2O=18$. The atomic weight of compounds, capable of existing in the form of vapour or gas, occupies the space of two volumes of our standard Hydrogen; consequently, if we desire to know the specific weight of a compound which occupies the space of two volumes, we must halve its atomic weight in order to compare it with $H=1$.

Thus, Hydrogen (1 vol.) unites with Chlorine (1 vol.) to constitute 2 volumes of hydrochloric acid: the symbol of that compound being HCl, and its atomic weight 36·5, its specific weight must be 18·25 when compared with 1 vol. of Hydrogen.

Again: Hydrogen (2 vols.) unites with Oxygen (1 vol.) to form 2 vols. of watery vapour H_2O: the specific weight of 1 vol. of H_2O must be 9.

Further: Hydrogen (3 vols.) unites with Nitrogen (1 vol.) to form 2 vols. of ammonia H_3N; the specific weight of 1 vol. of H_3N must be 8·5.

Lastly: Hydrogen (4 vols.) unites with Carbon (volume unknown, as it cannot be volatilized) to form 2 vols. of Marsh-gas H_4C; when 1 vol. of this gas is weighed, it is found to be 8 times heavier than 1 vol. of Hydrogen.

52. Whenever a Group of elements behaves like an element, it is called a radical, and the same term "atom" is applied to such a group. Thus, Carbon and Nitrogen unite to form Cyanogen, and the symbol for cyanogen expresses the weight of its atom: CN or Cy=26 expresses the atom, C_2N_2=52 represents the molecule. In combination, CN exists with 1 atom of Hydrogen, as prussic acid or hydro-cyanic acid HCN=27: but the volume-weight of its vapour must be half the atomic weight, therefore 13·5.

53. In the chemical notation adopted in this book, the formulæ of compounds, which can exist in the form of gas or vapour, always denote 2 volumes, unless expressly stated to the contrary.

54. The Elements themselves are called Monads, Dyads, Triads, Tetrads, or Hexads, according to the amount of Hydrogen they are capable of uniting with, or the character of the element which they can replace in combination. They are also called monatomic, diatomic, triatomic, tetratomic, and hexatomic elements.

The Table on the annexed page will be serviceable.

TABLE OF ATOMIC WEIGHTS.

Monads or Monatomic elements.

Hydrogen	H = 1	Bromine	Br = 80	Fluorine	F = 19
Chlorine	Cl = 35·5	Iodine	I = 127	Kalium	K = 39
Natrium	Na = 23	Rubidium	Rb = 85·5	Argentum	Ag = 108
Caesium	Cs = 133	Lithium	L = 7	Thallium	Tl = 204

Dyads or Diatomic elements.

Oxygen	O = 16	Ytrrium	Y = 68	Ferrum	Fe = 56
Sulphur	S = 32	Lanthanum	La = 92	Cobalt	Co = 59
Selenium	Se = 79·5	Cerium	Ce = 92	Nickel	Ni = 59
Tellurium	Te = 129	Didymium	D = 96	Zinc	Zn = 65
Barium	Ba = 137	Uranium	U = 120	Cadmium	Cd = 112
Strontium	Sr = 87·5	Palladium	Pd = 106·5	Cuprum	Cu = 63·5
Calcium	Ca = 40	Rhodium	R = 104·3	Plumbum	Pb = 207
Magnesium	Mg = 24	Manganesium	Mn = 55	Hydrargyrum	Hg = 200

Triads or Triatomic elements.

Nitrogen	N = 14	Boron	B = 11	Aluminum	Al = 27·5
Phosphorus	P = 31	Glucinum	G = 9·3	Bismuth	Bi = 210
Arsenic	As = 75	Chromium	Cr = 52·5	Aurum	Au = 196·6
Stibium	Sb = 122				

Tetrads or Tetratomic elements.

Carbon	C = 12	Zirconium	Zr = 90	Platinum	Pt = 197.5
Silicon	Si = 28	Thorinum	Th = 115·7	Iridium	Ir = 198
Titanium	Ti = 50	Niobium	Nb = 97·5	Osmium	Os = 199
Stannum	Sn = 118	Ruthenium	Ru = 104	Tantalum	Ta = 137·6

Hexads or Hexatomic elements.

Molybdenum	Mo = 96	Tungsten	W = 184	Vanadium	V = 137

CHAPTER IV.

ON THE CHEMISTRY OF FOUR TYPICAL ELEMENTS, HYDROGEN, OXYGEN, CARBON AND NITROGEN.

55. Hydrogen and its properties. 56. Monad element. 57. Found in combination. 58. Modes of preparation. 59. Oxygen. 60. Characters of Oxygen. 61. Preparation of Oxygen. 62. Basic Oxides and Acids. 63. Ozone. 64. Water. 65. Hydrogen Peroxide. 66. Carbon. 67. A tetrad element. 68. Compounds with Oxygen. 69. Carbonic Oxide. 70. Carbonic Anhydride. 71. Compounds of Carbon with Hydrogen. 72. Acetylene. 73. Methyl-hydride or Marsh Gas. 74. Ethylene. 75. Nitrogen. 76. Occurrence and Properties. 77. Ammonia. 78. Preparation and Properties. 79. Salts of Ammonium. 80. Laws of chemical combination. Law of definite proportions. Law of multiple proportions. Law of equivalent proportions. 81. Nitric Acid. 82. Nitrous Acid. 83. Nitrous Oxide. 84. Nitric oxide. 85. Nitric Peroxide. 86. Cyanogen. 87. Hydrocyanic or Prussic Acid. Cyanides.

I. Hydrogen H = 1.

55. HYDROGEN is a colourless gas, tasteless and, when pure, inodorous; it has never been liquefied. Cavendish discovered Hydrogen in 1766, and he prepared it from water. Owing to its lightness, hydrogen has frequently been employed for filling balloons. In water it is but little soluble, to the extent of only two per cent., but it is readily absorbed by various metals, especially Palladium. Although not poisonous, hydrogen cannot be breathed for more than two or three seconds; and, even then, care should be taken thoroughly to purify it. Hydrogen is inflammable in air, and burns at the expense

of the free Oxygen contained therein: the result of the burning is nothing but water, H_2O.

56. Hydrogen is a monad element, which is generally accepted by chemists as the standard of atomic weights, and of combining volume. One gramme weight (15·432 grains), at 0° C. and 760 mm. Bar., fills a space of 11·19 litres. Its atomic weight being = 1, its molecular weight = 2. Although considered as a non-metallic Element, in its behaviour Hydrogen comports itself as a metal.

57. Hydrogen always occurs in combination; not only as water, which is an oxide, but also as a constituent of nearly all vegetable and animal bodies. It derives its name from its leading character; it comes from the Greek ὕδωρ, hydor, water, and γεννάω, gennao, I generate.

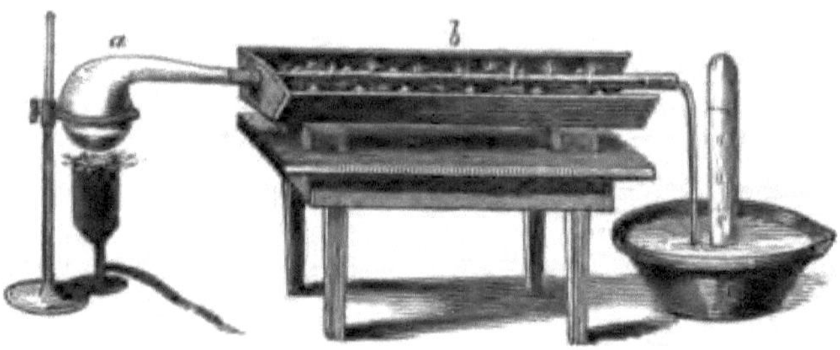

Fig. 25.

58. Hydrogen is most readily prepared from water, and by a variety of processes. If water be boiled in the retort *a*, Fig. 25, and its steam be passed over red-hot iron filings contained in the gun-barrel *b*, large quantities of hydrogen may be collected in the tube inverted over water. Water in this case gives up its oxygen to the iron, which becomes oxydized, and thus its hydrogen is set free. $3\ Fe + 4\ H_2O = Fe_3O_4 + 4\ H_2$.

Fig. 26.

The composition of water may, indeed, be at once determined. If the two electrodes of a galvanic battery terminating in platinum ends be plunged into water (acidulated, so as to render it a better conductor of electricity), bubbles of gas are seen to arise at each ·ctrode. If these gases be collected by placing tubes filled with water (Fig. 26) on the terminals of each electrode, twice as much gas will be found in one tube as in the other. At the positive (+) pole or platinode, Oxygen is collected; at the negative (—) pole or zincode, Hydrogen; and the latter is double the quantity of the former.

This experiment proves, that water is a compound of two volumes of Hydrogen with one volume of Oxygen; and if the two gases are mixed together in these proportions, and ignited, they unite with tremendous detonation, and produce nothing but Water. This combination may be readily effected in the so-called Eudiometer, (from εὔδιος, eudios, good, and μετρον, metron, a measure, because a supposed measurer of the goodness of air from the quantity of oxygen which it contains.) Fig. 27 represents the experiment in process of execution. The graduated tube standing in the mercurial trough, is supposed to be filled with the two gases in the proportions named. Two pieces of platinum wire, the ends of which do not quite touch, are melted into the upper end of the tube. When these are connected with the outer and the inner coating of a charged Leyden jar, a spark passes between them, the gases unite in consequence of the heat generated, the watery vapour condenses, and the mercury, pressed upon by the air, rises and fills the vacuum so formed.

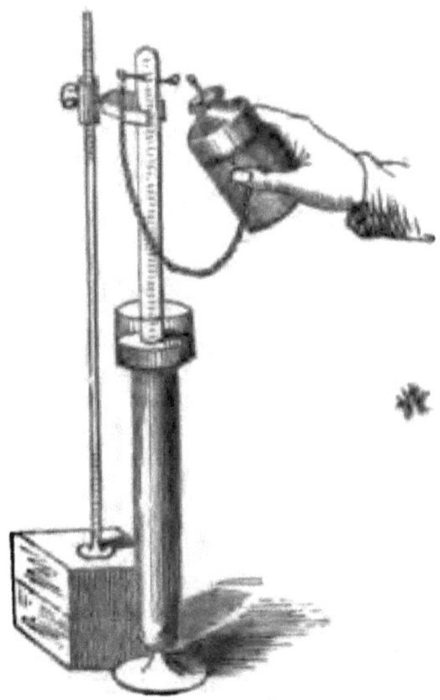
Fig. 27.

Very easily indeed is Hydrogen prepared from water, H_2O, by the action of the metal Sodium or Natrium Na. Hydrogen is set free, whilst the water is found to contain Sodium hydrate in solution.

$Na_2 + 2\ (H_2O) = 2\ (Na\,HO) + H_2$.

The mode of proceeding is well illustrated in Fig. 28. The Sodium, which has been fastened to the

Fig. 28.

wire held in the hand, rises, by its lightness, to the surface of the water contained in the test-tube. The hydrogen, from the decomposed water, collects at *b*, and speedily fills the tube, whilst the bulk of the water collects in the basin, *a*. Potassium, or Kalium K may be employed instead; but the heat evolved is so great, that the hydrogen is liable to be set on fire. The result is similar: $K_2 + 2(H_2O) = 2(KHO) + H_2$.

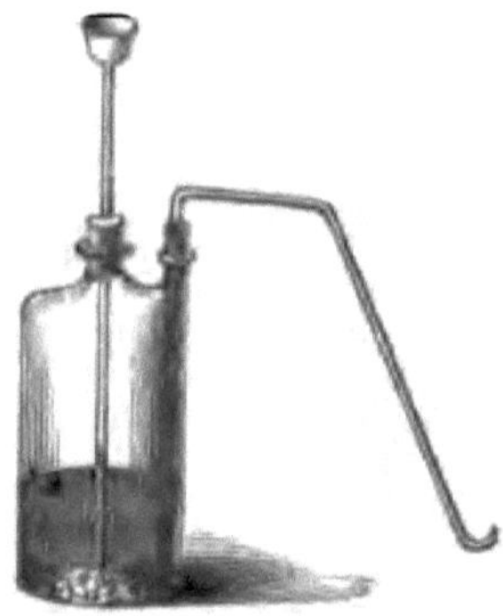

Fig. 29.

Hydrogen is, however, much more commonly prepared from water, with the aid of Zinc and sulphuric acid. The usual arrangement is such as is depicted in Fig. 29. The chemical action may be explained by the following statement: the zinc simply displaces the Hydrogen. Zinc is a Dyad element, and displaces 2 atoms, or the molecule of Hydrogen; Zinc sulphate, $ZnSO_4$ is found in solution. $Zn + H_2SO_4 = ZnSO_4 + H_2$. The gas thus evolved is by no means pure.

On account of its lightness, Hydrogen may be readily collected by displacement. (*See* Ammonia.)

II. Oxygen $O = 16$.

59. Oxygen was discovered by Priestley, in 1774. It is a colourless, tasteless and inodorous gas, a great supporter of combustion and uninflammable. It is magnetic; the diurnal variations of the compass needle are occasioned by the weakening and strengthening of its magnetism through the heating of the atmosphere and subsequent cooling.

Oxygen is essential to respiration; hence it is called "Vital air." It is also necessary to decay and various other processes. In water it is somewhat more soluble than Hydrogen; under favourable circumstances, water dissolves 3 per cent. at 15° C.

60. Oxygen is a Dyad element and takes its name from ὀξὺς, oxys, acid, and γεννάω, gennao, I generate, because it was supposed by

Lavoisier to be the great acidifying principle. It is sixteen times heavier than Hydrogen. 11·19 litres of Oxygen weigh 16 grammes at 0° C. and 760 mm. Bar. As the atomic weight of Oxygen $O=16$, its molecular weight is $O_2=32$.

61. Oxygen occurs both in the free state and in combination; indeed, it is the most abundant of all the elements. In the atmosphere Oxygen is found in the free state, constituting as nearly as possible 21 per cent. by volume or measure of dry air. No less than $\frac{8}{9}$ of the weight of water consists of oxygen, but here it is in combination. It is a component of nearly all animal and vegetable bodies, and the vast majority of the various minerals of which the earth is constructed.

Although so abundant, Oxygen is difficult to obtain from the air. This is owing to its powerful affinities. It is sufficiently easy to withdraw it from the air, but then it enters into combination with the substance employed.

For purposes of experiment, Oxygen is best prepared from such of its compounds as part with it most readily, or, in other words, have least affinity for it. Thus, all the oxides of the noble metals, Silver, Gold, Mercury, and Platinum, might be employed. The least expensive of these is best adapted, and so, when Mercuric oxide HgO, is heated strongly in a tube, as represented in Fig. 30, it resolves itself into Mercury and Oxygen: $2HgO=Hg_2+O_2$.

Fig. 30.

There are also certain oxides called Peroxides, all of which give off a portion of their Oxygen when very strongly heated, either by themselves, or with Sulphuric acid. The commonest peroxide is that of the metal Manganesium, which is found as an abundant mineral called Pyrolusite. When Manganesium peroxide MnO_2 is strongly heated in an iron retort, Manganous-Manganic Oxide, Mn_3O_4 remains, whilst Oxygen is set free. $3(MnO_2)=Mn_3O_4+O_2$.

If Manganesium peroxide be gently heated with Sulphuric acid, Manganous sulphate, $MnSO_4$, together with Water, H_2O, is formed, and Oxygen is set free. $2(MnO_2)+2(H_2SO_4)=2(MnSO_4)+2(H_2O)+O_2$. Barium peroxide, BaO_2, may also be employed; by heat it breaks up into Barium oxide, BaO and Oxygen. $2(BaO_2)=2(BaO)+O_2$.

There is, however, a salt of Potassium, which contains Oxygen so loosely, that soon after it has melted (420° C.), the whole of its Oxygen is set free. Potassium chlorate, $KClO_3$, becomes changed into Potassium chloride KCl, with evolution of all its Oxygen. $2(KClO_3)=2(KCl)+3O_2$. The experiment may be conducted as in Fig. 30. (As a compound is made up of the sum of its components, the atomic weight of Potassium chlorate must be 122·5. So that 122·5 grammes of the salt would furnish 48 grammes of Oxygen. And as 16 grammes of Oxygen are equal to 11·19 litres at 0° C. and 760 mm. Bar., it follows that 122·15 grammes of Potassium chlorate, will supply 3×16 grammes=33·57 litres of Oxygen).

If Manganese peroxide be mixed with four times its weight of Potassium chlorate, although the former undergoes no change, the latter gives up its Oxygen far more readily. It does so because the heat is concentrated, by separation from the mass of Potassium chlorate, and so liberates the Oxygen from each particle of the salt. Otherwise, not a portion of Oxygen can be separated, until the whole of the salt is melted.

The arrangement for making and collecting this gas, as well as most others, is represented in Fig. 31. The so-called Pneumatic trough, consists of a vessel of water made of japanned sheet-iron, and fitted with shelves, upon which stand the jars (filled with water and inverted) intended for the reception of the gas. The flask, containing the mixture from which the gas is generated, is fitted with a gas-delivery tube,

Fig. 31.

and placed upon a retort-stand: it is heated by means of a spirit-lamp placed under it. Florence flasks are most convenient.

Gases are readily transferred from jar to jar, by first filling the jar, intended to contain the gas, with water, and then bringing beneath it the aperture of the jar from which the gas is to be transferred. On gently inclining the latter, as in Fig. 32, the gas rises into the other jar and displaces the water.

Fig. 32.

That Oxygen may be obtained from the electrolysis (from ἤλεκτρον, electron, amber, in which the properties of electricity were first discovered, and λύσις, lysis, releasing,) of water has been already mentioned: the most electro-negative of the two elements, it separates at the platinode or + pole. Other methods of preparing Oxygen will be mentioned later.

62. It has been already stated that Oxygen is possessed of wide affinities: consequently there are many varieties among oxydized bodies. The metals unite with Oxygen and form "bases," or "basic oxides"; when soluble in water, they change many vegetable reds into blues, and restore the blue colours to reddened litmus paper. When a metal unites in only one proportion with oxygen, in its salts we speak simply of the metal. Thus: Silver nitrate; Lead nitrate; Potassium chlorate; because each of these metals has but one basic oxide. But, if a metal unites with oxygen in two proportions, to form two basic oxides, we distinguish them by the affixes, "ous" and "ic." Thus: Ferrous oxide and Ferric oxide; and, in combination in salts, Ferrous sulphate and Ferric sulphate.

63. Ozone $O_2O=48$ is a peculiar modification of Oxygen to which the term "allotropic" (ἄλλος, allos, another, and τρόπος, tropos, manner), has been given. By the term we understand the endowment of the same substance with different properties; or different

aspects of the same substance. It is called Ozone from ὄζω, ozo, I emit an odour). Although it has never been completely isolated, ozone is found possessed of wonderful oxydizing properties: it therefore both bleaches and disinfects. It is readily absorbed by oil of turpentine, and thus its density has been shown to be 24 times that of Hydrogen. When heated to 290° C., it is changed into common Oxygen gas. Ozone may be prepared by several methods. Mr. Siemens prepares it from Oxygen, by electric induction. Also, when sparks from an electrical machine pass into the air, ozone is produced and recognised by its odour. And, it may be made, by placing Phosphorus in moist Oxygen, or in moist air for a short time.

One of the best tests for Ozone is, that, whilst it liberates Iodine from Potassium iodide KI, and thus blues starch-papers containing this salt, this paper is found alkaline after the reaction, or with the property of turning red litmus paper blue. $6(KI)+3(H_2O)+O_3=6(KHO)+3I_2$.

64. Oxygen unites with Hydrogen in two proportions, to form: WATER OR HYDROGEN OXIDE H_2O, and HYDROGEN PEROXIDE H_2O_2.

WATER $H_2O=18$. When viewed in mass, water is blue; but in small quantities, it presents the appearance of a colourless liquid. Many of its properties have been already described. Under a pressure of 760 mm. Bar., water boils at 100° C. and freezes at 0°. Its point of maximum density is 4° C. The latent heat of steam at 100° C. is 537. One kilogramme of steam at 100° C would raise 5·37 kilogs. of water, from 0° C. to 100° C., or 537 kilogs. through 1°. Water is a great solvent of liquids and gases. Most of the oxides of the non-metallic elements when brought into water become changed into "acids," with the power, when soluble in water, of turning vegetable blues into reds. In a state of combination, water is found in many salts as water of crystallization: as in Zinc sulphate, $ZnSO_4.7H_2O$.

To demonstrate that the combustion of Hydrogen in air, is due solely to its combination with Oxygen, and not to the presence of atmospheric vapour, the gas should be previously dried. The ordinary apparatus for making hydrogen should then be arranged as in Fig. 33, in which the gas is dried by passing through a glass tube, *b*, filled with pumice-stone, soaked in sulphuric acid. By holding a bell-jar slightly inclined above the flame of hydrogen at *c*, the water formed by its combination with the oxygen of the air, trickles down the sides of the glass and collects in the plate.

Fig. 33.

On a small scale, the condensation of vapours, by distillation, is accomplished by a Liebig's condenser, Fig. 34. A glass tube, *b*, is placed in the axis of the metallic tube, *c*, and kept in its place by

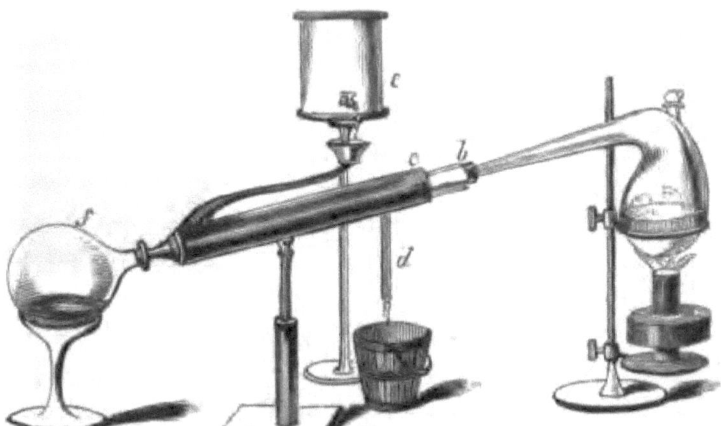

Fig. 34.

well-fitting perforated corks. The space between the two tubes is kept filled with cold water, entering from the cistern, *e*, and running off at *d*. The neck of the retort fits into the glass tube, *b*, and its receiver, *f*, receives the distilled water.

Such an arrangement is admirably adapted for distilling acids.

65. HYDROGEN PEROXIDE H_2O_2. A colourless, syrupy liquid, of sp. gr. 1·543. At 20° it begins to evolve Oxygen and then changes rapidly into water. It is a powerful oxydizing agent. Lead sulphide PbS, becomes Lead sulphate $PbSO_4$; Silver oxide Ag_2O is reduced to silver, with liberation of much oxygen. It also bleaches and disinfects.

Hydrogen peroxide is Dr. Frankland's so-called Hydroxyl. It is best prepared from Barium peroxide BaO_2 by Carbonic anhydride CO_2, and water H_2O: Barium carbonate $BaCO_3$ and Hydrogen peroxide H_2O_2 result. $BaO_2+H_2O+CO_2=BaCO_3+H_2O_2$.

III. Carbon C = 12.

66. Carbon, in its purest natural form, is known as the Diamond. This highly-prized gem, when found, has the appearance of a rounded pebble, enclosed in a thin, opaque crust; when freed from this coating, it is generally colourless. It crystallizes in regular forms, all allied to, or derived from, either the cube, Fig. 35, or the octohedron, Fig. 36, which therefore are called the typical forms of its crystallization. The diamond is the hardest known body; its density is from 3·5 to 3·55. It can only be cut by its own dust. When very strongly heated in absence of air, it becomes transformed into a substance like coke.

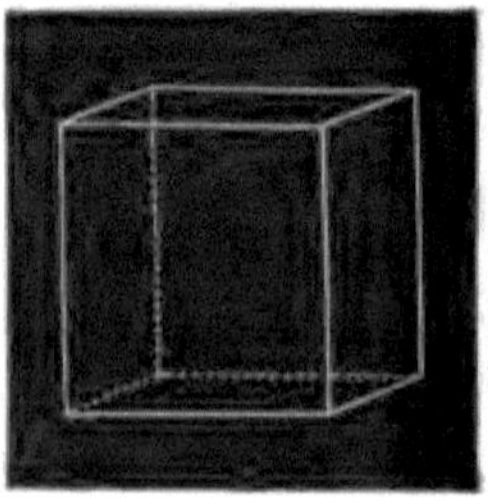

Fig. 35.

Fig. 36.

Carbon, like Oxygen, is allotropic (Par. 63). It occurs also as Graphite or Plumbago in the clay-slate at Borrowdale, in Cumberland; in Siberia, Ceylon, and other localities. It crystallizes in six-sided plates, and has a density of 2·2 to 2·35. Like its confrère, graphite is very hard and rapidly wears out the saws employed in cutting it. To the touch it is unctuous; it possesses a lustrous appearance and readily leaves a mark upon paper or the finger. It is quite unaffected by exposure to air, hence its use in "black-leading" iron. Both graphite and diamond burn in oxygen without flame and the result of the combustion is Carbonic anhydride CO_2. Unlike the diamond, Graphite is a conductor of electricity.

A third variety of Carbon is Lamp-black. It is entirely destitute of crystalline structure. It was formerly obtained by collecting the smoke or soot of ill-trimmed lamps. The mode of preparation now resorted to is interesting, as giving an insight into the Chemistry of vegetable matter, when submitted to the process of what is called destructive distillation. When resin is heated in an iron vessel, *a*, Fig. 37, and its vapours burnt in air, insufficient in quantity to produce complete combustion, lamp-black in a finely-divided state is produced in abundance. (The resin consists of Carbon, Hydrogen, and Oxygen. Hydrogen being most inflammable, and having a greater affinity for oxygen than has carbon, burns first; and as there is not sufficient oxygen to burn the carbon also, the latter is deposited in the solid form.) This is made to traverse a large chamber, *b*, furnished with a hood, *d*, on which the lamp-black collects. After being heated to redness in close vessels, from which the air is excluded, it becomes nearly pure carbon. Its chief use is in the manufacture of printers' ink.

Fig. 37.

Another variety of carbon, much used on the continent of Europe as fuel, is Charcoal. It is a black, brittle, and infusible substance, of well-known appearance. As it undergoes no change from exposure to air or moisture, it is a frequent custom to char the ends of posts and piles, with a view to their preservation. Thus, most of the houses in Venice are built on piles, all of which are deeply charred, and they have thus been able to resist the influence of centuries of exposure.

The quality of charcoal depends much upon the physical character of the wood from which it is prepared. Ebony affords a very hard, fir-wood a soft, and the vine the softest kind of charcoal; the last-named is highly esteemed by artists for sketching purposes, as lines made by it may be readily removed. For the manufacture of gunpowder, the willow, elder, and dog-wood are chiefly used, as they afford a very light, porous description of charcoal.

Charcoal is sometimes made by burning the wood in iron cylinders, or retorts, instead of in the open air, and is then called cylinder charcoal. But common charcoal is prepared by charring or carbonizing branches or trunks of trees. The following is an outline of the process. Logs of wood, not more than 6 feet long and 6 inches thick, are laid either horizontally or vertically, while stem and branches are employed to fill up the interstices. The whole is covered with, from three to five inches of earth, or, still better, with a mixture of earth and fine charcoal over a layer of leaves and small brushwood, and kept moistened with water. The heap is ignited by coals thrown into the chimney, when the fire draws to the sides, towards small openings, left around the base of the heap. A heavy, yellowish-grey smoke, and much watery vapour first appears, which condenses on the outer covering. During this sweating process, which lasts about sixteen hours, the fire should be brisk, to prevent explosions, and the heap carefully sweated off. The general shrinking of the wood opens cracks, when the coalman mounts the heap, rams the wood together, and replaces the covering. When the heap is fairly warmed, and no further explosions are to be feared, the openings are closed, and the heap suffered to burn from three to four days, while the collier only watches to secure equal combustion in all parts. A few openings are now and then made for the escape of the tarry matter, and a few others at the foot; and, after from four to eight days, others half-way up the heap, to clear the outside logs. If a blue flame rises, which shows that carbonic oxide is in process of formation, the openings are stopped and made lower down. When the fire gradually breaks out uniformly around the base, the charring is complete. The whole time required is from six days to four weeks. A heap of 3000 cubic feet requires about fifteen days. Fig. 38 affords an illustration of the arrangement most usual in France.

As no vegetable matter can be formed without the aid of certain mineral constituents, which are wholly unalterable by fire, it is clear, that charcoal (independently of the hydrogen which it always retains to some extent) is not carbon in a pure form. The purest charcoal leaves an ash when burnt with full access of air.

Fig. 38.

Coke is the carbon obtained from burning coal, by a kind of smothered combustion. It is an impure kind of Graphite, containing, in addition to Carbon, all the ash of coal. It is a good conductor of heat and of electricity, and is employed, among other purposes, in the construction of Bunsen's batteries.

Animal black is made by calcining bones in closed retorts: it is of course rich in Calcium phosphate and carbonate, both of which are removable by washing with hydrochloric acid. It is not as good a disinfectant as charcoal, as it does not so readily absorb oxygen, but it is far superior as a decolourizer.

67. Carbon is an element, which behaves generally as a Tetrad, but also as a Dyad. Its atomic weight is 12. Inasmuch as it has not been volatilized so as to contrast its vapour-volume with Hydrogen, its molecular weight is unknown. It is infusible and insoluble in all media. One common characteristic of all varieties of Carbon is, that when sufficiently heated in excess of Oxygen, they burn to Carbonic anhydride CO_2.

68. Carbon unites with Oxygen in several proportions. The two commonest compounds are: Carbonic oxide CO, and Carbonic anhydride CO_2.

69. Carbonic Oxide $CO=28$. A colourless, tasteless and inodorous gas, inflammable in air with blue flame, burning to Carbonic anhydride. It is but little soluble in water. Compared with Hydrogen, its density is 14; in other words, one volume of Carbonic oxide

is 14 times heavier than one volume of Hydrogen. 11·19 litres weigh 14 grammes. For complete combustion, two volumes require one volume of Oxygen, the result being two volumes of carbonic anhydride.

Carbonic oxide is always formed in the combustion of carbonaceous substances with a limited quantity of air. It may be made by passing Carbonic anhydride CO_2 over red-hot charcoal, or over red-hot iron. $CO_2+C=2CO$. Or: $3Fe+4CO_2=F_3O_4+4CO$. But it is most readily prepared by heating Oxalic acid $H_2C_2O_4.2H_2O$ with Sulphuric acid H_2SO_4. The arrangement of apparatus is depicted in Fig. 39. The first wash-bottle is supposed to contain water, the second a solu-

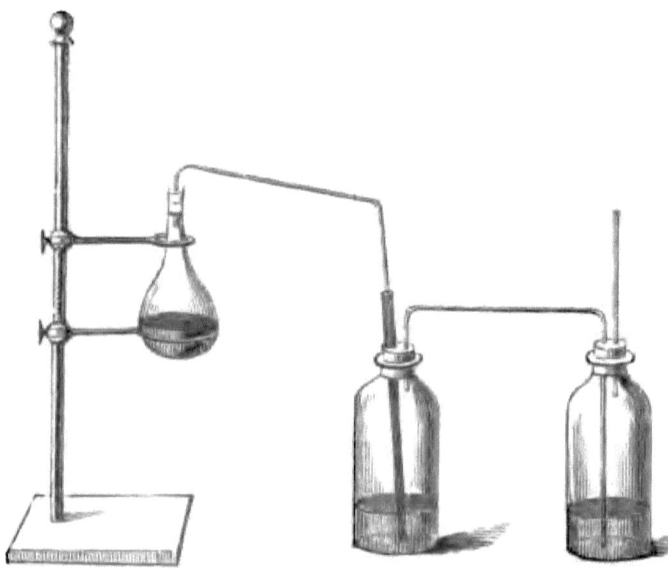

Fig. 39.

tion of Potassium hydrate. When Oxalic acid is thus heated, it breaks up into Water H_2O which unites with the sulphuric acid, and a mixture of Carbonic oxide CO, and Carbonic anhydride CO_2. By passing the mixed gases through Potassium hydrate, the Carbonic anhydride is absorbed.

70. Carbonic Anhydride $CO_2=44$. A colorless gas, of acidulous taste and odour. Not only does it not support combustion, but it furnishes one of the simplest means of extinguishing flame. It is

also irrespirable; the "Choke-damp" of miners is Carbonic anhydride. It is a very heavy gas; 22 times heavier than Hydrogen. 11·19 litres weigh 22 grammes. In water it is very soluble, especially if the pressure be increased. One volume of water dissolves its own volume of the gas under ordinary atmospheric pressure, and at 15° C and this quantity increases regularly for each additional atmosphere. By a pressure of 36 atmospheres, CO_2 becomes liquid, and when allowed to evaporate, a part immediately becomes solid. When solid carbonic anhydride is mixed with ether and placed under the receiver of an air-pump, the temperature falls to —100° C. Carbonic anhydride is contained in the atmosphere to the extent of 1 part in 2,500 parts of air. It is a product of respiration, decay, the combustion of carbonaceous matters, germination, fermentation and putrefaction.

Fig. 40.

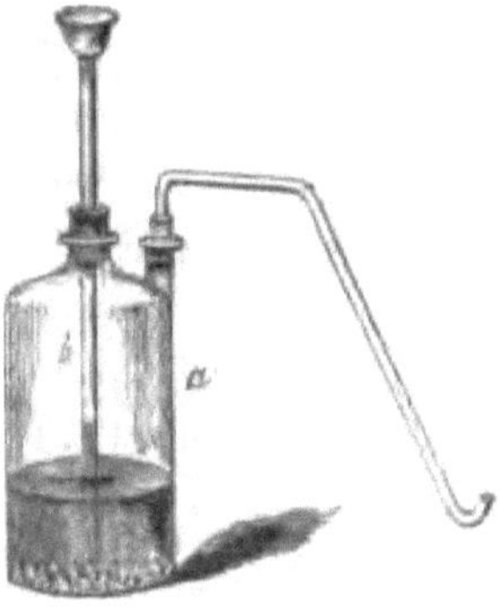

Fig. 41.

It is prepared by burning Charcoal in Oxygen: when heated in Oxygen, charcoal burns brilliantly and throws out white-hot sparks of ignited Carbon, Fig. 40. But the most easy method of obtaining Carbonic anhydride, is by the action of acids upon metallic carbonates. When marble (Calcium carbonate $CaCO_3$) is placed in a Wolfe's bottle, Fig. 41, and Hydrochloric acid HCl, poured upon it through the thistle-funnel *b*, the gas escapes through the gas-delivery tube and may be collected over water. $CaCO_3 + 2HCl = CaCl_2 + H_2O + CO_2$.

Fig. 42.

Carbonic anhydride may be easily proved to contain Carbon. If passed over the metal Potassium contained in the tube at *x*,

Fig. 42, and heat applied, the metal burns at the expense of the Oxygen of the CO_2, the Carbon being separated, $CO_2+2K_2=2K_2O+C$. The object of the wash-bottle *a*, is to dry the gas, by passing it through oil of vitriol.

Carbonic anhydride dissolves, but does not combine with water: Carbonic acid H_2CO_3 is hypothetical. It is di-basic. In its neutral salts, the two atoms of Hydrogen are replaced by two atoms of a Monad metal or by one atom of a Dyad; in acid carbonates, one atom of Hydrogen for the most part remains, or an additional atom of the anhydride is found in the salt. Sodium carbonate $Na_2CO_3.10H_2O$ contains 10 atoms of water of crystallization; this is the neutral carbonate. Sodium Hydrogen carbonate $NaHCO_3$ is the acid salt, or Sodium bi-carbonate. Carbonic anhydride contains its own volume of Oxygen. 100 parts consist of 72·73 parts of Oxygen and 27·27 parts of Carbon.

71. Carbon unites with Hydrogen in many proportions. Only three can be considered at this early stage. Acetylene C_2H_2, Light Carbonetted Hydrogen CH_4, and Ethylene C_2H_4: all three are gaseous.

72. Acetylene $C_2H_2=26$. Is a colourless gas, of peculiar odour, burning with a sooty flame. 11·19 litres weigh 13 grammes. It may be prepared by the direct union of the two elements at very high temperatures, and it is always produced by the incomplete combustion of Ethylene and of Light carbonetted hydrogen. Thus: $2C_2H_4+O_2=2H_2O+2C_2H_2$. Again: $4CH_4+3O_2=6H_2O+2C_2H_2$.

73. Light Carbonetted Hydrogen, Marsh Gas, or Methyl Hydride $CH_4=16$. A colourless, tasteless, inodorous gas. It constitutes the "fire-damp" or "wild-fire" of the coal mine. It is evolved in the decay of moist vegetable matter, and is consequently an occasional but frequent constituent of the air. This impure marsh-gas may be obtained by stirring the mud at the bottom of stagnant pools (whence it has the name of marsh gas), and collecting it as it escapes, in an inverted jar full of water (Fig. 43). When strongly heated in air the gas burns with pale flame to Carbonic anhydride and Water. Two

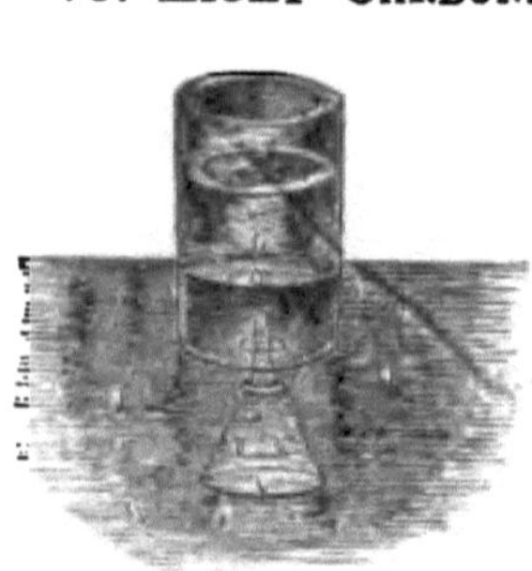

Fig. 43.

volumes of Methyl hydride CH_4 contain four volumes of Hydrogen and as much Carbon as two volumes of carbonic anhydride. For complete combustion, One volume requires two volumes of Oxygen or ten volumes of air. 11·19 litres of the gas weigh 8 grammes; its density is therefore 8 compared with Hydrogen. Methyl hydride is the most abundant constituent of coal-gas.

In a pure state, it is most easily prepared by heating Sodium acetate in a retort with Sodium hydrate. Thus :—$NaC_2H_3O_2+Na HO=Na_2CO_3+CH_4$. In Fig. 44 the wash-bottle is supposed to contain water.

Fig. 44.

74. Heavy Carbonetted Hydrogen or Ethylene $C_2H_4=28$. This is also a colourless gas, burning in air with a white flame to Carbonic anhydride and Water. 11·19 litres weigh 14 grammes. When mixed with three times its volume of Oxygen, or with fifteen volumes of air, and inflamed, it detonates. It is contained in coal-gas, and owes its luminosity to the fact of its containing twice as much Carbon as methyl hydride.

Ethylene is most readily prepared from Ethylic alcohol C_2H_6O (Spirit of wine), by heating it with six parts of sulphuric acid. Water separates and remains combined with the acid, whilst the gas is evolved. Thus: $C_2H_6O=H_2O+C_2H_4$.

IV. Nitrogen. N. = 14.

75. Nitrogen is a colourless, tasteless, and inodorous gas. It is not inflammable, neither does it support combustion, nor produce any effect upon lime-water. In water it is about as soluble, or as little

soluble as hydrogen. Although Nitrogen is not poisonous, it cannot be inhaled by itself. Nitrogen is a Triad element in its combination Ammonia NH_3, and a Pentad element in Sal-ammoniac or Ammonium chloride NH_4Cl. 14 grammes of Nitrogen fill the space of 11·19 litres. Its density is therefore 14.

76. Nitrogen occurs in an uncombined state in the atmosphere; therein it amounts to 21 per cent. by measure, after the removal of the watery vapour. From the very negative character of nitrogen, it remains behind after the abstraction of oxygen. Thus: Nitrogen may be prepared by burning Phosphorus in a bell-jar full of air (p. 78, Fig. 48). Phosphoric anhydride P_2O_5 results, which after a time unites with water, and dissolves as Meta-phosphoric acid HPO_3; Nitrogen remains. By a yet better method, Nitrogen is prepared by passing air over red-hot copper-turnings contained in a tube: the copper abstracts all the oxygen and remains in the tube in the form of Cupric oxide CuO. Nitrogen may also be obtained by decomposing Ammonia NH_3 through the agency of Chlorine gas: Hydro-chloric acid HCl is formed, which dissolves in water, and Nitrogen is set free. Thus: $2(NH_3) + 3Cl_2 = 6HCl + N_2$. Lastly, Nitrogen is obtained by heating Ammonium nitrite NH_4NO_2; it breaks up into: $2H_2O + N_2$.

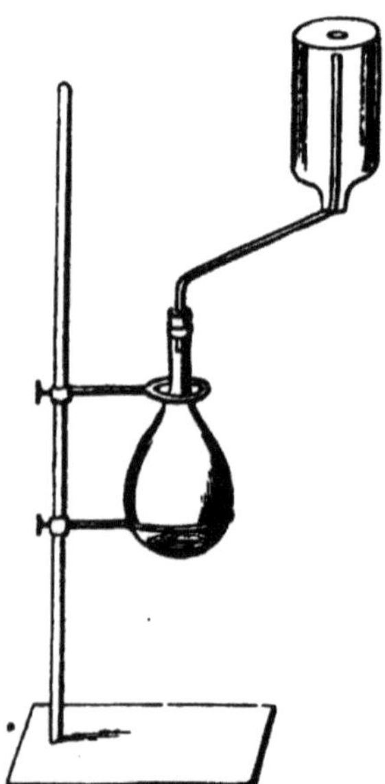

Fig. 45.

77. Nitrogen unites with Hydrogen in various proportions, the chief compound being: AMMONIA NH_3. It is a colourless gas, of highly pungent taste and odour as of Hartshorn. When pure, it is quite irrespirable: diluted, it acts as an admirable stimulant. It does not support combustion, but is itself somewhat inflammable in air or in Oxygen. It blues red litmus paper and possesses an alkaline* reaction. From its gaseous nature it is volatile, and is called

* Alkaline was the name given to Potash or Potassium hydrate (Kalium hydrate) to denote the superiority of the substance obtained, over the plant yielding it. The word is derived from the Arabic, *al*, the, and *kali*, the name of the plant. Potassium hydrate is a fixed alkali.

a Volatile Alkali. The solubility of Ammonia in water is most remarkable; at 15° C. water absorbs 727 volumes. By heating a solution of Liquor Ammoniae in a flask, Fig. 45, the gas is easily prepared for experiment: on account of its lightness (0·59 as compared with air) the gas may be collected by displacement. When a jar filled with the ammoniacal gas is plunged into water, the latter rises and fills it. In case the jar should not be made of stout glass, it is more safe to cover it with a cloth as it might be broken by the violence of the rush of water, (Fig. 46). Ammonia is a product of respiration, decay, putrefaction, and combustion of organic bodies containing Nitrogen and Hydrogen. It owes its name to Sal-ammoniac, a salt obtained by the Romans from Ammonia, a district in Lybia, in Africa, where an Egyptian God, Amun, was worshipped.

Fig. 46.

78. The density of the gas is 8·5, as 11·19 litres weigh 8·5 grammes. Two volumes of the gas may be detonated by admixture with 1½ volumes of Oxygen; the Nitrogen does not burn. When cooled down to—39° C, Ammonia condenses into a colourless liquid, of density 0·76, and becomes solid at—90° C.

Ammonia NH_3, is best prepared by heating Ammonium chloride NH_4Cl, with Calcium oxide or Quicklime CaO; Calcium chloride $CaCl_2$ with Water remains behind in the retort, whilst the gas passes over. Thus: $2(NH_4Cl) + CaO = CaCl_2 + H_2O + 2(NH_3)$.

A solution of ammonia in water behaves very much like a solution of Potassium hydrate: it is not only possessed of similar alkaline properties, but must be considered as a solution of Ammonium hydrate. Thus $NH_3 + H_2O = NH_4HO$.

79. Although AMMONIUM NH_4 is only known in combination, chemists regard it as a compound metal, which comports itself much like the alkaline metals Kalium and Natrium. The salts of ammonium are numerous. Ammonia is known to fume when brought together with hydrochloric acid, because Ammonium chloride NH_4Cl results.

AMMONIUM SESQUI-CARBONATE $2[(NH_4)_2CO_3]CO_2$ is the common carbonate or the smelling-salts of the shops. It forms a fibrous, semi-transparent mass, of strong, pungent odour as of hartshorn. It is made by heating Ammonium chloride with Calcium carbonate. Thus: $6(NH_4Cl)+3CaCO_3=3CaCl_2+H_2O+2NH_3+2[(NH_4)_2CO_3]$ CO_2. By exposure to air, it loses Carbonic ammonide $(NH_3)_2CO_2$ and becomes changed into a white porous mass of Ammonium hydrocarbonate NH_4HCO_3.

80. The combinations of NITROGEN with OXYGEN are both numerous and important. They offer a favourable opportunity for discussing in a few words the LAWS OF CHEMICAL COMBINATION.

These Laws are three in number. The first law is that of *Definite proportions*: in every compound, the nature and the proportions of the different elements are fixed, and invariable. Thus water is always a compound of two atoms of hydrogen with 1 atom of oxygen; its formula is H_2O.

The second law is the law of *Multiple proportions.* As is the case with Nitrogen and Oxygen, it frequently happens that the same elements unite in more than one proportion. When two elements N and O unite in more than one proportion, if we compare together quantities of each of the resulting compounds which contains the same amount of N, the quantities of O will bear a very simple relation to each other. Thus: N_2O; N_2O_2, N_2O_3; N_2O_4; N_2O_5. Again: H_2O and H_2O_2. Again: CO and CO_2.

The third law is known as the law of *Equivalent proportions.* Each element, in combining with other elements, or in displacing others from their combinations, does so in a fixed proportion which may be represented numerically. Thus, in 18 grammes of water, 39 grammes of Kalium or Potassium can displace one gramme of Hydrogen. In making Hydrogen by the action of Zinc upon Sulphuric acid, [redacted] grammes of Zinc take the place of 2 grammes of Hydrogen in the acid. When elements can be substituted for others, and form compounds which are analogous to them, they are said to be equivalent to each other.

81. NITRIC ACID $HNO_3 = 63$. Is a strong, fuming, volatile, liquid, of a density of 1·52 at 15° C., and known by the name of Aqua fortis. Even when highly diluted, it reddens litmus paper. It boils at 84°·5. With Ammonia, nitric acid unites with vehemence

and forms white vapours of Ammonium nitrate NH_4NO_3. In moist air, nitric acid forms dense white fumes, owing to its combination with water.

Nitric acid is called a mono-basic acid, because it contains only one atom of displaceable Hydrogen. Its salts are called Nitrates, and they are all, without exception, soluble in water, and they deflagrate or burn upon red-hot charcoal. Indeed, nitric acid may be considered as a salt of Hydrogen, and may be called "Hydrogen nitrate" or "Hydric nitrate."

The chief salts of nitric acid are: Potassium nitrate KNO_3, Nitre, or Saltpetre, and Sodium nitrate $NaNO_3$. From either of these salts Nitric acid may be prepared, through the direct action of sulphuric acid. When saltpetre is distilled with an equal weight of sulphuric acid, the potassium of the saltpetre displaces an atom of hydrogen from the stronger sulphuric acid; thus we obtain Potassium Hydrogen sulphate in the retort, whilst Nitric acid distils. In symbols, the change may be represented as follows: $KNO_3 + H_2SO_4 = KHSO_4 + HNO_3$.

The advantage of substituting Sodium nitrate is owing to the fact, that this salt is cheaper and affords more Nitric acid. The reason of this will be recognized from the statement, that, whereas the atomic weight of Potassium is 39, that of Sodium is only 23: now, as a compound is made up of the sum of its components, the atomic weights of these two salts must be respectively 85 and 101. Consequently, 85 parts of Sodium nitrate will afford the same quantity of Nitric acid (63 parts), as 101 parts of saltpetre.

Nitric acid is slowly formed by the transmission of electric sparks through moist atmospheric air. This accounts for its presence in the atmosphere during severe storms accompanied by lightning.

The best tests for Nitric acid are various of the metals, as will be understood by reference to sections 84 and 85. But Ferrous sulphate is also an excellent test, owing to the production of an olive-brown colour when mixed with free nitric acid.

It remains to be mentioned that, when Nitric acid HNO_3 is boiled, it gradually decomposes: oxygen and orange-red fumes are evolved,—the boiling-point rises to 121° C., and a weaker acid distils of a density 1·42, with the composition; $2HNO_3.3H_2O$. This hydrate is far more stable than the former, and is the ordinary acid of the shops.

"NITRIC ANHYDRIDE" N_2O_5. Crystallizes in brilliant, right-rhombic prisms. It melts at 29°·5 C., and boils at 45°. It cannot be obtained from nitric acid, but may be prepared by passing Chlorine gas over Silver nitrate: Silver chloride results, Oxygen is evolved and passes off, together with the Nitric anhydride. $4(AgNO_3)+2Cl_2=4(AgCl)+O_2+2(N_2O_5)$. When brought into contact with water, Nitric acid is formed: $N_2O_5+H_2O=2(HNO_3)$.

82. NITROUS ACID $HNO_2=47$. In the free state, nitrous acid is very unstable. Above 0° C., it forms red vapours, which may be obtained as a blue liquid, when cooled down to—20° C. When nitric acid is exposed to the sun-light, the vapours given off are those of nitrous acid. Salts of nitrous acid are called Nitrites: Potassium Nitrite KNO_2 is obtained from the nitrate by strongly heating it.

Nitrous anhydride N_2O_3 is formed by mixing Nitrogen peroxide NO_2 with $\frac{1}{4}$ of its volume of Oxygen. When condensed into a liquid, and mixed with the smallest needful quantity of water, nitrous acid is produced: $N_2O_3+H_2O=2(HNO_2)$. In presence of excess of water, Nitrous acid is decomposed into: Water, Nitric acid and Nitric oxide; $6(HNO_2)=2H_2O+2(HNO_3)+4(NO)$. When the metal silver is brought together with nitric acid, Nitrous anhydride is also formed: $4Ag+6(HNO_3)=4AgNO_3+3H_2O+N_2O_3$.

83. NITROUS OXIDE $N_2O=44$. Is known as Laughing Gas, and possesses, when inhaled, valuable anaesthetic properties, which are now turned to account. It is a colourless, tasteless and inodorous gas, supporting with brilliancy the combustion of ignited wood, and phosphorus. 11.19 litres of the gas weigh 22 grammes. Under a pressure of 36 atmospheres at 0°C., Nitrous oxide becomes a liquid. In water, it is soluble to the extent of nearly its own volume at 15° C.

Nitrous oxide is most readily made from Ammonium nitrate NH_4NO_3, by heating the salt carefully at 249° C.: it resolves itself into water and nitrous oxide. Thus; $NH_4NO_3=2H_2O+N_2O$. It may also be prepared by dissolving Zinc in very dilute nitric acid: Zinc nitrate $Zn2NO_3$ is formed, together with Water and Nitrous oxide. $4Zn+10(HNO_3)+nH_2O=4(Zn2NO_3)+5H_2O+nH_2O+N_2O$.

84. NITROGEN OXIDE or NITRIC OXIDE $NO=30$. This compound

forms a colourless gas, utterly irrespirable and suffocating. Into a liquid, it has never yet been condensed. 11·19 litres of nitric oxide weigh 15 grammes. In water it is but little soluble. The chief characteristics of the gas are shown in its affinity for oxygen, with which it unites in two other proportions. When one volume is mixed with half a volume of Oxygen, Nitrogen peroxide NO_2 is formed; but when mixed with $\frac{1}{4}$ its volume of Oxygen, Nitrous anhydride N_2O_3 results. The behaviour of nitric acid towards Ferrous sulphate has been mentioned: the formation of the olive-brown compound is due to Nitric oxide, which unites with some unchanged Ferrous sulphate ($2[FeSO_4].NO$). Nitric oxide is best made from nitric acid by the action of copper or mercury: cupric or mercuric nitrate is formed, together with water and the gas. $8(HNO_3)+3Cu=3(Cu2NO_3)+4H_2O+2NO$. Inasmuch as nitric oxide changes from a colourless gas, to one of a deep orange-red colour, when in contact with air (owing to the Oxygen which the air contains), it constitutes an admirable test for Oxygen. And for the same reasons, the metals are tests for nitric acid, but only when in a free state; if in combination, the nitric acid must be separated by means of sulphuric acid.

85. Nitrogen peroxide or Nitric peroxide $NO_2=46$. It is most generally known as a deep orange-red vapour, because it is a liquid which boils at 21°·7 C. Below 0°C., Nitric peroxide forms a colourless liquid which may be crystallized, by cooling down with a mixture of snow and salt. Although capable of entering into combinations and, in various organic substances, of displacing Hydrogen atom for atom, Nitrogen peroxide is most unstable in water; it is decomposed into nitric acid and nitric oxide: $3(NO_2)+H_2O=2(HNO_3)+NO$. Nitrogen peroxide may be prepared by mixing nitric oxide with half its volume of Oxygen, or, 2 volumes with one volume. But the easiest mode is by strongly heating Lead nitrate $Pb2NO_3$; it breaks up into: Lead oxide PbO, Oxygen and Nitrogen peroxide. Thus: $2(Pb2NO_3)=2PbO+O_2+4NO_2$. When tin is acted upon by nitric acid, nitrogen peroxide also results.

So much for the compounds of nitrogen with oxygen. The formulæ given in some works, as well for Nitrogen oxide N_2O_2 as for Nitrogen peroxide N_2O_4, denote 4 volumes instead of two volumes, as accepted in the notation here adopted.

86. CARBON unites with NITROGEN to form : CYANOGEN $C_2N_2=52$. It is a gas, of pungent odour, reminding of prussic acid. It is inflammable in air and burns with a peach-blossom-coloured flame. Water dissolves four times its volume at 15° C. Cyanogen is poisonous; its name is derived from the Greek (*κύανος*, kyanos, blue and *γεννάω*, gennao, I produce) because it is essential to the formation of Prussian blue. Under a pressure of four atmospheres, it condenses into a liquid.

Although a compound, Cyanogen behaves like an element, and is as much a salt-radical as Chlorine, Bromine, Iodine and Fluorine: it unites with the metals to form salts, called Cyanides. In combination with Hydrogen, we have Hydrocyanic acid HCN; Potassium cyanide is KCN. Cyanogen is a heavy gas; 11·19 litres weigh 26 grammes. If Nitrogen be passed over a mixture of Potassium carbonate and charcoal contained in a red-hot tube, Carbonic oxide is evolved and Potassium cyanide is formed: $K_2CO_3+4C+N_2=3CO+2(KCN)$. Cyanogen is best prepared by heating Mercuric cyanide HgC_2N_2 to redness: it breaks up principally into Mercury and Cyanogen.

PROBLEMS.

f. What volume of Carbonic anhydride can be obtained from 1000 grammes of Calcium carbonate? Bar. 760 mm., Thermometer 0° C.

$$100 : 1000 = 44 ; \therefore x = 440 \text{ Grammes.}$$

11·19 Litres of Hydrogen weigh 1 Gramme; ∴ 11·19 Litres of Carbonic anhydride weigh 22 grammes.

$$\therefore 22 : 440 = 11{\cdot}19 \; x = 223{\cdot}8 \text{ litres.}$$

g. The capacity of a gas-holder is 3357 Litres. What weight of Potassium chlorate will be required to fill it with Oxygen at the standard temperature and pressure, supposing the Potassium chlorate to give up all its Oxygen?

11·19 Litres of Oxygen weigh 16 Grammes.

$$\therefore 11{\cdot}19 : 3357 = 16 \therefore x = 4800 \text{ Grammes.}$$

There are 48 Grammes of Oxygen in 122·5 Grammes of Potassium chlorate.

$$\therefore 48 : 4800 = 122{\cdot}5 ; \; x = 12250 \text{ Grammes.}$$

CHAPTER V.

THE ATMOSPHERE IN RELATION TO ANIMAL AND VEGETABLE LIFE.

87. The atmosphere not an element. 88. Consists of two distinct kinds of matter. 89. Consists of 21 per cent. of Oxygen and 79 per cent. of Nitrogen. 90. Properties of Oxygen. 91. Nitrogen. 92. Watery vapour. 93. Carbonic anhydride, always present. 94. It is the source of Carbon to plants and they restore the Oxygen to the air. 95. Carbonic anhydride produced in respiration of animals. 96. Ammonia in the air: as ammonium hydrogen carbonate. 97. Nitric acid also in the air. 98. Constituents of the air enumerated. 99. Light carbonetted hydrogen. 100. Ozone also in the air. 101. Impurities in the air. 102. Uniform composition of air. 103. Diffusion of gases. 104. All power derived from the Great Creator.

87. It is no long time since the Atmosphere was included among the Elements, (Chapter III. p. 49, par. 44,) a term applied by Chemists to those bodies which have hitherto resisted all attempts to resolve them into simpler forms. It is now, however, well-known to be a mixture of several substances, in very different proportions and of widely-dissimilar properties. All of them are found in animal and vegetable bodies, and bear a direct relation to organic life.

By far the greater number of the substances of which the vegetable world is composed, are contained in an elementary form in the atmosphere; besides everything that goes to the formation of animals, with the exception of the bones, and those minuter quantities of earthy matter which enter into the composition of every liquid and solid portion of the frame.

What, then, is this atmosphere, without which neither life, nor that very important source of heat and light—combustion, can be supported? Of what is it made, that it can furnish so many constituents to the animal and vegetable worlds? We have seen it to be a transparent vapour or gas, a material substance held by the attraction of gravitation to the surface of the globe, with which it revolves, and which it encloses.

88. That the atmosphere does not consist wholly of one kind of matter, is readily shown by the following experiment:—A float, Fig. 47 (made of cork, into which is fitted a small porcelain capsule), containing a piece of Phosphorus about the size of a small pea, is placed upon a basin of water, and over it is inverted a glass jar, open at the bottom and carefully stoppered. On removing the stopper, the water will stand on a level inside and outside the jar. The phosphorus is then inflamed by means of a heated wire introduced through the neck of the bottle, and the stopper immediately replaced. The Phosphorus burns brilliantly for a time, emitting white vapours of Phosphoric anhydride which soon dissolve in the water, which then rises in the jar, Fig. 48, till it fills one-fifth of the space previously occupied by the air. By the combustion, therefore, of the phosphorus, something has been removed from the air in the jar, equivalent to one-fifth of its whole bulk, or, more accurately, to 21 per cent.

Fig. 47.

Fig. 48.

89. That it is not simply one-fifth of the air that has been removed is shown by the fact, that what remains, although clear and colourless as the air, is possessed of different properties. A lighted taper introduced into the jar is at once extinguished, proving that what is now there, will not support combustion.

In fact, that portion of the atmosphere, forming one-fifth of its bulk, that will alone support combustion, has been consumed by the burning phosphorus and has entered into union with it.

The matter so removed is Oxygen; what is left behind is called Nitrogen; and of these two invisible gases, oxygen and nitrogen, atmospheric air is mainly composed, in the proportions of 21 per cent. of oxygen and 79 per cent. of nitrogen.

90. Oxygen is a colourless, tasteless, and inodorous gas, heavier than air, and very widely diffused throughout nature. For, besides forming one-fifth of the air, it is a constituent of earth and of water, and indeed of most created things. It is the most abundant of all the elements, and its importance can scarcely be overrated. It is the chief supporter of combustion: all substances that burn well in air, burning with greatly-increased brilliancy in Oxygen (p. 56, par. 59).

So necessary is it to life, that oxygen used to be called "Vital air." And not only to life and combustion, but also to germination. No vegetation is possible without it; indeed, a seed may lie buried for centuries, without giving any signs of life, if placed beyond the reach of atmospheric air.

Nor is this all. Without oxygen there would be no decay, none of that change by means of which dead matter is got rid of. Heaps of fallen leaves and other refuse, would accumulate upon the earth, were it not for this busy, active oxygen: under its magic touch they assume new and simpler forms of existence, being for the most part changed into invisible gases. In those grand processes of nature, decay, respiration, and combustion, oxygen plays the chief part. It enters into combinations with the decaying matter and the living body, and is either removed from the atmosphere, or re-enters it in some new form.

According to the late illustrious Faraday, about 7,124,824 tons of oxygen are daily removed from the air.

Yet, necessary as it is, we could not live in an atmosphere of pure oxygen. If a rabbit is made to breathe pure oxygen, although it does not appear to suffer immediately, yet its respiration soon becomes very rapid, the circulation of blood is accelerated, and the animal shows signs of excitement. Symptoms of weakness appear, which are followed by insensibility, and death takes place in a few hours.

This result is ascribed to the excessive rapidity of the destruction of the tissues, exceeding their powers of repair.

91. Mixed with oxygen, the atmosphere, when dry, contains 79 per cent., by measure, of Nitrogen, (p. 69, par. 75). Like Oxygen, it is coolurless, tasteless, and inodorous; unlike oxygen, it does not support either respiration or combustion. Its presence in such large proportion in the atmosphere is a proof that it is not injurious to res-

piration, while it serves to moderate the activity of both respiration and combustion. From the difficulty with which it enters into combination, nitrogen is admirably adapted for this purpose; indeed. there is no other gas which could be substituted for it.

92. Next to oxygen and nitrogen, the most important constituent of the atmosphere is Watery Vapour. Its amount varies greatly. The quantity with us in England may be said to average 1·4 per cent., by measure.

93. Far behind these three in quantity, but always present, is Carbonic anhydride. According to the most recent experiments, one hundred volumes of inland air contain 0·04 per cent. of this compound; that is, there is one volume in 2500 volumes of air. The maximum quantity of carbonic anhydride is found in the air towards morning; the minimum at the close of day. Its sources are most numerous, the great processes of respiration, decay, combustion, germination, and fermentation, being each and all productive of carbonic anhydride.

Carbonic anhydride is a compound of one atom of Carbon, with two atoms of Oxygen, (p. 66, par. 70). As Carbon is contained in all combustible matter, it is easy to conceive how every species of combustion will introduce a certain quantity of carbonic acid into the atmosphere. The blood, also contains carbon in union with oxygen, hydrogen, and nitrogen. Every time a breath is drawn, a portion of the inspired air, but especially of its oxygen, is taken up by the blood circulating through the lungs, and the air which is returned by expiration is considerably changed. In place of the oxygen which it gave up, it contains a large excess of carbonic anhydride: instead of 0·04 per cent. which it took in, it brings out 4 per cent.

This air, therefore, will neither support combustion nor serve any longer for respiration. Thus is a certain portion of air spoilt by every breath drawn. Germination is, in this respect, akin to respiration. If some seeds be moistened and placed under a shade containing common air, all the signs of vegetation will soon commence. An examination of the air after the seeds have sprouted, will give evidence of a change in its composition; a portion of its oxygen has disappeared, and a corresponding volume of carbonic anhydride been generated. This fact explains the necessity of oxygen in water, to the germination of aquatic plants. The presence of oxygen is as favourable to germination, as that of large quantities of carbonic anhydride is unfavourable; hence, the process is hastened by the introduction into the soil of Lime, which absorbs the carbonic anhydride as fast as it is developed by the plant.

That Carbonic anhydride is evolved in the process of decay, is easily proved. Its deleterious effects are manifest on a person sleeping upon dead leaves; and it is found that the air in a confined space, over decaying matter, soon ceases to have the power of supporting combustion. If this were not so, and did not leaves and plants, in their decay, resolve themselves into gases, what would become of all the refuse vegetable and animal matter, the accumulation of ages of living and dead organized beings?

94. This question at once provokes another. Since carbonic anhydride is always present in the air, being continually produced, by respiration, combustion, and the like, at the expense of the oxygen, how does it happen that the oxygen does not unduly diminish, and the carbonic anhydride increase, to an injurious or even fatal extent? It suggests an inquiry into the manner in which plants obtain the Carbon which they require for their development.

A very simple experiment will prove the presence of Carbon in every part of the vegetable kingdom. The charring of a leaf, stalk, or piece of wood, is due to the presence of carbon, which separates as charcoal, and which is not burnt so readily as those other constituents of plants which contain Hydrogen. Now as this carbon is an elementary substance, and, therefore, cannot be produced by the plant itself, it must be derived from some external source. Now, the only source from which the constituents of plants are derived, are the soil, water, and the air. From the soil come all its mineral ingredients; all that which remains as solid matter after the plant is completely burnt—its Ash, which, however, rarely exceeds 1½ per cent. of its weight. However great may be the influence of water on the plant, both as forming one of its chief constituents, and as contributing largely in other ways to its growth, it brings to it no Carbon. This must come from the air, and indeed only from the Carbonic anhydride which is always there to be found, and which plants have the singular power of decomposing under the influence of solar light, taking for their own use the Carbon, and restoring the Oxygen to the air. The researches of De Saussure have proved, that the presence of Light is essential to this action of the plant. He found that young plants, under the influence of solar light, throve in an atmosphere containing one-twelfth of its bulk of carbonic anhydride and that their growth was evidently more vigorous than in common air; while at the conclusion of one of his experiments, he found that almost the whole of the carbonic anhydride had been replaced by oxygen.

On the other hand, if the plants were kept in the shade, he found that the smallest addition of carbonic anhydride to the atmosphere was injurious to their growth. For the above reason, this appropriation, or, as it is called, assimilation, of carbonic anhydride ceases at night-time; oxygen is no longer exhaled from the leaves of the plants, but the reverse chemical process takes place, similar to that which occurs in animals. Oxygen is inhaled by the leaves, and carbonic anhydride formed at the expense of the carbon contained in them; and the carbonic anhydride so formed is retained, partially or wholly, according as the green part of the leaf is more or less plentifully supplied with water. This absorption of oxygen (respiration) appears to take place equally during the day, but it is then overpowered by the opposite process—the absorption of carbonic anhydride.

Owing to its solubility in water, some of the carbonic anhydride of the atmosphere is washed by every shower into the soil, and is thus conveyed into plants by another channel, viz., through their roots. This solubility of carbonic anhydride enables aquatic plants to get their due supply of it, which they require as much as land-plants.

95. It is evident that the carbonic anhydride of the air would rapidly be exhausted by the constant processes of vegetation, were there not some equally continuous source of supply. Such a source, however, is found in the respiration of animals, and in the decay of both animal and vegetable matter. Thus, there is a mutual interchange of kind offices between animals and vegetables, by which the purity and uniformity of the air is maintained, the one supplying to the air what the other requires for its existence.

96. To the constituents of the air already enumerated, viz., Oxygen, Nitrogen, Watery vapour, and Carbonic anhydride, must be added Ammonia, which, in combination, as Ammonium Hydrogen

carbonate, is always present, although only in minute proportions. According to Fresenius, one million parts of air contain only 0·283 parts of Ammonium Hydrogen carbonate by day, and about twice that quantity by night. The comparative smallness of the quantity present by day is due, partly to its consumption in the nutrition of plants, and partly to the fact, that the ammonia which accumulates in the day and night, is dissolved and deposited by the morning mists. At all times, and from many sources, ammonia is given out to the air. There can be neither decay nor putrefaction without its development, nor can any material containing nitrogen be burnt without its formation: it is also evolved in respiration and perspiration. Not that it is ever generated in any of these processes in equal quantity with carbonic anhydride, inasmuch as, under the circumstances most favourable to its production, as in respiration, it is never produced in sufficient quantity to allow of recognition by its familiar odour of hartshorn. Nor can it ever accumulate like carbonic anhydride, as its very ready solubility occasions its constant removal from the air by every passing shower, or falls of dew. From it (and from nitric acid) plants derive their Nitrogen, without which they would be useless as food for animals, whose flesh and blood contain Nitrogen in abundance. Though surrounded by air, which contains so vast a proportion of nitrogen, all plants would languish for want of it, were it not for the ammonia also present. Of nitrogen in its simple form the plant can make no use; it can only obtain it by decomposing ammonia. Hence the value of manures, is, for the most part, in proportion to the ammonia which they contain or afford.

How admirable, therefore, is the provision, that decay and putrefaction should thus become the great ministrants to the life and vigorous growth of fresh vegetation!

97. There is yet a third form in which nitrogen is, sometimes at least, present in the air. In rain-water, after a thunder-storm, Nitric acid (p. 72, par. 81) is nearly always discoverable; and it may be thence inferred that it also is a constituent of the atmosphere during storms. It is well ascertained, that the mere transition of sparks from the electrical machine through the air, is sufficient to cause some of the nitrogen to unite with oxygen, in infinitesimal quantities, and to produce nitric acid. It was thus, indeed, that the composition of nitric acid was first proved by Cavendish in 1785.

Nitric acid has been found to be as useful a manure to plants, as ammonia: and it is likely, that one among many causes of the greater fertility of tropical climates, is due to the presence of so much nitric acid in the rains falling during the frequent electrical states of the atmosphere in those regions.

98. It must not be forgotten that while Watery Vapour is always present in the air, and Carbonic anhydride is also to be found in small, but thoroughly-appreciable quantities, Oxygen and Nitrogen form by far the greater bulk of it. The other constituents mentioned, Ammonium Hydrogen carbonate and Nitric acid, are present in such minute proportions, that they are only discoverable by careful experiments upon large volumes of air.

99. The latter observation holds good of numerous occasional constituents of the air, which, when present, impart to it distinctive qualities. Thus, whenever decay of vegetable matter takes place in presence of much moisture, Methyl hydride, Marsh-gas, or Light Carbonetted Hydrogen (p. 68, par. 73) is always generated; so that, by some observers, this gas is added to the number of constant ingredients of the atmosphere.

100. Among the most curious occasional constituents of air, is a peculiar modification of oxygen which is known by the name of Ozone (p. 59, par. 63). Although possessed of very different properties, it is only Oxygen under different physical aspects. It was discovered by Schoenbein, in 1840, and was called ozone on account of its marked odour. Its best recognised property is that of a disinfectant, hence it is supposed by some to be one of the great natural purifying agents: whether this be so or not, it has, undoubtedly, all the attributes of a scent-destroyer. Although it is never present in the air in anything like sufficient quantity to be observable by its odour, yet is it often largely produced during storms; and we have good reason for believing that its presence sometimes gives rise to affections of the mucous membrane, to colds and catarrhs. Perfectly insoluble in water, the air which traverses seas does not lose its ozone; while, on the contrary, when it meets with putrescent matter, every vestige of ozone is lost.

101. In large cities, the air is always more or less contaminated by the various gases evolved from putrefying or decaying animal or

vegetable matter, from combustion of fuel, and from the numerous chemical processes carried on in such places.

The name of such gases is legion. In the atmosphere of London, Manchester, and other cities, both Sulphurous and Sulphuric acid are present in the air; from drains and sewers, Carbonic anhydride, Sulphuretted hydrogen, Methyl hydride, and Ammonium hydro-sulphide, add their contaminating influences. The two sulphur compounds last-named are exceedingly noxious, but they are, fortunately, attended with such offensive odours as to be readily detected, so as to allow of the removal of the immediate source from which they emanate. So fatal is sulphuretted hydrogen (or rotten-egg gas, as it is sometimes called, from being evolved from putrid eggs), that its presence in the air, to the extent of $\frac{1}{1500}$ of its bulk, will destroy the life of small birds, while a horse will die where the air contains $\frac{1}{250}$ of its volume of this gas.

Everything volatile, of necessity, finds its way to the atmosphere, as does everything soluble to the ocean.

102. Setting aside such minor differences as are occasioned by the presence of slight casual impurities, how does it happen that, as regards its chief constituents, the air maintains its composition so uniformly in all open places? Samples brought from distances most remote from each other, exhibit an absolute identity of composition; and to such an extent is this the case, that the oxygen and nitrogen of the air were long believed to be chemically united. This supposition is now ascertained to be incorrect, the air being merely a mixture of gases, in which all their characteristic features are retained; and this, too, in spite of the fact, that all the gases contained in the air are of different densities,* and that there is a far greater

* Specific gravity of air	1·000	Specific gravity of Nitrous oxide	1·529
" " nitrogen	0·972	" " Carbonic anhydride	1·529
" " ethylene	0·972	" " Hydrochloric acid	1·250
" " carbonic oxide	0·972	" " Hydrogen sulphide	1·191
" " ammonia	0·589	" " Nitric oxide	1·038
" " methyl hydride	0·560		
" " chlorine	2·470		

difference between the respective weights of Nitrogen and Carbonic anhydride, than between oil and water.

103. The subject of this intermixture of gases of different specific gravities, has been thoroughly investigated by Professor Graham. Owing to the absence of cohesion between their particles, vapours and gases mingle very freely and completely—a peculiarity of such mixture being that they never again separate into their component parts. To such an extent is this tendency to mingle carried into effect, that if two gases—say Hydrogen and Carbonic anhydride (the latter being 22 times heavier than the former)—be allowed access to each other through a tube several feet in length, as is represented in Fig. 48, they will be found to have commingled thoroughly in the course of a few hours.

This process of intermingling of gases is called "Diffusion." The rapidity with which such Diffusion takes place, is inversely proportional to the square root of the densities of the different gases. Oxygen being 16 times heavier than Hydrogen, it follows, that only 1 volume of Oxygen would diffuse itself in the same time that 4 volumes of Hydrogen would be able to do so.

Fig. 48.

This diffusive power of gases is of the utmost importance. Were it not for this law, the carbonic acid evolved in such vast quantities in our large towns would, from its intrinsic weight, collect and speedily destroy the inhabitants. The foul and noxious gases, constantly arising from the numerous operations of a large city, would spread disease and pestilence around it. But for this law, the perfume of flowers and the sweet scent of newly-mown hay would not rise to fill our hearts with a sense of gladness, but would fall from their own weights on the senseless earth. Were it not for this law, the oxygen would separate from the nitrogen, and a universal conflagration, which no water could subdue, would arise on the ignition of the first fire. In fact, but for this law there could be no life. Vegetation must languish and die for want of a supply of carbonic acid, which, by this provision, is wonderfully and completely diffused throughout all the particles of air. The watery vapour, of no less importance, would not keep flexible the skins of animals, and supply parched vegetation with rain and dew. Lastly, in the process of respiration, the minute cells of the lungs, in which the aëration of the blood takes place, would not be emptied at each exhalation of the heavy carbonic acid which they contain: but the operation would, after a few minutes, be suspended, and death inevitably ensue.

104. Modern science has done much to clear away the mystery that surrounds the operations of nature, and we have learned to know something of the wonders that encompass our daily path. But the more we know, the more we feel the imperfection of our knowledge; and however clearly we may trace the laws by which the world is governed, true wisdom consists in acknowledging that these laws are but secondary causes in the hands of Him who "ruleth by His power for ever." While we thankfully accept such glimpses into the mysteries of creation as scientific research has procured for us, let us never forget in theories as to the law of storms, or laws of diffusion, and the like, that it is God, "at whose word the stormy wind ariseth," and "who maketh the storm to cease;" that it is He "who covereth the heaven with clouds, who prepareth rain for the earth, who maketh grass to grow upon the mountains;" who "giveth snow like wool, and scattereth the hoar-frost like ashes;" and that the phenomena whose causes we have been investigating—"fire and hail, snow and vapours, stormy winds"—are not blind instruments of arbitrary laws, but in their errands of mercy and judgment are but "fulfilling His Word" "who stretcheth forth the heavens alone, and spreadeth abroad the earth by Himself."

CHAPTER VI.

THE CHEMISTRY OF THE DYAD ELEMENTS, SULPHUR, SELENIUM AND TELLURIUM.

105. V. Sulphur; a dyad element. Brimstone; roll sulphur; flowers of Sulphur. 106. Allotropic conditions of Sulphur. 107. Native compounds of sulphur with other elements: metallic sulphides. 108. Effects of roasting sulphides in air or in absence of air. 109. Hydrogen sulphide or Hydrosulphuric acid: a test for the metals. 110. Hydrogen persulphide. 111. Sulphurous anhydride: a bleaching and disinfecting agent. Used in making Sulphuric acid. Di-basic acid. Sulphites. 112. Sulphuric anhydride. 113. Sulphuric acid: oil of vitriol. Di-basic. Sulphates. Manufacture of the acid in leaden chambers. Tests for Sulphuric acid. 114. Nordhausen Sulphuric acid. 115. Hyposulphuric acid. [116. Trithionic acid. 117. Tetrathionic acid. 118. Pentathionic acid. 119. Nitro-sulphurous acid.] 120. Carbon di-sulphide. 121. VI. Selenium. Derivation and Properties. 122. Selenous anhydride and acid. Selenites. 123. Selenic acid. 124. Hydrogen selenide. 125. VII. Tellurium. 126. Tellurous anhydride. Tellurous acid and tellurites. 127. Telluric acid. 128. Hydrogen telluride. 129. Tellurous and telluric sulphide. 130. Analogy between Sulphur, Selenium and Tellurium. 131. Isomorphism.

V. Sulphur S = 32.

105. The non-metallic element Sulphur is a solid substance of a pale-yellow colour. It is met with in the uncombined state or "native," in the blue clay of Sicily, as well as generally on the coasts of the Mediterranean. Sometimes Sulphur is met with in large, transparent, yellow crystals, derived from the Octohedron with Rhombic base (Fig. 49.) When mixed with clay, it is easily separated by the aid of heat, inasmuch as Sulphur is volatile and may be sublimed in absence of air. The process may be conducted as represented in Fig. 50.

Fig. 49.

In commerce, Sulphur is known by the names "Brimstone" or

"Roll Sulphur" cast in moulds, Fig. 51. Also, as "Flowers of

Fig. 50.

Sulphur," so much used in medicine. Indeed, its medicinal properties have been long known and valued; its old name is from the Greek, θεῖον, theion, sulphur.

Fig.51.

106. Sulphur affords a notable illustration of the allotropic condition. There are few distinct varieties of this element. Native sulphur, or $S\alpha$, is of an amber-yellow colour, semi-transparent, fusible at 115° C., and of sp. gr. 2·05. It is insoluble in water, soluble in Carbon disulphide CS_2, in oil of turpentine, and other media. It is a non-conductor of heat and electricity, and very brittle : it crystallizes as shewn in Fig. 49.

The second variety, or $S\beta$ is obtained by carefully melting Sulphur at 115° C., and allowing it to solidify on the surface: on piercing the crust, and pouring out the yet liquid mass, it is obtained in prismatic crystals, Fig. 52, of bright yellow colour. It melts at 120° C., and has a sp. gr. of 1·98. In solubility, and other respects, it much resembles the first variety, into which it becomes changed after a time, by mere exposure, or by dipping the crystals into Carbon di-sulphide.

Fig. 52.

The third variety, or $S\gamma$, is also called "viscid sulphur." When Sulphur is melted, the amber-coloured liquid deepens at 140° C., and turns from brown to deep red-brown at 160°. From being thin-fluid, at 230° C. its tenacity is such, that the containing vessel may be inverted without the Sulphur running out. If, now, the Sulphur be further heated to its boiling point (440° C.) and then suddenly cooled, by pouring it into water or otherwise, this third variety results. It is plastic, and may be cast into moulds or pulled out into threads resembling barley-sugar, but soft. It is insoluble in all media, and has a sp. gr. of 1·957. In a few hours it returns in great part to the condition of common sulphur, with evolution of considerable heat; or it may be at once re-converted by heating to 100° C. The vapour density of this Sulphur would give 192 or S_6 as the molecular weight of this variety.

A fourth variety is pulverulent and of sp. gr. 1·95. It is obtained from Hydrogen Sulphide or other Sulphides by the aid of Chlorine.

11·19 litres of Sulphur-vapour weigh 32 Grammes. In its chief combination with Hydrogen, Sulphur behaves like a Dyad.

107. Many combinations of Sulphur are met with in nature. One of its compounds with Hydrogen, Hydrogen Sulphide H_2S, is a very common emanation from sewers, and from decomposing animal matters which contain Sulphur. With the metals, Sulphur unites to form SULPHIDES, many of which correspond to those of the Dyad element "Oxygen": thus, H_2S; H_2O: H_2S_2; H_2O_2: CS_2; CO_2. Lead, or Plumbum is found native as Galena, or Lead Sulphide PbS; Iron or Ferrum Fe, frequently as Iron-pyrites, or Ferrous sulphide FeS_2. Copper, as Copper pyrites $FeCuS_2$. Zinc, as Blende, or Zinc Sulphide ZnS. Silver or Argentum, as Silver-glance or Silver-Sulphide Ag_2S. Quicksilver or Hydrargyrum, as Cinnabar or Mercuric sulphide HgS. Antimony or Stibium as Antimony-Glance or Antimonous Sulphide Sb_2S_3, &c.

108. As a rule, when a Sulphide is heated strongly in the air, the metal and the Sulphur become oxydized; the metal remains in combination as Oxide, and the Sulphur volatilizes in the form of a gas, the well-known SULPHUROUS ANHYDRIDE SO_2. If Sulphur is heated either in air or in Oxygen, it also oxydizes: at a temperature of 260° C, it inflames and burns with lilac colour to Sulphurous anhydride, the vapours of which are most suffocating and recall the smell of the burning brimstone match.

Many sulphides undergo no change when strongly heated in absence of air, except perhaps as to fusion. But, if a higher sulphide, like Iron-pyrites FeS_2, be strongly heated in a retort in absence of air, a portion of the Sulphur may be obtained by condensation in a receiver. ($3FeS_2=Fe_3S_4+S_2$.)

109. Combinations of Sulphur with Hydrogen. Sulphur unites with Hydrogen in two proportions, analogous to those of the Dyad Oxygen. The most important compound is, HYDROGEN SULPHIDE $H_2S=34$. Also called SULPHURETTED HYDROGEN, HYDRO-SULPHURIC ACID, and "Rotten-egg gas." It is a colourless gas, of most offensive odour, irrespirable, poisonous. It is a powerful narcotic. Water dissolves three volumes of the gas, and the latter acquires all its properties: Harrogate and other "hepatic" waters contain it in solution. In air, it may be burnt with flame, forming Sulphurous anhydride and water ($2H_2S+3O_2=2SO_2+2H_2O$.) When exposed to air in solution in water, its Hydrogen oxydizes and Sulphur is deposited ($2H_2S+O_2=2H_2O+S_2$). Chlorine immediately decomposes it.

11·19 litres of Hydrogen Sulphide weigh 17 grammes. Under a pressure of 17 atmospheres, it may be liquefied; the liquid has a sp. gr. of 0·9 compared with water.

Hydrogen sulphide is a first-rate test for the metals. Many Sulphides are black. The best test for the gas is "lead-paper," or filtering paper which has been soaked in a solution of Lead acetate; by the formation of Lead Sulphide, which is black, the presence of the gas is made known. ($Pb2[C_2H_3O_2]+H_2S=2(HC_2H_3O_2)+PbS$.)

Hydrogen sulphide is best prepared by two methods. 1. By the addition of dilute sulphuric acid to Ferrous sulphide, FeS: Ferrous sulphate will remain in solution and Hydrogen sulphide will be evolved. $FeS+H_2SO_4=FeSO_4+H_2S$. 2. By the action of Hydrochloric acid upon Antimonous sulphide: Antimonous chloride is formed in the solution and Hydrogen sulphide is set free. $Sb_2S_3+6HCl=2SbCl_3+3H_2S$.

[110. HYDROGEN PERSULPHIDE H_2S_2. Possesses properties similar to those of Hydrogen peroxide H_2O_2. It is obtained by dropping a solution of Calcium di-sulphide, CaS_2, into Hydrochloric acid HCl: Calcium chloride $CaCl_2$ and Hydrogen persulphide H_2S_2 result ($CaS_2+2HCl=CaCl_2+H_2S_2$.)]

111. Sulphur unites with Oxygen. It forms two anhydrides, viz., SULPHUROUS ANHYDRIDE, SO_2, and SULPHURIC ANHYDRIDE SO_3. There are, however, seven acids, but only three of importance

Sulphurous anhydride $SO_2=64$. A colourless gas, of pungent, suffocating odour, with the well-known smell of the burning brimstone match. Water dissolves 68·8 times its bulk at 0° C., and 43·5 times its bulk at 15° C. In presence of water, it acquires the properties of Sulphurous acid H_2SO_3, by combining with one atom of it, and when this solution is exposed to the air, it rapidly absorbs Oxygen, and becomes changed into sulphuric acid H_2SO_4. At —10° C., Sulphurous anhydride liquefies and may be readily condensed by a mixture of snow and salt.

Sulphurous anhydride is a bleaching agent, and is employed in bleaching straw and flannels. It is also a great reducing agent, from the affinity which it possesses for Oxygen. For this reason it acts as an admirable disinfectant, both in the way of decomposing Hydrogen compounds and in the prevention of decomposition. By its affinity for Oxygen, so long as Sulphurous anhydride is present, articles of food may be prevented from undergoing putrefaction. Hydrogen sulphide decomposes in contact with Sulphurous anhydride, forming Water and Pentathionic acid, with separation of Sulphur, ($3H_2S+4SO_2=2H_2O+H_2S_5O_6+S_2$), 11·19 litres of Sulphurous anhydride weigh 32 grammes. With water it forms crystalline compounds containing respectively 9 and 14 atoms of water.

Sulphurous anhydride is best prepared by four methods. *a.* By burning Sulphur in Oxygen. *b.* By roasting Ferrous di-sulphide FeS_2, or Iron pyrites, on the bed of a reverberatory furnace: Ferric oxide Fe_2O_3, and Sulphurous anhydride SO_2, result ($4FeS_2+11O_2=2Fe_2O_3+8SO_2$). *c.* By boiling Copper in sulphuric acid: Cupric sulphate $CuSO_4$, water H_2O and Sulphurous anhydride SO_2 are produced ($Cu+2H_2SO_4=CuSO_4+2H_2O+SO_2$.) If Mercury is employed, Mercuric sulphate $HgSO_4$, is formed instead of Cupric sulphate. *d.* When impure Sulphurous anhydride is wanted in a Laboratory, pounded charcoal may be boiled with sulphuric acid: a mixture of carbonic and sulphurous anhydrides results ($2H_2SO_4+C=2H_2O+CO_2+2SO_2$.)

Sulphurous acid H_2SO_3, is a di-basic acid. It therefore forms two classes of salts, neutral and acid. Sodium sulphite Na_2SO_3 is the neutral salt of Sodium. Sodium Hydrogen sulphite NaH SO_3 is the acid salt. The great use of Sulphurous anhydride is in the preparation of Sulphuric acid: it can decompose both

Nitric and Nitrous acids. When Sulphurous anhydride is brought together with Nitric acid, Nitrogen peroxide and Sulphuric acid result: $SO_2 + 2HNO_3 = 2NO_2 + H_2SO_4$. And, when Nitrous anhydride and Sulphurous acid are brought together, they yield Nitric oxide and Sulphuric acid: $N_2O_3 + H_2SO_3 = 2NO + H_2SO_4$.

112. SULPHURIC ANHYDRIDE $SO_3 = 80$. A solid substance, crystallizing in white, silken needles. It melts at 29° C., and boils at 46° C. At a red-heat it decomposes into: Oxygen and Sulphurous anhydride ($2SO_3 = O_2 + 2SO_2$.) Sulphuric anhydride has no acid properties. For water, it possesses the most greedy affinity, hissing with it like a red-hot iron: SULPHURIC ACID H_2SO_4 is produced.

Sulphuric anhydride may be prepared by passing a mixture of Sulphurous anhydride and Oxygen over red-hot spongy platinum. Or, by distilling Nordhausen Sulphuric acid,—a solution of Sulphuric anhydride in Sulphuric acid. And again, by distilling Sodium acid sulphate Na_2SO_4,SO_3 at a red-heat ($Na_2SO_4,SO_3 = Na_2SO_4 + SO_3$).

113. SULPHURIC ACID $H_2SO_4 = 98$. Is the "oil of vitriol" of commerce, the most important of all acids, and, at common temperatures, the most powerful. Sulphuric acid is a dense, oil-like liquid, of sp. gr. 1·842. It boils at 338° C., and distils unchanged. It chars most organic substances, such as sugar, wood, starch, &c. With water, it unites in two other proportions; the first hydrate H_2SO_4,H_2O, has a density of 1·78, crystallizes at 7° C., and boils at 224° C.; the second hydrate $H_2SO_4,2H_2O$, has a density of 1·632 and boils at 176° C. Indeed, its affinity for water is the cause of its destructive action; when mixed with water, care should be taken to introduce the acid in a thin stream into the water, and to stir well during the whole time on account of the heat occasioned. Sulphuric acid is a di-basic acid; its salts are called SULPHATES. Many of them contain water of crystallization.

Ferrous sulphate $FeSO_4,7H_2O$ is the common "Green Vitriol." Zinc sulphate $ZnSO_4.7H_2O$ is "White Vitriol." Magnesium Sulphate $MgSO_4.7H_2O$ is "Epsom Salts." Cupric sulphate $CuSO_4.5H_2O$ is "Blue Vitriol." Calcium Sulphate $CaSO_4.2H_2O$ is likewise known as "Selenite," and "Alabaster." Sodium sulphate Na_2SO_4 $10H_2O$ is "Glauber Salts."

The manufacture of Sulphuric acid is conducted in leaden chambers, from 150 to 300 feet long, 15 to 20 feet wide, and 12 to 15 feet high. The necessary Sulphurous anhydride is supplied, either by the burning of Sulphur, or of Ferrous di-sulphide (Iron pyrites FeS_2); whilst the Nitric acid is obtained by heating Sodium nitrate with Sulphuric acid. Water is placed on the chamber-floor; steam is admitted by a boiler from without and the requisite Oxygen, from the atmospheric air accompanying the Sulphurous anhydride. The action has been in great part explained, but in few words is as follows.

a. Nitric acid gives up Oxygen and Hydrogen to Sulphurous anhydride and becomes reduced to Nitrogen peroxide, whilst it raises Sulphurous to Sulphuric acid. $2(HNO_3)+SO_2=2(NO_2)+H_2SO_4$. *b.* Nitrogen peroxide unites with an atom of Water and of Sulphurous anhydride to a crystalline body, which, by more water is decomposed into Nitric oxide, and Sulphuric acid. NO_2,SO_2,H_2O, by water, becomes: $NO+H_2SO_4+H_2O$. *c.* Nitric oxide by union with Oxygen in the chamber becomes Nitrogen peroxide. $2(NO)+O_2=2(NO_2)$. And thus, a limited quantity of Nitric acid converts a large amount of Sulphurous anhydride into Sulphuric acid.

The Sulphuric acid at the bottom of the chamber corresponds with the second hydrate already mentioned; it is evaporated in shallow, leaden pans until it becomes nearly the first hydrate ($H_2SO_4.H_2O$) and it is then concentrated in retorts of glass or platinum until the boiling point becomes stationary and the acid or "oil of vitriol" is obtained. If the process were continued, the Sulphuric acid would volatilize in dense, white, suffocating fumes.

It only remains to be stated that ordinary Sulphuric acid always contains Lead Sulphate in solution; such an acid deposits the salt of lead as a white powder, when it is poured into water and allowed to cool.

The chief test for sulphuric acid and for all its soluble salts is Barium Nitrate, or Barium Chloride; a white, insoluble Barium sulphate results, whilst nitric or hydrochloric acid is set free $Ba2NO_3+H_2SO_4=BaSO_4+2(HNO_3)$. Or: $BaCl_2+H_2SO_4=BaSO_4+2(HCl)$ Barium Sulphate is insoluble in free nitric and hydrochloric acids, as well as in water, and so, also, are Strontium and Lead sulphates.

If sulphuric acid be passed through a red-hot porcelain tube, it breaks up into Water, Sulphurous anhydride and Oxygen; the Sulphurous anhydride may be removed from the Oxygen and held back by means of Calcium hydrate. $2(H_2SO_4)=2H_2O+2SO_2+O_2$.

114. "NORDHAUSEN SULPHURIC ACID" is made from Green Vitriol or Ferrous sulphate. The salt is dried so as to free it from nearly all its water of crystallization, and it is then distilled in earthen retorts. It is a brown, fuming liquor of sp. gr. 1·9, and consists mainly of Sulphuric acid containing a solution of Sulphuric anhydride $H_2SO_4.SO_3$.

115. HYPOSULPHUROUS ACID $H_2S_2H_2O_4=132$. Is also called *dithionous acid.* Only in combination is it known to exist, for any attempt to separate the acid from its salts, breaks it up into Water, Sulphurous anhydride and Sulphur. Its chief salt is: SODIUM HYPOSULPHITE $Na_2S_2H_2O_4.4H_2O$. This salt, which is so much used in Photography, crystallizes in striated, oblique-rhombic prisms. It is prepared by digesting Sodium sulphite with Flowers of Sulphur: $Na_2SO_3+S+H_2O=Na_2S_2H_2O_4$. A solution of Sodium hyposulphite dissolves Silver chloride and is immediately decomposed by Hydrochloric acid. Mercurous nitrate is also at once changed into black Mercurous sulphide by it.

[116. TRITHIONIC ACID $H_2S_3O_6$. In combination with Potassium, as Potassium trithionate, is made, by digesting Flowers of Sulphur in a saturated solution of Potassium Hydrogen Sulphite. The acid, in a free state, gradually changes into Sulphur, Sulphurous and Sulphuric acids.

117. TETRATHIONIC ACID $H_2S_4O_6$. Unstable and unimportant.

118. PENTATHIONIC ACID $H_2S_5O_6$. Is also very unstable. It is best prepared by transmitting Hydrogen sulphide through a saturated solution of Sulphurous acid. $10(H_2SO_3)+10H_2S=18H_2O+5S_2+2(H_2S_5O_6)$.

119. NITRO-SULPHUROUS ACID $H_2SO_3.2NO$. Is only known in combination with metals.]

120. Sulphur unites with Carbon. CARBONIC SULPHIDE $CS_2=76$. Is the common Bi-sulphide of carbon, so much used as a solvent of essential oils, caoutchouc, sulphur, phosphorus. It is a colourless liquid of sp. gr. 1·272 at 15° C., and boils at 48° C. Volatile, of great refractive power, insoluble in water, it is well recognized by its odour of cabbage-water. It is very poisonous and destructive of

insect life. In air it is very inflammable, and burns with blue flame to Sulphurous and Carbonic anhydrides: two volumes of the vapour, require 6 vols. of Oxygen for complete combustion.

Carbonic sulphide is very easily made by passing the vapour of Sulphur over red-hot charcoal contained in a porcelain tube.

[VI. Selenium Se = 79·5.

121. Selenium is a rare element, discovered by Berzelius in 1817, in the refuse of a manufactory of Sulphuric acid at Fahlun, its name being derived from σελήνη, selene, the moon. It is obtained from certain *Selenides*.

Selenium forms a brittle solid, of dark brown semi-metallic hue and glassy fracture. Its density is from 4·3 to 4·8. Above 100° C. it melts, and undergoes changes similar to those of the element Sulphur. Indeed, it comports itself like Sulphur, and is also a Dyad.

122. Selenium burns in air to: SELENOUS ANHYDRIDE SeO_2. This anhydride crystallizes in snow-white prisms. With water it unites to: SELENOUS ACID H_2SeO_3. Its salts are called *Selenites*. When this acid is brought into contact with Sulphurous anhydride, Selenium is obtained: $2H_2SeO_3+2H_2O+4SO_2=4H_2SO_4+Se_2$.

123. SELENIC ANHYDRIDE SeO_3; is unknown. SELENIC ACID H_2SeO_4, very much resembles Sulphuric acid. Its salts are called: *Selenates*. Those of Barium, Strontium, and Lead are insoluble. Selenic acid is best obtained from Lead selenate through the action of Hydrogen sulphide: $PbSeO_4+H_2S=PbS+H_2SeO_4$.

124. HYDROGEN SELENIDE H_2Se. A colourless gas, of most offensive odour. Like its analogue, H_2S, it is inflammable. In water, it is somewhat soluble and forms Selenides, when added to many metallic salts.]

[VII. Tellurium Te = 129.

Tellurium derives its name from *Tellus*, the earth. It was examined by Klaproth in 1798. It is found native, as well as in combination as a Telluride with Gold, Silver, &c.

In its properties, Tellurium closely resembles Sulphur and Selenium: like them, it is a Dyad. In density, 6·26; in lustre, like Bismuth, appearance and other properties, Tellurium is, however, more allied to the metals. It fuses at about 495° C.

126. When strongly heated in air, Tellurium burns with a blue flame edged with green, and emits white fumes of TELLUROUS ANHYDRIDE TeO_2. Nitric acid oxydizes and dissolves Tellurium; the solution, when poured into water, deposits a white powder. This is TELLUROUS ACID H_2TeO_3. Salts of the acid are called *tellurites*.

127. TELLURIC ACID H_2TeO_4. Can be obtained as a *tellurate*, by oxydizing Tellurium with Potassium nitrate. Also in crystals, as an acid with two atoms of water. $H_2TeO_4.2H_2O$.

128. HYDROGEN TELLURIDE H_2Te. An inflammable gas; burns with blue flame and with an odour resembling Hydrogen sulphide. It precipitates most of the metals as tellurides.

129. Tellurium also combines with Sulphur, forming: TELLUROUS SULPHIDE TeS_2 and: TELLURIC SULPHIDE TeS_3. These compounds are quite unimportant.]

130. The analogy existing between Sulphur, Selenium and Tellurium may be well seen, by comparing the formulæ of their chief compounds:—

H_2S.	SO_2.	H_2SO_3.	SO_3.	H_2SO_4.
H_2Se.	SeO_2.	H_2SeO_3.	SeO_3.	H_2SeO_4.
H_2Te	TeO_2.	H_2TeO_3.	TeO_3.	H_2TeO_4.

131. Many of the sulphates, selenates and tellurates are not only similar in crystalline form, but also analogous in composition. Such similarity in form, with analogy of composition, is called: ISOMORPHISM, (from ἴσος, isos, equal, and μορφή, morphe, form). It is a common feature of many chemical compounds. Magnetic oxide of iron FeO,Fe_2O_3 and Chrome-iron-ore FeO,Cr_2O_3 afford also excellent illustrations of Isomorphism.

CHAPTER VII.

ON COAL AND COAL GAS.

132. Oxygen necessary to combustion. 133. Composition of the atmosphere. An oxydized body is really burnt. Matter indestructible. Shakspere on indestructibility. 134. Combustion attended by light and heat: takes place at a fixed temperature. 135. Combustible bodies contain as a rule Carbon and Hydrogen. 136. Charcoal, not pure Carbon. 137. Coal: its formation and origin. 138. Varieties of coal: Cubical, Fat Bituminous, Anthracite, Steam and Cannel Coal. Quantity raised and exported from the United Kingdom in 1868. 139. Lignite or Brown Coal. 140. Peat. 141. Wood. 142. Necessity for previously heating coals, before they will inflame. Smoke. 143. Temperature of a fire. 144. Carbonic anhydride, Watery Vapour, Ammonia and Sulphurous anhydride evolved in combustion of coal: the gases pass up the chimney. 145. On the waste of fuel by smoke. 146. Formation of Coke by heat in absence of air. 147. The value of coal for heating purposes. 148. The volatile matters constitute the differences between coal and coke. 149. Discovery of coal-gas by Dr. R. Clayton. First introduction in 1792 by Mr. Murdoch. 150. Production of coal-gas. 151. Amount of gas from varieties of coal. 152. Amount of gas made on a single night in London. 153. Coal-gas, a mixture of certain gases. 154. Hydrogen. 155. Light and Heavy Carbonetted Hydrogen. 156. Hydride of Methyle is Light Carbonetted Hydrogen: also called Marsh-gas. 157. Ethylene or Olefiant gas, the other names for Heavy Carbonetted hydrogen. 158. Butylene and Acetylene assist in the illuminating powers of gas. 159. Carbonic oxide in gas. 160. The odours of gas due to Benzole, Naphthaline, Carbonic sulphide, &c. The removal of all undesirable; only of injurious Sulphur compounds.

132. It has been already stated that the great supporter of combustion is Oxygen; that under ordinary circumstances, burning bodies obtain their Oxygen from the air; that the intensity of combustion

is in proportion to the amount of Oxygen present; that, therefore, it is most energetic in pure Oxygen—so energetic, in fact, that for the ordinary operations of nature, it requires to be restrained, by the dilution of the Oxygen in the atmosphere with four times its volume of NITROGEN.

The exact composition of the air, as far as its most important constituents are concerned, is, then, as follows:—

Oxygen	20·61
Nitrogen	77·95
Watery vapour . . .	1·40
Carbonic anhydride . .	0·04

One cubic metre of dry air, measured at 0° C. and 760 mm. Bar., weighs 1·2932 kilogrammes.

133. Combustion is a simple chemical process, consisting in the union of the body to be burnt, with Oxygen. After the union, the body so burned is said to be oxydized. And when any substance has once entered into as full combination with Oxygen as is possible, it cannot be either set on fire, or burnt: it is both uninflammable and incombustible. It has been already burnt as much as is possible. Now, as combustion consists in the union of any substance with oxygen, it is clear that the substance, after it is burnt, must be increased in weight by just so much as the weight of the oxygen which it has absorbed. To speak of the destruction of a body by burning, cannot, therefore, be correct, for all matter is indestructible; the body is not destroyed, it has only changed its form. This is clear enough in such a case as the burning of Phosphorus; there we see and handle, and easily collect and weigh, the Phosphoric anhydride produced. But it is not so clear in the case of a candle, or piece of coal, because the new substances into which they are transformed are gaseous and invisible, and are dissipated in the air. But by a simple experiment we can collect and weigh these gases, and prove that in their case, also, the burnt substances have actually increased in weight.

The grand truth of the indestructibility of matter has been happily enunciated by Shakspere, and aptly put into the mouth of the philosophic Hamlet:—

"HAMLET. To what base uses we may return, Horatio! Why

may not imagination trace the noble dust of Alexander, till he find it stopping a bung-hole?

"HORATIO. 'Twere to consider too curiously, to consider so.

"HAMLET. No, faith, not a jot; but to follow him thither with modesty enough, and likelihood to lead it. As thus: Alexander died, Alexander was buried, Alexander returneth into dust; the dust is earth; of earth we make loam; and why of that loam, whereto he was converted, might they not stop a beer-barrel?

"Imperial Cæsar dead, and turn'd to clay,
Might stop a hole to keep the wind away.
Oh! that that earth, which kept the world in awe,
Should patch a wall t' expel the winter's flaw."

Hamlet, Act V. Scene II.

134. Combustion is always accompanied by the development of heat, and generally by that of light also. Indeed, the two tend to produce, and are probably convertible into, each other.

Every combustible substance has a certain fixed temperature at which it burns or inflames in air, but this temperature differs greatly for different substances. Some, such as phosphorus, lead, iron, &c., when in a fine state of division, are so eager for oxygen, so anxious to exchange their condition of singleness, that they take fire the moment they are exposed to the action of the air, at any temperature; others, at a temperature but slightly above summer-heat; while others, again, can only be induced to change their condition by the urgent persuasion of great heat. Neither of these classes of substances could answer the purposes of fuel: the one would burn the housemaid's fingers while she was lighting the fire, and would be burnt out before breakfast-time; the other would allow us to be frozen, before it warmed the room. The one would burn too rapidly, the other too slowly to meet the requirements of life. But there is a large class of bodies, principally of vegetable, partly of animal origin, which possesses the proper characteristics.

135. The only point in which all these bodies agree, whatever may be their visible properties, is, in containing both HYDROGEN and CARBON, the predominant feature being the very large proportion of Carbon. This, then, is the fundamental constituent of all kinds of fuel, whether used for the sake of their heat, or light.

136. As no vegetable matter can be formed without the aid of certain mineral constituents, which are wholly unalterable by fire, it is clear, that CHARCOAL, (independently of the Hydrogen which it always retains to some extent) is not Carbon in a pure form. The purest charcoal leaves an ash when burnt with full access of air.

137. It is, however, with COAL that we have to deal; a substance now well ascertained to be entirely of vegetable origin. In it we see the remains of a vegetation, which covered the earth before it was inhabited by man. The former existence of land has been proved by the occurrence of numerous upright fossil trees, with their roots terminating downwards in seams of Coal. Our Coal-mines furnish almost a complete fossil Flora; a history of many of the now lost species, which once decorated the surface of the earth. There is a great resemblance between the plants of the Coal-formation and the Flora of New Zealand. The botany of those islands is characterized by abundance of Ferns, one hundred and forty species of which are already known, some of them attaining the size of trees. Another point of resemblance is the prevalence of the Fir tribe. A large proportion of the purest kind of Coal has been formed from plants which grew on the spot, by a process similar to that which is still seen at the bottoms of marshes, lakes, and rivers, in the formation of peat. The vegetable matter, submerged in water, undergoes decomposition, losing its Water and some of its Hydrogen, and leaving its Carbon behind: this is subsequently covered by accumulations of clay and sand, which, in process of time, harden into slate and gritstone. The pressure of the deep overlying strata has not only prevented the evolution of gas, thus giving the Coal the property of burning with flame, but has destroyed most traces of vegetable structure, and given to the Pit-coal the close and compact quality of stone.

138. The several varieties of Coal differ greatly in appearance, but not more so than in composition. All coal consists chiefly of Carbon, with a somewhat varying percentage of Hydrogen, Oxygen, and Nitrogen, and with very variable quantities of Sulphur and of Ash.

The principal varieties of Coal in common use are five—cubical coal, fat bituminous coal, anthracite, steam coal, and cannel coal.

Cubical, or "dry" Coal, is dark and shiny, but containing very little bituminous matter, does not cake while burning, and hence is the best adapted for blast-furnaces. The "splint" coal of Scotland, so famous for the smelting of ores, is of this kind.

Fat bituminous Coal, is the coal of the North of England. Owing to the quantity of bituminous matter it contains, it softens when heated, swelling up and "caking," so as to require much poking. It leaves a good deal of cinder. It is a pleasant and the most useful coal for household purposes, making cheerful fires.

Anthracite is a black, brittle, lustrous coal. It contains more Carbon than any other kind; it never smokes, because it contains less bituminous matter. For the same reason it burns without flame, and, when heated in a close vessel, undergoes scarcely any change, leaving a Coke differing but little from its original self; it gives out a great deal of heat, and leaves but little ash. Anthracite abounds in Wales and North America; also in Ireland, where it is known as Kilkenny coal; and in Scotland as "Blind" coal, from the absence of gas.

Steam, or "Smithy" coal, is intermediate between the two last kinds in quality, containing more Hydrogen than anthracite, and more Carbon than bituminous coal: it inflames readily and makes but little smoke, hence it is much prized for steam vessels.

Cannel, or gas-coal, is greyish black or dark brown. It comes principally from Lancashire and Scotland, is very compact, ignites readily and burns with a brilliant flame. It derives its name from the practice among the Scotch farmers, of burning it for its light instead of candles, or, as they pronounce it, "cannels." It is the best coal for gas-making.

According to information received from Mr. R. Hunt, F.R.S., we find that in the year 1868, 104,500,480 tons of coal were raised and used in the United Kingdom, and 10,837,513 tons were exported to foreign countries! So great, indeed, is the consumption, that considerable alarm existed among well-informed persons lest the supply should fail us.

The composition of Coal may be inferred from the following table :—

	Wigan Cannel.	Newcastle Caking coal.	Newport Steam coal.	S. Wales. Anthracite.
Carbon	80·07	86·75	81·47	90·39
Hydrogen	5·53	5·24	4·97	3·28
Nitrogen	2·12	} 6·61	1·63	0·83
Oxygen	8·08		5·32	2·98
Sulphur	1·50		1·10	0·91
Ash	2·70	1·40	5·51	1·61
	100·00	100·00	100·00	100·00

139. Lignite, or brown coal, which is used as fuel in some parts of the world, rarely occurs in extensive fields, as does pit coal. It has usually a brown colour, burns with a dark smoky flame, and a very disagreeable odour; it consists, in fact, of vegetable matter but little altered.

140. Peat, so much employed in Ireland and in other peat countries, consists chiefly of the remains of plants which have undergone comparatively little change.

141. Wood, may be said to contain, on an average, nearly equal weights of carbon and the elements of water. From this it may be inferred, that wood fires will not only burn with much flame and little smoke, but will give out intense heat and consume very fast. There is also a very peculiar odour which usually accompanies the burning of wood. Among the products of its destructive distillation, which differ in many respects from those of coal, are Wood-naphtha, Vinegar, and Kreasote; and it may, therefore, be expected that the vapours of the burning wood will, according to the low or high temperature of the fire, partake more or less of the odour of these.

142. Combustible as are all kinds of coal, they must, nevertheless, be heated to some extent, before they will inflame. For domestic purposes, this heating is accomplished by the use of the more readily combustible materials, paper and wood. These are laid in the grate and the coal heaped on them in such a manner as to allow the air free access through them. When the coal is heated below redness by the lighted wood, its Hydrogen unites partly with its own Oxygen and partly with that of the air, and escapes up the chimney in the form of steam. The heat being at first below that at which Carbon

ignites, a quantity of it passes off unconsumed as Smoke. Soon, however, the heat increases by continued chemical action, gases are formed which kindle on contact with atmospheric oxygen, and burn as long as the coals continue to supply volatile matter. After the flame has ceased, the coals retain their glow as long as any Carbon is left, and at length nothing remains in the grate but the incombustible ash.

143. As to the temperature of a fire, a dull red heat, visible in daylight, is probably equal to 525° C., a full red heat to 1000° C , an orange heat to 1100° C., and a white heat to 1400° C. The latter is never attainable in a common fire-place.

144. The chief results of the full combustion of coal are Carbonic anhydride and Water, with small quantities of Sulphurous anhydride and Ammonia. In a good grate, therefore, all the gases and vapours produced, are carried up the chimney, upon the draught or current in which, the more or less perfect combustion of the coal depends. A fire thus serves the purpose both of ventilation and warming; and the larger the fire in a room, the more necessary it becomes to keep it supplied by fresh air from without; the draught in the chimney depending entirely on the replacing with colder air the rarefied air that ascends the flue.

145. The waste of fuel that takes place when much of it escapes in the form of smoke, is mainly due to the insufficient supply of air, and may be prevented by judicious management. The very common practice of heaping up a fire with fresh coal, when it has burnt low, is clearly objectionable, because by this it is so much chilled, that the fuel, instead of burning, evaporates in the form of unconsumed smoke and gases.

146. The effect of heat on any substance, depends on the presence or absence of air, that is, of Oxygen. The results of heating coal in air have been sufficiently explained: but if it be heated in the absence, or with very slight access, of air, by a kind of smothered combustion, as in the manufacture of charcoal, it is much altered, but not burnt. The gas is driven off, and the greater part of the carbon remains unburnt as Coke,—a kind of graphite, containing, in addition to the Carbon, all the ash of the coal. It is a good conductor heat, and well adapted for fuel, as it burns without smoke, and

with an intense, steady heat. Coke, for use in locomotive engines, is largely manufactured, by burning the coal in ovens constructed expressly for the purpose.

147. The value of coal, as, indeed, of all fuel for commercial purposes, depends on the amount of heat generated by its combustion; and this is expressed by the quantity of boiling water which a pound of coal, while burning, will evaporate or boil away. Thus, a pound of bituminous coal will boil away about 11·18 pounds of water.

One pound of hydrogen will boil off		. .	43·94 lbs	of water.
,, ,,	pure charcoal	,,	14·41 lbs	,,
,, ,,	common charcoal	,,	13·95 lbs	,,
,, ,,	oil	,,	17·75 lbs	,,
,, ,,	turf	,,	4·65 to 5·58 lbs	,,

148. The great difference, then, between coal and coke, is the presence of volatile matter in the former, and its absence in the latter. The power of coal to burn with flame, depends upon this volatile matter, which, on the application of heat, is converted into gas.

149. Of the numerous services which chemistry has rendered to household economy, there is certainly none which possesses more general importance and interest, and none that has been more fully developed, than that of lighting by means of gas, obtained by the destructive distillation of coal. Although its application as a means of procuring artificial light is of modern invention, yet the germ of it may be traced back nearly two hundred years. Dr. Robert Clayton, Bishop of Cork and Ossory, first distilled coals in a retort and obtained gas. But the first application of coal-gas to illumination was made in 1792, by Mr. Murdoch. Novel as was this mode of lighting his own dwelling, it did not attract public attention until the year 1802, when the front of the Soho manufactory of Messrs. Boulton and Watt was lighted by him with a public display of gas-lights, on occasion of the national illumination in the spring of that year, at the peace of Amiens. In 1810 the National Light and Heat Company obtained an Act of Parliament for the purpose of lighting the streets of London; and though the new system was vehemently opposed at first, gas had, by the year 1825, superseded every oil-lamp in the metropolitan thoroughfares, with the exception of one or two benighted parishes.

150. The manufacture of COAL-GAS is at present a trade of great magnitude and importance, employing an immense amount of capital and a great number of persons.

For the production of Gas, the Coal is placed in closed oblong cylinders of cast iron, called Retorts, which are ranged in furnaces to keep them at a red-heat, the volatile products being conveyed by pipes in connection with the retorts, to a condensing vessel kept cold by immersion in water. Here all the condensible vapours, such as Water, Tar, and Ammonia, as well as some other impurities, are retained; and the gas, still very impure, from the presence of Carbonic anhydride and Hydrogen Sulphide, is then passed over Lime, or Calcium oxide,* in vessels called Purifiers. In this process, the lime unites with and retains the Sulphur and Carbonic anhydride, but does not act on the gas, which, still further purified, is then passed into Gas-holders where it is stored, for subsequent distribution by means of iron-pipes to the places where it is required for use.

151. For gas illumination, the value of coal is proportional to the quantity of volatile matters which it disengages at a red-heat. Different coals vary greatly in this respect: for gas-making, a ton of coal should yield not less than 9,500 cubic feet of gas of specific gravity 0·42.

Boghead cannel coal	yields	14 to 15,000	cubic feet of gas.
Newcastle Wallsend	„	12,000	„ „ „
Wigan coal	„	10,850	„ „ „
West Hartley	„	10,500	„ „ „
Newcastle cannel coal	„	9,833	„ „ „

In actual practice, however, some of this yield is lost by leakage.

152. Some idea of the enormous extent to which gas is manufactured, may be formed from the statement, that several of the London gas works supply more than 1,500,000 cubic feet, in a night in the middle of winter.

153. Coal-gas is a mixture of several gases, the most important

---

* When Calcium oxide CaO is brought together with Carbonic anhydride CO_2, Calcium Carbonate $CaCO_3$ results. So, also, when Hydrogen Sulphide H_2S is placed in contact with Lime; Calcium Sulphide CaS and Water H_2O are produced. Thus both compounds are held back by the Calcium oxide.

of which are certain compounds of Carbon with Hydrogen, commonly known as Hydro-carbons, Hydrogen, Carbonic oxide and Nitrogen.

154. Hydrogen, which has already been so frequently mentioned, is to be met with in all ordinary flames. This gas is the lightest substance in nature, and on this account is often employed for filling small balloons, an illustration of the mode of filling which is given in Fig. 53. It does not support combustion, but is itself inflammable, burning with a pale flame at the expense of the Oxygen of the air to Hydrogen oxide or Water. Its light is, indeed, so feeble, that were it not for the affinity it displays for Carbon, with which it forms light-giving gases, it would be quite useless for purposes of illumination. In coal-gas, Hydrogen exists in considerable quantities.

Fig. 53.

155. The two chief compounds of Carbon and Hydrogen found in coal-gas, are known as Light and Heavy Carbonetted Hydrogen. They are the main sources of its illuminating power.

156. LIGHT CARBONETTED HYDROGEN, OR HYDRIDE OF METHYL, is the principal component of coal-gas. In many coal mines it is also found abundantly collected between the seams of coal, escaping from which, it mingles with the air of the mine, and is the common cause of fatal accidents by its explosive inflammation. It is known to miners as Fire-damp and Wild-fire. It is likewise evolved in the decay of moist vegetable matter; and as decay, like life, is in perpetual activity on the earth, this gas is generally contained in the air (p. 84, par. 99). It may be readily obtained by stirring the mud at the bottom of stagnant pools (whence it has the name of marsh gas), and collecting it, as it escapes, in an inverted jar full of water (Fig. 54.)

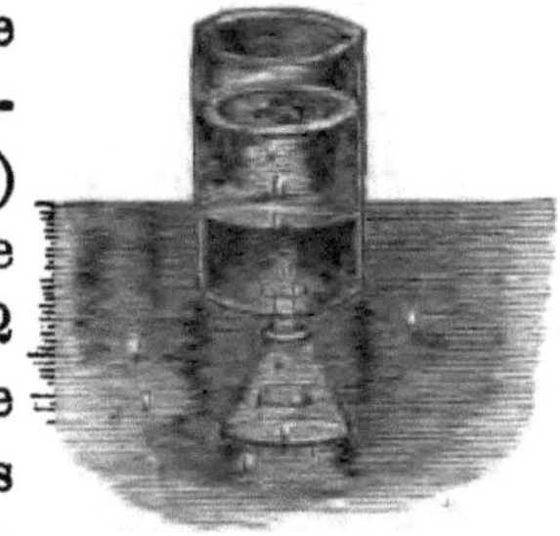

Fig. 54.

This marsh gas burns with a yellowish-white flame. If mixed with 10 measures of air or 2 measures of Oxygen, it explodes with violence on the application of flame. The sole products of its combustion in air, are Carbonic anhydride and Water.

157. HEAVY CARBONETTED HYDROGEN, OLEFIANT GAS, OR ETHY-

LENE, is likewise contained in coal-gas, and is, indeed its most important constituent for the purpose of illumination. It burns quietly with a white light; but, when one measure of Ethylene is mixed with 15 measures of air, or with 3 measures of Oxygen, it detonates with violence on the approach of flame.

158. These two gases are the chief HYDROCARBONS,† but the illuminating power of gas is much assisted by the presence of others. BUTYLENE, or oil-gas, has twice the light giving power of ethylene, and is found sometimes to the extent of 2·38 per cent. in coal-gas. ACETYLENE is also present. But coal-gas contains also NITROGEN and CARBONIC OXIDE.

159. CARBONIC OXIDE is not only a constituent of coal-gas, but is produced abundantly in the open fire, and is often seen as a blue flame flickering over the hot coals. In this case Carbonic anhydride is always formed first, but as it passes over the red-hot coals, it unites with more Carbon, and becomes changed into Carbonic oxide. The heated gas takes fire when it mixes with the air, and passes once more into the condition of Carbonic acid.

Carbonic oxide is also abundantly produced in the combustion of charcoal in stoves. It is a very poisonous gas, and is very often the cause of fatal accidents when respired.

---

† HYDRIDE OF METHYL $CH_4=16$. A colourless gas, eight times as heavy as its own volume of Hydrogen. One atom of the gas occupies the space of two atoms of Hydrogen. When one volume is mixed with two volumes of Oxygen and detonated, we obtain one volume of Carbonic anhydride and two volumes of watery vapour: the latter condenses. It is obtained by heating Sodium acetate with Sodium hydrate. Thus :—$NaC_2H_3O_2+NaHO=Na_2CO_3+CH_4$.

ETHYLENE $C_2H_4=28$. A colourless gas, 14 times heavier than Hydrogen. Requires 3 vols. of Oxygen for complete combustion: we obtain 2 vols. of Carbonic anhydride, and 2 vols. of watery vapour which condenses. It is made by heating Sulph-ethylic acid to the point of decomposition; $C_2H_5HSO_4=H_2SO_4+C_2H_4$.

ACETYLENE $C_2H_2=26$. A colourless gas. May be prepared by synthesis of the elements.

BUTYLENE $C_4H_8=56$. A colourless gas. 28 times heavier than Hydrogen. One volume requires six volumes of Oxygen for combustion and affords four volumes of Carbonic anhydride and four volumes of watery vapour, which condenses.

160. But as all the gases hitherto mentioned as present in coal gas are inodorous, whence does it derive the peculiarly offensive smell by which it may always be recognized?

This odour is partly owing to the presence of minute quantities of BENZOLE and NAPHTHALINE†; partly also to that of the vapour of CARBONIC SULPHIDE (p. 95. par. 120). If it be asked, whence does the coal-gas obtain its Carbon di-sulphide, the answer is readily given. Nearly all kinds of coal contain Iron pyrites. At a red-heat, this mineral loses a portion of the Sulphur which is in combination with the Iron, and the Sulphur unites with Carbon to form Carbonic sulphide, with Hydrogen to form Hydrogen Sulphide, and with Ammonium to Ammonium Sulphide. The two latter compounds are retained by Water and by Calcium oxide, whilst the Carbonic sulphide volatiles and defiles the Coal-gas.

The removal of the odorous principles in coal-gas offers special difficulties, and it may well be doubted whether it is desirable, considering that the strong smell is an excellent safe-guard against accidents, and that well-made gas-fittings, in good order, are amply sufficient to prevent its escape. Apart from this, however, it may be safely asserted, that the purification of coal-gas could with propriety be carried considerably beyond the present practice.

---

† BENZOLE C_6H_6=78. Is a colourless, mobile liquid, of sp. gr. 0·85. It is very volatile and has a very peculiar gas-like smell. Benzole is one of the best solvents of fats, oils, and caoutchouc, and it is prepared from oil of coal-tar. It is insoluble in water and boils at 80° C. Nitric acid changes benzole into Nitro-benzole $C_6H_5NO_2$, a very fragrant oil-like compound; thus: $C_6H_6 + HNO_3 = H_2O +$ Nitro-benzole $C_6H_5NO_2$.

NAPHTHALINE $C_{10}H^8$=128. A solid, crystalline substance, in shiny plates. Its odour is aromatic, reminding of gas. It melts at 79° C., and boils at 220° C. It is best obtained from the last portions of the distillation of coal-tar. Although a solid, Naphthaline is often deposited in considerable quantities in gas-pipes when Coal-gas is made and distributed rapidly after its manufacture.

CHAPTER VIII.

ON FLAME, AND THE CHEMISTRY OF A CANDLE.

161. What is flame? The hollow structure thereof demonstrated. 162. Oxy-hydrogen flame. 163. Light of flames due to heated Carbon. Proofs. 164. Mechanism of flame. 165. Bunsen's burners. 166. Heat increased by Blowpipe. 167. Composition of candles. Brunnen's respirator. 168. High temperature of flame. Hemming's jet. Davy Lamp. 169. Illuminating flames principally Hydro-Carbons. 170. Employment of different kinds of burners to obtain different amounts of light. 171. Length of flame in an Argand burner. Chimney glasses. 172. Arrangements of jets. 173. Bude Light. 174. The rate at which candles burn. 175. Burning of Coal-gas in inhabited rooms. 176. Nevertheless, there are advantages in the use of Coal-gas. 177. Amount of Carbonic anhydride generated per hour from different flames. 178. Photometers. 179. Manufacture of candles. 180. Use of wicks. 181. Capillary attraction. Chemical composition of wicks. 182. Dips and moulds. 183. Tallow. 184. Palm candles. 185. Wax candles. 186. Spermaceti. 187. Paraffine.

161. What is flame? It has been seen already, that the difference between coal and coke, the one burning with flame, the other without, depends on the presence or absence of gas. The removal of the gas prevents the formation of flame. The inference is simple, that flame is the burning gas. This, however, might be, and for a long time was, thought inapplicable to the flame from solid or liquid bodies, as candles or oil. But oil, when heated, is readily decomposed and changed into gas; and that the same is true of a candle, is proved by a very simple experiment, with which every one is familiar. If a candle, with a glowing wick, is blown out, there rises from it, for a few seconds, a small cloud of smoke, which may be lighted even at a distance from the candle. What is this but gas? Flame, then, is burning

gas. And inasmuch as combustion is the result of the union of the gas with the oxygen of the air, it is clear that the gas can only burn where it comes into contact with this oxygen—that is, at its surface.

Hence, flame is hollow: it is a luminous envelope to a quantity of gas. The hollow structure of flame may be readily shown by experiment. By inserting into the flame of a candle, *a*, one end of the glass tube, *b*, the unburnt gases of the interior may be drawn off, and inflamed at the other end of the tube, Fig 55.

Fig. 55.

162. The simple burning of a gas would not, however, account for the light it gives. The excessively feeble light of burning Hydrogen has already been noticed. It is scarcely visible in daylight. And even the flame of Oxy-hydrogen, though possessing the greatest heat of all artificial flames, is in no degree more luminous.

Oxy-hydrogen is a mixture of Oxygen and Hydrogen in the proportions to form water. As has been fully explained, Water is a compound of Hydrogen and Oxygen: it consists, in fact, of one atom of Oxygen and two atoms of Hydrogen. If these two gases be mixed together in the proportions of two volumes of Hydrogen with one volume of Oxygen we obtain Oxy-hydrogen: it is only on combustion that water is formed. Fig. 56 represents the method of filling the bag, *d*, with the mixed gases. *f* represents the pneumatic trough, and *a*, the water in the trough. The tap, *e*, cuts off communication from *b*, when the bag is filled; and, if any gas should remain in the bell jar, *c*, it is preserved by closing the stop-cock, *b*.

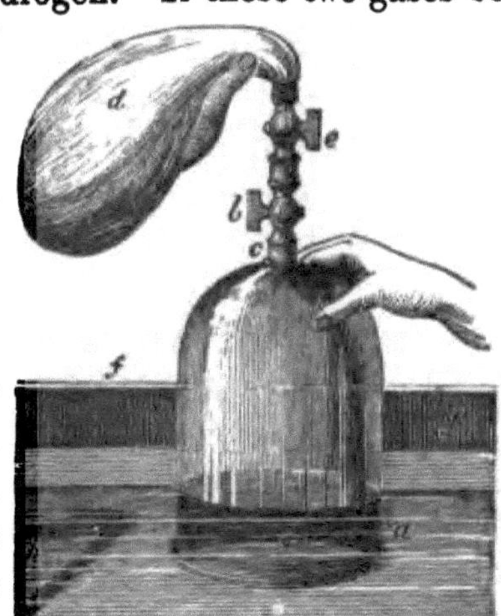

Fig. 56.

The flame of Oxy-hydrogen is solid. When fired in considerable quantity, it detonates loudly: the water formed, issues as highly-heated steam, which is instantly condensed.

and its place supplied by air,—the collision of the particles of which occasions the noise.

The introduction of any solid object, such as wood, or Platinum, into such a flame, immediately renders it luminous. One of the most intensely brilliant lights known, the Lime-light or Oxy-calcium light, is obtained by introducing a piece of Lime into the feeble Oxy-hydrogen flame; when its rays of light are concentrated by a parabolic mirror, the light has been visible at a distance of 112 miles.

163. The solid matter which gives LIGHT to the flame of a candle, or of gas, is FINELY DIVIDED CARBON. The gases, consisting of Hydrogen, Hydride of Methyle and Ethylene, under the influence of heat, are at first decomposed. In the portion of the flame where this takes place, there is only enough Oxygen to burn the Hydrogen, which inflames on account of its greater affinity for Oxygen. The Hydrogen, in burning, forms Water, and gives out so much heat, that the liberated Carbon, instead of being deposited in the form of soot, is heated to whiteness, but does not burn until it reaches the very outermost part of the flame, where, finding a free supply of oxygen, it is converted into Carbonic anhydride.

The presence of free Carbon in every light-giving flame is easily shown, by depressing a plate, or other cold object, upon it; the Carbon is immediately deposited as lamp-black. It is often unpleasantly manifested by a smoky flame, and the deposit of soot on the ceiling. A flame carried rapidly through the air always smokes, because the quick accession of air cools it below the point necessary for the complete combustion of its Carbon.

Fig. 57.

164. This, then, is the mechanism of Flame. It may be seen by the eye to consist of three parts. A quantity of undecomposed gas, ready to burn, forms the centre of the flame, as a dark cone, Fig. 57: in the bright, or light-giving part, this gas is decomposed, and the burning Hydrogen ignites the particles of Carbon; whilst in the outermost portion of the flame this Carbon itself burns, forming Carbonic anhydride which, with the water from the Hydrogen, forms a thin atmosphere round and above the flame. The hottest part of the flame is near the top, where combustion is complete, while the

temperature in the centre is so low, that gunpowder may be held there without being ignited.

The different characters of these three parts of flame may be well demonstrated by an ingenious experiment suggested by the late Professor Faraday. If a glass syphon, A, be held in the dark centre of the candle flame, D, a quantity of white vapour rises in the tube, which may be collected in the glass (placed on the stand C), Fig. 58, and inflamed. If it be held at the point of the flame, water is collected, and a colourless gas, found to be Carbonic anhydride, fills the glass. But, if the syphon be depressed into the light-giving portion of the flame, the tube and reservoir become filled with blacks, or Carbon.

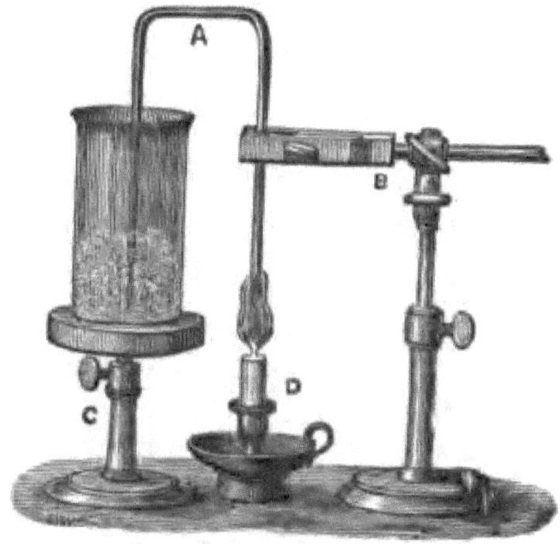

Fig. 58.

165. At the very bottom of the flame, as well as at the outside, the gas is in perfect contact with the air, and here, also, complete combustion takes place, so that in this portion of the gas-flame, a colourless semi-circle may always be seen. If a plate be depressed into this part of flame, no soot is deposited on it.

Advantage is taken of this fact, in the various gas-burners employed by professional chemists for heating purposes. Fig. 59 represents a "Bunsen burner," so constructed that the coal-gas mixes in the tube with air, which is admitted through small holes near the base. The mixture burns with a pale flame and with very slight luminosity,

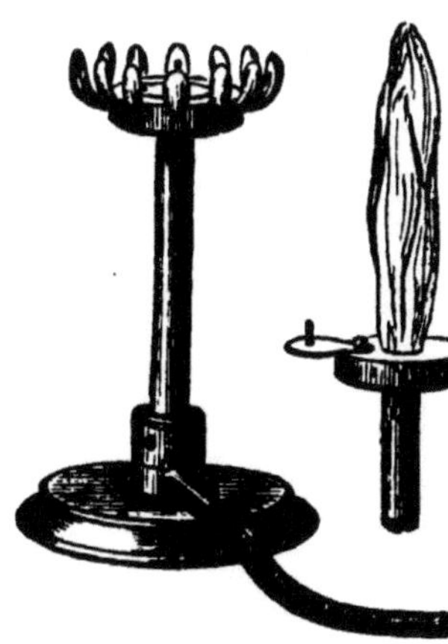

Fig. 59.

owing to the complete oxidation of the Carbon at the same time as the Hydrogen.

166. If the combustion is accelerated by a jet of air, the heat of a flame may be considerably increased; upon this principle depends the action of the BLOW-PIPE. A current of air is projected, by means of the tube, *a*, along the surface of the wick, Fig. 60, and the flame

Fig. 60.

blown out of its vertical position into the form of a pointed cone, in which the three parts of flame are still distinctly visible. In the luminous part are enclosed the yet unburnt combustible gases, possessing such great affinity for Oxygen, as immediately to deprive of that element any Oxide that may be brought into them. Hence, this portion is called the REDUCING OR DE-OXYDIZING FLAME. At the summit of the flame, because fully oxydized, the opposite effect is produced, so that a metal, such as Tin, introduced into it, immediately becomes converted into an Oxide. This part of the flame is hence called, the OXYDIZING FLAME.

167. All the substances from which candles are made, consist of Carbon, Hydrogen, and Oxygen, unless expressly stated to the contrary, in various proportions. The results of their combustion are therefore only Carbonic anhydride and Water. The components of a candle, instead of being destroyed, increase in weight by just that amount of Oxygen with which its Hydrogen and Carbon combine.

The following experiment is derived from Dr. Miller's admirable work on Chemistry. A glass tube, *a*, *b*, Fig. 62, is divided at *c* by a diaphragm of wire-gauze, and the upper part filled with Potassium hydrate, *d*, which will retain the whole of the Carbonic anhydride and Water given off in the combustion of the taper, *b*, placed on a perforated cork, fitted to the lower end of the tube. Fig. 61 represents an ASPIRATOR filled with water, and capable of closure by a stop-cock: its upper end being connected by the bent tube, *a c*, with the principal tube, *a b*. Fig. 61, at *e*. Now, as the water can-

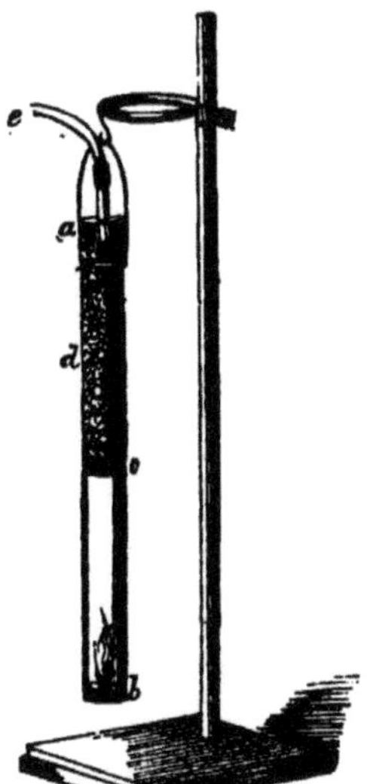

Fig. 62.

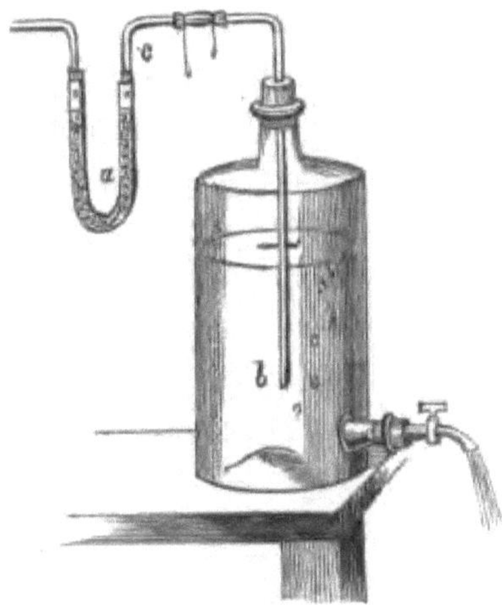

Fig. 61.

not escape from the open tap, unless communication is made with the outer air, on account of the upward pressure of the latter; and as the air can only enter through the perforations of the cork upon which the taper is placed, it follows, that when the tap is turned, a current of air is established through *a b*, and the water flows. When all arrangements for the experiment are made, the tube is weighed, connected at *e* by means of vulcanized india-rubber-tubing with the aspirator, the taper lighted, and the tap opened. The air at *b* supplies the taper with the Oxygen necessary to its combustion; and, if the tube be weighed after the taper has burnt out and the whole cooled down, it will be found to have increased several grains in weight.

168. Flame requires a very high temperature, so that if it be cooled down below a certain point, it is extinguished. The deposit of soot from a flame on a cold plate, or from a burning candle when wafted through the air, already mentioned, are illustrations of the effects of cooling.

169. If Hydride of Methyle or Light Carbonetted Hydrogen,—

the Fire-damp of the Coal-mines, is mixed with more than sixteen times its bulk of air, it can no longer be inflamed, much less exploded, by reason of the cooling power possessed by so large a volume of air. Sir Humphry Davy found that a mixture of explosive gases could not be inflamed through a narrow tube, owing to the cooling influence exerted by the tube on the flame. If wire gauze, with about 400 meshes to the square inch, be depressed into a gas flame, the upper part of the flame will be cut off, because the unburnt gas is reduced in temperature below the point of combustion. It may be kindled by the application of a light above the gauze, and the gas will then continue to burn above and below. The experiment may be varied, by turning on an unlighted jet of gas beneath the gauze; on the application of a light above the gauze, the gas will inflame, but the cooling power of the wire will prevent the flame from passing below.

Fig. 63.

Hemming's Safety-jet, Fig. 63, depends for its efficacy upon this cooling power of metals. The brass tube, which, in the woodcut, is represented as unscrewed from the jet, is filled longitudinally with brass wires. The interstices form metallic tubes, which, by their cooling power, prevent the passage of flame. If, therefore, the tube be screwed on to a bag containing a mixture of the most explosive gases, such as Oxy-hydrogen (p. 111, par. 162), they may be burnt at its extremity quietly, and without danger of explosion.

This principle finds its most perfect and useful application in the Davy Safety-Lamp. It consists of an oil-lamp, Fig. 64, enclosed within a cylinder of fine wire-gauze, *a*, *b*, the lowest part, *c*, of which is glazed. A small movable wire, *w*, passing through a closely-fitting tube, and bent at its upper end, reaches the wick, and can be used to trim it without removing the cage.

When such a lamp is carried into a coal-mine, containing an explosive gas, the flame is sometimes seen to fill the whole interior of the gauze. When that is the case, the miner has notice of danger, and should at once withdraw; for, if the cage should become oxydized, as it would speedily do at that high temperature, several meshes might break, so as to form a hole in the gauze, and an explosion would arise, fatal to the miner either from the flash of the flame, or by the resulting Carbonic anhydride, or from both causes combined.

Fig. 64.

The chief use of the Davy-lamp ought to be, to enable the viewer of the coal-mine to examine it carefully before the miners are allowed to enter in the mine. An explosive atmosphere is, of necessity, unfit for respiration; so that good ventilation, by a double shaft, is the miner's only complete protection from danger to health or life.

169. The above principles find their application in the varieties of flame used for illumination. In the first place, as the light depends on Carbon at a white heat, it follows, that the richer a gas is in Carbon, the higher is its illuminating power. It has been already stated, that the substance on which this power depends in coal-gas, is chiefly Ethylene, or Heavy Carbonetted Hydrogen, which contains just twice as much Carbon, in proportion to its volume, as Hydride of Methyl. The illuminating power of coal-gas depends, therefore, on the relative amount of this heavier gas: consequently, the lighting power may be increased by artificially raising the proportion of Carbon. This may be done by the employment of cannel-coal, in admixture with the ordinary coal used in the manufacture of coal-gas.

170. The degree of light given out by a flame depends on the amount of air supplied to it. If it receives too little air, the combustion is incomplete, the flame smokes, and gives out only an imperfect light. If, on the other hand, it receives too much, it is cooled down, and again its light is diminished. Hence, the best burner will be that, which admits the air to the flame in just sufficient quantity to obtain the greatest light.

By the Bat's-wing and Swallow-tail burners (in the former of which the gas issues through a narrow slit, and in the latter through two), the flame is made as thin and wide as possible, so as to present a large surface to the air. These, accordingly, give considerable light, for a small quantity of gas.

In the Argand burner, the air is made to pass through the interior of the flame, as well as round it. This flame, therefore, condensing, as it were, a large surface in a small space, is the brightest and most economical.

171. The length of the flame is a point worthy of consideration. The most economical length of flame to burn, is about 4½ or 5 inches. In other words, a flame of this length gives most light, in proportion to the amount of coal-gas it consumes. For example, in an Argand burner, a flame only half an inch in length, is six times as expensive, in proportion to the amount of light it gives, as one 5 inches long. The small flame is cooled down by an excessive supply of air.

An Argand burner requires a chimney, partly to increase and direct the draught from below, and partly to regulate the relative supplies of air to the outside and inside of the flame. If the chimney is too large, the flame contracts at the top; if too small, it expands. The best light is obtained when the external and internal supply of air being equal, the flame is cylindrical; as large at the top as the bottom. There should be a slight shoulder to the chimney, to direct the current of air against the flame.

172. Where several jets are directed against one another, a certain amount of light is lost; so that, however beautiful such combinations of jets may be made, they are far from economical.

173. Up to a certain point, an increase in the Oxygen of the air, by accelerating combustion, increases the light of the flame. The original Bude light, introduced by Mr. Gurney, was obtained by passing a jet of oxygen into the centre of the flame of an Argand burner. The light now known by that name, however, is no more than the ordinary flame from a number of Argand burners, arranged concentrically, and fitted with chimneys and reflectors.

We say, an increased supply of oxygen is desirable only up to a certain point. For a gas-flame, surrounded by pure Oxygen, gives very little light, in consequence of the combustion being too rapid. So, too, if the gas be made to pass through the air for a short distance before it is lighted, so as to mix with it, as in the Bunsen burner, the combustion is perfect in all parts of the flame at the same time, and intense heat is obtained, but scarcely any light.

174. The rate at which candles burn is not altered by variations in the atmospheric pressure, but the light emanating from them diminishes directly as the pressure. On the other hand, the light-

giving power of a flame is proportionately increased as the atmospheric pressure is raised. For these facts we are indebted to Dr. Frankland.

175. Many persons object to the burning of coal-gas in inhabited rooms; and it does, undoubtedly, render the air impure more than any other artificial source of light. One volume of good coal-gas produces three volumes of Carbonic anhydride and a large supply of watery vapour. And all gas contains some Sulphur-compounds (p. 109, par. 160), which, when burnt, still further contaminate the air by the Sulphurous anhydride which they contribute to it. These several impurities produce very injurious consequences on the health and comfort of many who breathe them. The deleterious effects of gas-burning have been sometimes strikingly shown by the injury it has inflicted on articles in rooms, such as silver-ornaments, curtains, book-binding, and so forth. These effects can only be wholly remedied by perfect ventilation. Every room in which gas is burnt should have a ventilator in the ceiling over the gas-burner, communicating by a pipe with the outer air or chimney. But the only perfect mode of ventilation, is to have the hot and impure air carried up to this ventilator by a pipe dipping down over each gas-burner: by that means, the air of the room may be kept perfectly pure. Other inconveniences of gas-burning in rooms are, its heat, the unsteadiness of its flame, and the imperfection of its light from a super-abundance of yellow and red rays. This last defect, however, may be greatly mitigated, by the use of shades or chimneys of a pale-blue tint.

176. In spite, however, of all these drawbacks, Coal-gas has numerous advantages over all other sources of illumination. If people would be contented with the same amount of light they were accustomed to, when they burnt candles or oil, then the advantage of coal-gas becomes obvious; not only is much less heat, but less Carbonic anhydride produced. Of course, the greatest advantages of coal-gas consist in its high illuminating power, its relative cheapness and the facility with which the flame can be raised and depressed to any desired point.

177. The following table (taken from Ure's Dictionary) exhibits the amount of Carbonic anhydride and heat, generated per hour,

from various sources of light; the heat generated by tallow being assumed to be 100, for the purpose of comparison.

	Carbonic Anhydride in Cubic Feet.	Heat.
Tallow	10·1	100
Wax and spermaceti	8 3	82
Sperm oil	6·4	63
Coal-gas	5·0	47

178. When light diverges from a luminous source, its intensity diminishes directly as the square of the distance increases. Suppose the flame of a candle to be placed in the centre of a hollow sphere two square inches in diameter, its light will fall upon the whole internal surface of the sphere, and the candle will be one square inch distance from each point. The same candle, placed in the middle of a globe four square inches in diameter, will be two square inches distant from each point of the surface: it will be at double the distance that it was in the first globe, but its light will still illuminate the whole interior. The surface of the second globe, however, is four times greater than that of the first, because the surfaces of spheres are to each other as the squares of their radii; in this case as $1^2 : 2^2$, or as 1 to 4. Consequently, each point of the surface will only receive one-fourth of the light that fell on an equal space in the smaller globe, and yet the candle is only twice as far from it. (Miller's Physics).

An application of this law affords a ready means of determining the relative intensities of two lights. The illuminating power of different flames is estimated by instruments called PHOTOMETERS, a word derived from the Greek φῶς, phos, and μετρέω, metreo, I measure. The simplest photometer is Count Rumford's, in which the two flames to be compared, are made to cast each a shadow of the same vertical post on a screen. The shadow from each light is illuminated by the rays coming from the other light. If the shadows be sensibly equal, the amount of light falling upon the screen from each source, must, at that distance, be equal also: the relative intensities of the two lights are then found, by squaring the distances of each light from the screen.

The standard of comparison, used by different experimenters, has not always been the same; but a spermaceti candle, consuming 120 grains per hour, is now generally adopted. Taking this standard, the relative expense of burning different lights may be easily reckoned. Oil will be found to cost five times as much as gas, tallow candles seven times as much, and wax 19 times as much.

179. It remains now to consider the manufacture of the candle. Nearly all the substances from which our candles are manufactured, are, as has been stated, composed of the three elements, Carbon, Hydrogen, and Oxygen. All of these inflame spontaneously at a high temperature in air, and burn with a bright flame; but the temperature which converts them into gas, a necessary preliminary to the production of flame, decomposes them and they emit a most suffocating and nauseous vapour. It is evident, therefore, that unless by some contrivance the melted materials are made to undergo complete combustion, such substances would be utterly unfit for domestic use.

180. As all materials suitable for candle-making melt at a comparatively low temperature, and become perfectly liquid, it is easy, by the use of Wicks, to accomplish the combustion of a candle without smell. By the term "Wick," is meant the solid body, of whatever material, which forms the axis of the candle, and from which, when lighted, the flame apparently issues. Wicks are generally made of fine cotton threads twisted, plaited, or otherwise combined. In a well-made candle, the Wick should be perfectly free from knots or loose fibres, as their presence always tends to produce the well-known phenomenon of guttering. This effect is caused by the extension of the flame beyond the cup-like hollow in the top of the candle, which ordinarily holds the whole of the melted material, by which the outside edge of the candle is melted and the grease allowed to run down.

The finer the threads that form the wick, the more perfect is the absorption of the material of the candle, and, consequently, the more brilliant the combustion.

181. The difference in the action of a fine and coarse wick depends upon Capillary attraction, a term intended to express an adhesive property, peculiar to liquids, by which they rise, especially in tubes of narrow bore, to a considerable height above the surface of the

fluid in which the tubes are placed. Since adhesion is a surface-power, the more nearly the sides of the tube approach, the higher will be the level to which the liquid rises. The word "Capillary" is derived from the Latin word Capillus, a hair, because the finer and more hair-like the tube, the higher will be the ascent of the liquid in which it stands. Though the fibres of the wick are not themselves actual tubes, yet the spaces between the fibres have precisely the same action, the melted material of the candle being drawn, by capillary attraction, to the top of the wick, where it is inflamed. In finer wicks, the ascent of the liquid is higher, and the flame being further removed from the candle, a smaller portion becomes melted by it; and the combustion being slower, and therefore more complete, the candle burns longer.

In chemical composition, most wicks have the same character as dry wood, and require, for complete combustion, more Oxygen than they can obtain in their position in the centre of the flame; and as, for want of it, the whole of their Carbon cannot be consumed, they need occasional SNUFFING. With a view to the removal of this incomplete combustion of the wick, which not only necessitates the use of snuffers, but interferes materially with the uniformity in the light of the flame, some wicks are so plaited as to untwist in burning, and to bring their ends into the outer flame, and so into contact with the air, where they find the Oxygen necessary for their complete combustion. The same effect is produced in some wicks, by steeping them in solutions of certain Salts containing much Oxygen, which, at the high temperature of the ignited wick, oxydizes and consumes it. In some candles a double or treble wick is used, in order to give a larger illuminating surface. In those known as Palmer's candles, these wicks are twisted spirally, and the gradual burning of the candle causes them to untwist and spread laterally beyond the limits of the flame.

182. Common candles are made by the processes of DIPPING and MOULDING. The former process consists in immersing the prepared wicks repeatedly in a bath of the melted material, until a sufficient quantity has adhered to them, and in this case the candles are called "Dips." In the latter process, the wicks are fixed in the centre of

metallic moulds, of the dimensions of the intended candles, into which the melted fat is poured. These are sold by the name of "Mould candles."

Wax candles cannot be moulded, on account of their liability to contract in cooling, and of the tenacity with which they adhere to the mould. They are commonly made by dipping and basting, and they are then, while still warm, subjected to a process of rolling between marble slabs, to render them of uniform thickness.

183. The commonest kind of candles are, as is well known, made from TALLOW, the fat of oxen and sheep, melted down and freed from its natural impurities. When fresh it has but little smell, and fuses at about 38° C. Its most disagreeable property is the greasy stain which it leaves when handled. Tallow contains about 75 per cent. of a hard, lustrous fat, which melts at about 62° C., and is called STEARINE, from στέαρ, stear, tallow. Stearine itself is a compound of Stearic acid with Glycerine; when this latter is removed, STEARIC ACID remains, a substance which can be made to yield candles of a very superior description, as the melting point is as high as 70°·5 C.

184. But the most important substance in use for candle-making, is Palm oil, from which the candles of Messrs. Wilson, at Vauxhall, are manufactured. The Palm oil of commerce contains about 30 per cent. of a solid fat, which is known by the name of PALMITINE. Like most of the materials employed in making common candles, Palmitine melts at too low a temperature to be generally serviceable. Like Stearine, Palmitine is a compound of Palmitic acid with Glycerine. When PALMITIC ACID is separated from the Glycerine with which it is combined, it presents the appearance of a perfectly white substance, of considerable hardness, melting at 62° C.

185. The WAX used for making candles is derived entirely from bees. This substance is collected by these insects, in very small quantities, from the pollen and other parts of flowers, but is chiefly produced by them from the sugar contained in the nectaries of flowers, or from sugar artificially prepared. As collected from the honeycomb, wax is of various shades of colour, but, after the first purification, is generally yellow. It is melted in hot water, and drawn out into the thinnest possible sheets, which are exposed to the bleaching action of the air, light, and moisture. When pure, Wax

is a very beautiful substance. It is white, transparent, and inodorous, leaving no greasy stain upon the fingers, and not melting below 63° C. It softens, however, between 29° C. and 32° C., and in this state it can be kneaded and moulded. Chemically considered, wax is an exceedingly complex substance, though its constituents are only the very general ones, Carbon, Hydrogen, and Oxygen.

186. SPERMACETI is another material well adapted for candle-making. It is a fat, found in certain cavities in the heads of various species of Delphinus, but chiefly in the Spermaceti whale, Physeter macrocephalus. During the life of the fish, it is found dissolved in spermaceti oil, but, after death, it separates from the oil in hard crystalline masses. It is afterwards further purified by pressure. Spermaceti melts at 49° C. It consists almost entirely of CETINE, a compound of Carbon, Hydrogen, and Oxygen.

187. But if, among the various materials of Candle-manufacture, Palm oil is the most important substance, the most interesting is undoubtedly PARAFFINE, It is obtained in very considerable quantities from Leeswood and Boghead cannel-coal, in which it cannot be said to exist ready formed, but is found among the products of its destructive distillation. It may also be prepared from the products of the distillation of peat and tar.

Chemically speaking, Paraffine may be considered as solid Gas, it being a compound containing the same elements as Ethylene (p. 69, par. 74) in the same relative proportions, but in a more condensed state.

In the candles of this description manufactured by Messrs. Field, of Lambeth, the Paraffine assumes the beautiful appearance of the best wax. They burn with a bright white light, and are in every respect more economical than wax.

We are all, more or less, slaves to fashion, and the prevailing taste for artistic decoration has extended even to the manufacture of candles. Fortunately, in this instance, an addition to beauty is not incompatible with economy—the grooved and coloured paraffine candles are beyond all comparison the prettiest candles made.

CHAPTER IX.

THE CHEMISTRY OF THE FOUR MONAD ELEMENTS, CHLORINE, BROMINE, IODINE AND FLUORINE.

188. The Halogens. 189. Chlorine. 190. Chlorides. 191. Chlorine a bleaching and disinfecting agent. 192. Preparation of Chlorine. 193. Hydrochloric acid. Aqua regia. 194. Oxy-acids of Chlorine. 195. Hypochlorous anhydride and Hypochlorous acid. 196. Chlorous anhydride and Chlorous acid. 197. Chloric peroxide. 198. Chloric anhydride and Chloric acid. 199. Perchloric acid. 200. Nitrogen chloride. 201. Carbon tetrachloride and oxy-chloride. 202. Sulphurous and Sulphuric chloride. 203. Bromine. 204. Preparation of Bromine. 205. Hydro-bromic acid. 206. Combinations of Bromine with Oxygen. 207. Hypobromous anhydride and Hypobromous acid. 208. Bromic acid. 209. Perbromic acid. 210. Nitrogen bromide. 211. Iodine. 212. Preparation of Iodine. 213. Hydriodic acid. 214. Iodine and Oxygen. 215. Iodic anhydride and Iodic acid. 216. Per-iodic anhydride and Per-iodic acid. 217. Nitrogen iodide. 218. Sulphur, Chlorine, and Nitrogen with Iodine. 219. Fluorine. 220. No compound with Oxygen. 221. Hydrofluoric acid. Used in etching glass. 222. Preparation of the acid. 223. Combining number of Fluorine.

VIII. Chlorine $Cl=35{\cdot}5$, $Cl_2=71$.

188. The group of elements which comprises Chlorine, Bromine, Iodine and Fluorine is characterized by their great affinity for Hydrogen and the metals generally. These elements are never found native, or in the free state; but as Chlorides, Bromides, Iodides and Fluorides. Because they produce compounds analogous to Sodium chloride (common salt), they are sometimes called Halogens, or Salt-formers, from the Greek words, ἅλς, *als*, sea-salt, and γεννάω, *gennao*, I generate.

189. CHLORINE $Cl=35{\cdot}5$, was discovered by Scheele, in 1774. It is a gas, possessed of a yellowish-green colour, whence its name, χλωρὸς, *chloros*, yellowish-green; it is suffocating and can only be breathed when largely diluted with air. The specific weight of Chlorine, compared with Hydrogen, is 35·5; in other words, 11·19 litres of the gas weigh 35·5 Grammes: it is consequently 2·47 times heavier than its own bulk of air. Under a pressure of four atmospheres, and at 15°·5 C., Chlorine forms a yellow liquid. Chlorine is also soluble in water, which takes up more than twice its volume at 15°·5 C. A little above the freezing-point of water, a crystalline compound of Chlorine and Water $Cl,5H_2O$ separates in large scales.

190. Chlorine unites with other elements and forms CHLORIDES. For Hydrogen its affinity is so great, that it decomposes water in the Sun-light, setting Oxygen free. $2Cl_2+2H_2O=4HCl+O_2$. And the very same action is brought about when Chlorine and Steam are passed together through a red-hot porcelain tube; the resulting Oxygen may be purified from the Hydrochloric acid which accompanies it, by means of Calcium hydrate.

A taper burns in Chlorine with a reddish light and much smoke, because Chlorine has no affinity for Carbon at common temperatures. Turpentine $C_{10}H_{16}$ bursts into flame when introduced into the gas; Hydrochloric is formed amidst the separation of nearly all the Carbon. Ethylene C_2H_4 when mixed with twice its volume of Chlorine, burns on the application of a light with a beautiful white flame, and with a heavy deposit of soot.

Many metals take fire when immersed in Chlorine, such as Dutch-leaf, Antimony and Arsenicum, and the highest Chlorides of the metals result. Sodium, when heated, burns with a bright orange light to Sodium chloride NaCl, the chief salt of Chlorine.

Nearly all the metallic chlorides are soluble in water: the exceptions are SILVER CHLORIDE AgCl, MERCUROUS CHLORIDE Hg_2Cl_2, and, with difficulty, LEAD CHLORIDE $PbCl_2$. Silver chloride is very soluble in Ammonia, but insoluble in Nitric acid. Silver nitrate is a chief test for Chlorine and soluble chlorides.

191. Chlorine is a great bleaching and disinfecting agent. All Hydrogen compounds are decomposed by it. Inasmuch as the most injurious gases are Hydrogen-compounds, Chlorine unites with the

Hydrogen and decomposes them. Thus, Hydrogen sulphide is decomposed by Chlorine, with deposition of the Sulphur.

192. Chlorine is never found native. It is, however, most easily prepared from its compound with Hydrogen, Hydrochloric acid, by the aid of Manganese peroxide, MnO_2. When heated together, the peroxide oxydizes the Hydrogen of the acid, and forms water; at the same moment a portion of the Chlorine unites with the metal Manganesium, whilst another portion is set free. Manganous chloride $MnCl_2$, remains in the retort in which the operation is conducted.

Fig. 65.

On a small scale, Chlorine may be produced in a Flask as in Fig. 65, and collected, owing to its weight, by displacement. It is, however, more safe to collect it in the pneumatic trough over water.

In symbols, the action is as follows:—$MnO_2 + 4HCl = MnCl_2 + 2H_2O + Cl_2$.

Chlorine may be also prepared by heating Sodium Chloride with Sulphuric acid and Manganese Peroxide: a mixture of Sodium and Manganous Sulphate remains in the retort, whilst Water and Chlorine pass over. In symbols: $2NaCl + 2H_2SO_4 + MnO_2 = Na_2SO_4 + MnSO_4 + 2H_2O + Cl_2$.

193. Hydrogen unites with Chlorine to form Hydro-chloric acid. When equal volumes of Chlorine and Hydrogen are mixed together, they unite spontaneously in the actinic rays of light, and also at the approach of flame. The gases unite with detonation, but, when dry, without condensation.

Hydrochloric acid $HCl = 36{\cdot}5$. Is a colourless gas, irrespirable, uninflammable, and possessed of all the properties of an acid. For water it possesses great affinity: it fumes in air owing to the moisture contained therein. Water at 0° C. absorbs 500 times its measure of the gas: the saturated solution has a sp. gr. of 1·21, and contains 43 per cent. of the gas. When Hydrochloric acid is brought together with Ammonia, it combines with the latter, forming dense white fumes of Ammonium chloride NH_4Cl.

Hydrochloric acid gas reddens litmus paper. 11·19 Litres weigh 18·25 Grammes; it is, therefore, 18¼ times heavier than its own volume of Hydrogen.

The solution of the gas in water is also called Hydrochloric acid, as well as Hydric chloride and Hydrogen chloride. The pure acid has a sp. gr. of 1·1 at 15°·5 C., and may be distilled unchanged at 111°·7 C.: it is not a hydrate, but a solution containing 20 per cent. of the gas. This acid is generally prepared by carefully re-distilling the common *Muriatic acid* of the shops, a strong, fuming, impure solution, coloured yellow by Ferric chloride, and containing also Sulphuric acid, as well as small quantities of common salt.

When pure and diluted with water, Hydrochloric acid does not precipitate Barium chloride; but, if Sulphuric acid is present, a white, insoluble, Barium sulphate $BaSO_4$ results. Hydrochloric acid is easily prepared by distilling Sodium chloride with Sulphuric acid, and collecting the gas by absorption in water. $2NaCl + H_2SO_4 = Na_2SO_4 + 2(HCl)$. Upon a gigantic scale, muriatic acid is obtained in the manufacture of Sodium carbonate; in the first stage of the process, Sodium chloride is changed into Sodium sulphate, and the gas evolved is absorbed in towers filled with moist coke.

The behaviour of Hydrochloric acid towards many metals, consists in the formation of a metallic chloride, and the liberation of Hydrogen. Thus: $Zn + 2HCl = ZnCl_2 + H_2$. When it is brought to a metallic oxide, a metallic chloride is formed, together with water. Thus, with Silver oxide Ag_2O, we obtain Water H_2O and Silver chloride $AgCl$. $Ag_2O + 2HCl = H_2O + 2AgCl$. With Ferric oxide, Hydrochloric acid produces water and Ferric chloride: $Fe_2O_3 + 6HCl = 3H_2O + Fe_2Cl_6$. Its behaviour towards a hydrate will be equally well understood by three examples. $KHO + HCl = H_2O + KCl$. $BaH_2O_2 + 2HCl = 2H_2O + BaCl_2$. $Fe_2H_6O_6 + 6HCl = 6H_2O + Fe_2Cl_6$.

Hydrochloric acid is employed for making Carbonic anhydride from marble: $CaCO_3 + 2HCl = CaCl_2 + H_2O + CO_2$. Also for preparing Hydrogen sulphide from Ferrous sulphide: $FeS+2HCl=FeCl_2+H_2S$. And from Antimonous sulphide: $Sb_2S_3+6HCl=Sb_2Cl_6+3H_2S$. It is also used in making Chlorine, by its action upon Manganese peroxide: $4HCl+MnO_2=MnCl_2+2H_2O+Cl_2$. A mixture of Hydrochloric acid and Nitric acid, constitutes AQUA REGIA, the great solvent of Gold and Platinum. The two acids decompose each other, Chlorine is evolved together with red fumes of a gas called NITRIC OXY-CHLORIDE $NOCl_2$, and Water separates at the same time. $6HCl+2HNO_3=4H_2O+2(NOCl_2)+Cl_2$. The metals dissolve as chlorides.

194. The Oxy-acids of Chlorine are four in number. They are all mono-basic and may be regarded as oxides of Hydrochloric acid.

1. Hypochlorous acid $HClO$.
2. Chlorous acid $HClO_2$.
3. Chloric acid $HClO_3$.
4. Perchloric acid $HClO_4$.

[195. HYPOCHLOROUS ANHYDRIDE $Cl_2O=87$. A pale-yellow gas of peculiar odour. 11·19 litres weigh 43·5 grammes. It is prepared by the action of Chlorine upon excess of Mercuric oxide: Mercuric oxy-chloride remains in an insoluble form. Thus: $2HgO+2Cl_2=HgO,HgCl_2+Cl_2O$. It can be condensed into a deep-red liquid.

HYPOCHLOROUS ACID $HClO$ is readily formed when the anhydride is put into water. The latter absorbs 200 volumes of the gas. $Cl_2O+H_2O=2HClO$. The acid is pale-yellow and acrid. It is a great bleaching and oxydizing agent. Hydrochloric acid decomposes it with formation of Water and Chlorine. $HCl+HClO=H_2O+Cl_2$. A solution of Hypochlorous acid is readily made by passing Chlorine into Calcium carbonate stirred up in water. Thus: $2Cl_2+CaCO_3+H_2O=CaCl_2+CO_2+2HClO$. Salts are called Hypochlorites. From Calcium Oxy-chloride, or Chloride of lime, the acid may be obtained by distillation with half the quantity of acid necessary to neutralize the Calcium: $2CaOCl_2+H_2SO_4=CaSO_4+CaCl_2+2HClO$.

196. CHLOROUS ANHYDRIDE $Cl_2O_3=119$. Is a gas of a deep yellowish-green colour. It decomposes with loud detonation at 55° C., and is, therefore, dangerous to prepare. It is anomalous in composition as the molecule occupies three volumes; 11·19 Litres weigh 39·66 Grammes. Chlorous anhydride is best prepared by the action of Nitrous acid upon Potassium chlorate. $2HNO_2+2KClO_3=2KNO_3+H_2O+Cl_2O_3$.

Chlorous acid $HClO_2$. Is formed when Chlorous anhydride is absorbed by water: the latter dissolves 6 times its volume of the gas. It possesses great bleaching properties. Salts are called Chlorites.

197. Chloric peroxide $ClO_2=67\cdot5$. A gas of a very deep-yellow-colour, far less irritating than Chlorine. It explodes at about 63° C., and is consequently dangerous to make. Water dissolves 20 volumes of the gas, but the solution possesses no acid properties. At—20° C., it condenses to a red liquid. The gas is best made by the action of sulphuric acid upon Potassium chlorate. $3KClO_3+2H_2SO_4=2(KHSO_4)+H_2O+KClO_4+2ClO_2$.

198. Chloric anhydride Cl_2O_5. Has not yet been prepared.]

Chloric acid $HClO_3=84\cdot5$. Is a well-known acid, especially in its salts, the Chlorates. It forms a syrupy liquid, powerfully acid. At a little above 38° C., it decomposes into Perchloric acid, Water, Oxygen and Chlorine: $8(HClO_3)=4(HClO_4)+2H_2O+3O_2+2Cl_2$. Chloric acid is made from Potassium chlorate by means of a solution of Silico-fluoric acid: insoluble Potassium Silico-fluoride separates from the solution of Chloric acid. $2(KClO_3)+H_2SiF_6=K_2SiF_6+2HClO_3$. A chlorate is readily formed when Chlorine is passed into a hot solution of Potassium-hydrate; 3 molecules of Chlorine and 6 molecules of Potassium-hydrate furnish 5 molecules of Potassium chloride, 3 molecules of water and 1 molecule of Potassium chlorate: $3Cl_2+6(KHO)=5KCl+3H_2O+KClO_3$. All chlorates are decomposed by heat: Oxygen is expelled and a chloride generally remains. All chlorates are soluble in water. They detonate on red-hot charcoal. They explode violently when rubbed together with Sulphur or Phosphorus.

[199. Perchloric acid $HClO_4=100\cdot5$. An oily liquid, which forms with water a crystalline hydrate. It distils at 110° C., readily attracts moisture, and explodes vehemently under conditions not always intelligible. It is best prepared by distilling one part of Potassium chlorate with 4 parts of Sulphuric acid. $8KClO_3+4H_2SO_4=4K_2SO_4+2H_2O+3O_2+2Cl_2+4HClO_4$. Nitric acid also decomposes Potassium chlorate, and the resulting mixture of Potassium nitrate and perchlorate can be separated by crystallization, as the Potassium perchlorate is so little soluble in water. $8KClO_3+4HNO_3=4KClO_4+4KNO_3+2H_2O+3O_2+2Cl_2$. Potassium chlorate when heated, in making Oxygen, also affords Potassium perchlorate if the operation be broken off when the mass becomes pasty. $2KClO_3 = KCl + KClO_4 + O_2$.

200. Chlorine unites with NITROGEN. NITROGEN CHLORIDE NCl_3, $NHCl_2$ is a liquid of sp. gr. 1·653, the most detonating of all agents. It explodes spontaneously below 100° C., as also at common temperatures, when brought together with inflammable bodies like Phosphorus, the fixed oils, oil of turpentine; also with the alkalies. It is prepared by placing a jar full of Chlorine over a solution of Ammonium chloride.

That Chlorine decomposes ammonia, with formation of Hydrochloric acid and Nitrogen, should be remembered. $2NH_3 + 3Cl_2 = 6HCl + N_2$.

201. Chlorine in a free state has no affinity for CARBON. But indirectly, by the decomposition of Ethylene C_2H_4, four compounds can be prepared. C_2Cl_2; C_2Cl_4; C_2Cl_6 and CCl_4. The latter, or CARBON TETRACHLORIDE CCl_4 is a colourless liquid: it is the true Carbonic chloride.

CARBONIC OXY-CHLORIDE $COCl_2$ is a colourless gas, suffocating, irrespirable. By water it is decomposed: $COCl_2 + H_2O = 2HCl + CO_2$. Equal measures of Chlorine and Carbonic oxide in the sunlight condense into half their volume of Carbonic Oxy-chloride.

202. Chlorine forms two compounds with SULPHUR. SULPHUROUS CHLORIDE S_2Cl_2 is a yellow, volatile liquid, fuming on exposure to air. It is prepared by passing Chlorine into excess of melted Sulphur. SULPHURIC CHLORIDE SCl_2 is a deep-red fluid, prepared from the former by saturation with Chlorine.]

IX. Bromine Br = 80. Br_2 = 160.

203. Bromine is a monad element never found native. It was discovered by Balard, in the mother-liquor of sea-water, wherein it is found, in quantities not exceeding one grain per gallon, as Magnesium bromide $MgBr_2$.

Bromine is liquid at temperatures above —12°·5 C., and of deep reddish-brown colour. Its vapour, which is of the colour of Nitrogen peroxide, is very irritating and more injurious even than Chlorine:

indeed, it derives its name from *βρῶμος, bromos*, a bad smell. Bromine boils at 64° C., volatilizes and condenses in blood red drops. Its sp. gr. is 2·97. In water it is very little soluble; in ether, however, it is so very soluble that it can be extracted even from its solution in water. With starch, Bromine unites to form an orange-coloured Bromide of starch. Bromine, in the condition of vapour is 80 times heavier than Hydrogen; 11·19 Litres weigh 80 Grammes.

Like its analogue Chlorine, it forms with the Metals, Salts, called BROMIDES. SILVER BROMIDE $AgBr$ is yellowish, insoluble in water, insoluble in nitric acid, and requires a large amount of Ammonia for solution.

204. Chlorine displaces Bromine from all its combinations, except those with Oxygen. The first step in obtaining Bromine, from the Magnesium bromide, is by means of Chlorine; $MgBr_2+Cl_2=MgCl_2+Br_2$. The Bromine is then dissolved out by means of ether and Potassium hydrate added to the ethereal solution: $6KHO+3Br_2=5KBr+KBrO_3+3H_2O$. Ether is distilled off; the Potassium bromate is changed by heat into Potassium bromide, with loss of its Oxygen, and then the Bromine itself is prepared in the same way as Chlorine: $2KBr+MnO_2+2H_2SO_4=K_2SO_4+MnSO_4+2H_2O+Br_2$.

[205. HYDRO-BROMIC ACID $HBr=81$. A colourless gas, soluble in water, with acid and other properties resembling Hydrochloric acid; but its solution oxydizes in part on exposure to air, and its Hydrogen is completely removed by Chlorine. 11·19 Litres of the gas weigh 40·5 Grammes.

When Hydrogen Sulphide H_2S is passed into Bromine, Hydrobromic acid is formed with separation of Sulphur: $2H_2S+2Br_2=S_2+4HBr$. It is best made by causing water to act upon Phosphorous bromide $PBr_3+3H_2O=H_2PHO_3+3HBr$.

206. Bromine unites with Oxygen, but its combinations are unimportant.

207. HYPOBROMOUS ANHYDRIDE Br_2O is formed when the vapour of Bromine is passed over dry Mercuric oxide: $HgO+2Br_2=HgBr_2+Br_2O$. When, however, Mercuric oxide is shaken with Bromine diffused through water, we obtain HYPOBROMOUS ACID $HBrO$. Thus: $2HgO+H_2O+2Br_2=HgO.HgBr_2+2HBrO$.

208. BROMIC ACID $HBrO_3$ is similar to Chloric acid, but more stable. It can be made from Barium bromate $Ba2BrO_3$ by the aid of Sulphuric acid H_2SO_4: insoluble Barium sulphate $BaSO_4$ is then readily separated from the solution of Bromic acid. As has been already stated, a mixture of a Bromate with a Bromide results, when Bromine and Potassium hydrate are brought together; by crystallization the Bromate is separated.

209. PERBROMIC ACID $HBrO_4$. Is made from Perchloric acid $HClO_4$ by means of Bromine.

210. NITROGEN BROMIDE NBr_3. An oily fluid, resembling the chloride in detonating properties.]

X. Iodine I = 127. I_2 = 254.

211. Iodine, like Chlorine and Bromine, is a Monad element never found native. It was discovered in 1811, by Courtois, in the waste liquors from the manufacture of Sodium carbonate from the ashes of sea-weed. The chief source is Sodium iodide, which is obtained from Kelp, the ash of various kinds of sea-weed.

Iodine resembles Graphite in appearance. It is a solid, with lead-grey, metallic lustre. In water, it is but little soluble, but it is very soluble in alcohol—*Tincture of Iodine*, and in ether. Sp. gr. 4·95. It fuses at 107° C., and boils at 175° C. The vapour of Iodine is deep-violet, whence the name, *ἰώδης*, *iodes*, violet-coloured. Externally applied, Iodine is most valuable in glandular swellings; internally, it acts as an irritant poison. It stains and corrodes the skin.

The vapour of Iodine is 127 × heavier than Hydrogen; 11·19 Litres weigh 127 Grammes.

Like its analogues Chlorine and Bromine, it unites with the metals and forms salts, called IODIDES. POTASSIUM IODIDE KI is much used in medicine. SILVER IODIDE AgI is yellow, and insoluble both in Nitric acid and in Ammonia. MERCURIC IODIDE HgI_2 is scarlet: MERCUROUS IODIDE Hg_2I_2 is green. LEAD IODIDE PbI_2 is bright yellow. Iodides are decomposed by Chlorine and by Nitric acid, but Potassium iodide decomposes Silver chloride. The chief test for Iodine is Starch, with which it forms purple Iodide of starch, rich in colour, but evanescent when heated.

212. Iodine is prepared from kelp, after the removal of certain salts, such as Potassium chloride and Plate Sulphate K_3Na2SO_4, in the same manner as Chlorine and Bromine: $2NaI + MnO_2 + 2H_2SO_4 = Na_2SO_4 + MnSO_4 + 2H_2O + I_2$.

[213. HYDRIODIC ACID $HI = 128$. Closely resembles Hydrochloric and Hydro-bromic acids, but it is much less stable than the latter. Its solution in water oxydizes: $4HI + O_2 = 2H_2O + 2I_2$. The solution becomes deep-brown. Hydriodic acid, however, decomposes both Chloride and Bromide of silver, although both Chlorine and Bromine decompose Hydriodic acid. It may be prepared in the same manner as Hydrobromic acid. Thus: $2H_2S + 2I_2 = S_2 + 4HI$. Again: $PI_3 + 3H_2O = H_2PHO_3 + 3HI$. Hydriodic acid is decomposed in contact with Mercury, into Mercuric iodide HgI_2 and Hydrogen; it is, therefore, best collected by displacement, on account of its greedy absorption by water.

214. Iodine unites with Oxygen. Two anhydrides are known as well as their respective acids.

215. IODIC ANHYDRIDE I_2O_5. May be obtained by carefully heating Iodic acid to 170° C., when the latter breaks up into Water and Iodic anhydride: $2HIO_3 = H_2O + I_2O_5$. IODIC ACID HIO_3, is the analogue of Chloric acid $HClO_3$ and Bromic acid $HBrO_3$. It may be obtained in crystals; its solution is very acid and astringent. Its salts are called Iodates, all of which are sparingly soluble except those of the Alkaline metals. Hydrochloric acid immediately decomposes the free acid as well as iodates.

The action of Potassium hydrate upon Iodine, is similar to that of Bromine. $6KHO + 3I_2 = 5KI + 3H_2O + KIO_3$. From Barium iodate $Ba2IO_3$, Iodic acid is readily obtained by means of Sulphuric acid, as Barium sulphate is quite insoluble in water. But it may also be prepared by oxydizing Iodine with Nitric acid, or, better still, by acting upon Iodine and Water with Chlorine. $5Cl_2 + 6H_2O + I_2 = 10HCl + 2HIO_3$.

216. PER-IODIC ANHYDRIDE I_2O_7. By heating Per-iodic acid to 160° C. $2HIO_4 = H_2O + I_2O_7$. PER-IODIC ACID HIO_4. May be obtained in crystals. It may be prepared from its Lead salt by Sulphuric acid; but it can also be made by the action of Iodine upon Perchloric acid: $2HClO_4 + I_2 = Cl_2 + 2HIO_4$.

217. Iodine unites with Nitrogen. NITROGEN IODIDE NI_3 and

NHI_2. A black powder, detonating often spontaneously and with great violence. It is easily made by digesting Iodine in a strong solution of Ammonia: $3NH_3+2I_2=2NH_4I+NHI_2$.

218. The compounds of Sulphur, Chlorine and Bromine with Iodine are quite unimportant.]

XI. Fluorine F = 19.

219. FLUORINE, in its elementary form, is almost unknown, and this because of its powerful affinities. It occurs mainly in combination with Calcium, as CALCIUM FLUORIDE CaF_2. Fluorine is a colourless gas.

220. No compound of Fluorine with Oxygen is known.

221. HYDROFLUORIC ACID $HF=20$. Is a densely fuming acid, very volatile and colourless. It occasions most painful burns, and is dangerous in the highest degree, except when diluted. It boils at 15°·5 C, and has a sp. gr. of 1·06.

The hydrate $HF,2H_2O$ has a density of 1·15, boils at 120° C., and distils unchanged. It must be preserved in vessels of Silver, Lead or Platinum.

Hydrofluoric acid is much employed for ETCHING glass. It attacks the Silica SiO_2, which is contained in glass in combination with Potassium or Sodium, and forms with its Oxygen, WATER, and with its Silicon, SILICON FLUORIDE. Thus: $SiO_2+4HF=2H_2O+SiF_4$.

Fig. 66.

222. Hydrofluoric acid is easily made by distilling one part of Calcium fluoride or Fluorspar with 2 parts of Sulphuric acid. $CaF_2+H_2SO_4=CaSO_4+2HF$. The still and receiver must be of Lead, Fig. 66.

223. One hundred parts of Calcium Fluoride are composed of 51·25 parts of Calcium and 48·75 parts of Fluorine. As the combining number of Calcium is 20, the combining number of Fluorine is found as follows:—

Calcium per cent.		Comb. number of Calcium.		Fluorine per cent.		Comb. number of Fluorine.
51·25	:	20	::	48·75	:	19

CHAPTER X.

ON THE TRIAD ELEMENT PHOSPHORUS, ITS COMBINATIONS WITH PRECEDING ELEMENTS AND THE CHEMISTRY OF A LUCIFER-MATCH.

224. Phosphorus never native. Origin of the name. Occurrence. 225. Properties of common or vitreous Phosphorus. 226. Atomic weight 31: molecular volume 124. 227. Phosphorus burns in Oxygen, Chlorine and Bromine. 228. Illustrations of the allotropic condition afforded by Phosphorus. Vitreous, white, black and red varieties. Red phosphorus not poisonous. 229. Use of Phosphorus: cast in sticks. 230. Hydrogen phosphide H_3P, 231. Phosphorous chloride PCl_3. Phosphoric chloride PCl_5. Phosphoric oxy-chloride $POCl_3$. 232. Phosphorous bromide PBr_3. Phosphoric bromide PBr_5. Phosphorous iodide PI_3. 233. Phosphorous anhydride P_2O_3: Phosphorous acid H_2PHO_3. Phosphoric anhydride P_2O_5. Meta-phosphoric acid HPO_3. Pyro-phosphoric acid $H_4P_2O_7$. Ortho-phosphoric acid H_3PO_4. Ortho-phosphates. The making of Phosphorus from Calcium ortho-phosphate. Hypo-phosphorous acid HPH_2O_2. Phosphorous oxide P_4O. 234. Phosphorous sulphide P_2S_3. Phosphoric sulphide P_2S_5. 235. Phospham N_2HP. 236. Chemistry of Lucifer-matches. Bryant and May's Safety-matches.

XII. Phosphorus $P = 31$. $P_4 = 124$.

224. THE element Phosphorus is never found native. It was accidentally discovered by Brandt, in 1669. Gahn, in 1769, pointed out the existence of this acid in bones, and Scheele devised a process for extracting it. It is called Phosphorus from its luminosity in air, from φῶς, *phos*, light, and φέρειν, *pherein*, to carry. In an oxydized form, Phosphorus is an abundant constituent of the animal and getable kingdoms, and also, in smaller quantities, of all fertile ls.

225. Phosphorus is a pale, wax-like, lustrous solid, of sp. gr. 1·83. It is insoluble in water, slightly soluble in alcohol and ether, more so in turpentine and benzole, and very soluble in Carbonic sulphide CS_2, from which it may be crystallized in rhombic dodecahedra. At 44° C., it melts, and boils at 290° C. On account of its inflammability, it must be preserved under water, and for this reason also it requires careful handling, always under water. It is very poisonous.

226. Although the atomic weight of Phosphorus is 31, its molecular weight is 124. P_4 consequently represents the molecule. 11·19 Litres of the vapour of Phosphorus weigh 62 Grammes. The vapour-density of Phosphorus is 4·32.

227. Phosphorus burns in air to PHOSPHORIC ANHYDRIDE P_2O_5, when the Oxygen is in excess; otherwise to PHOSPHOROUS ANHYDRIDE P_2O_3. In Chlorine, it burns spontaneously to PHOSPHOROUS CHLORIDE PCl_3, and in Bromine, to PHOSPHOROUS BROMIDE PBr_3. Owing to its attraction for Oxygen, Phosphorus is a great reducing agent.

228. Phosphorus is a splendid illustration of the allotropic condition of matter. When common or vitreous Phosphorus is kept for some time under water, and exposed to light, it becomes *white*. Its sp. gr. is now 1·515, and it is much more dangerous to handle, from its inflammability at a lower temperature. Again, when Phosphorus is frequently melted, and suddenly cooled, it becomes *black*. But the most important variety is the *red* or *amorphous* Phosphorus. This is made by heating common Phosphorus from 235°—250° C., in tubes from which the air has been expelled by a stream of Carbonic anhydride. Red phosphorus is a powder, of sp. gr. 2·14. It is not soluble in any medium and not poisonous. It may be heated in air to 260° C., when it bursts into bright inflammation and burns to Phosphoric anhydride. Rubbed together, even gently, with Potassium chlorate, it detonates.

229. The chief use of Phosphorus of either kinds, the vitreous or the red, is in making Lucifer-matches. Vitreous phosphorus is generally sold in sticks. When melted, and after it has been strained through wash-leather, the melted Phosphorus is sucked into tubes under water and then transferred to cold water, wherein it is left until it solidifies.

230. Phosphorus unites with Hydrogen. Its chief compound is the analogue of Ammonia, HYDROGEN PHOSPHIDE $H_3P=34$. A colourless gas, of most offensive odour, as of putrid fish, insoluble in water, soluble in alcohol and ether. It is very inflammable, and has a density of 1·185, when compared with air. It burns to Water and Phosphoric anhydride, but generally deposits a portion of its Phosphorus as red Phosphorus. [In a pure state, it is best prepared by distilling Phosphorous acid $\overset{\prime}{H}_2PHO_3$, which breaks up into Orthophosphoric acid H_3PO_4 and Hydrogen Phosphide H_3P. Thus:— $4H_2PHO_3=3H_3PO_4+H_3P$. Chlorine immediately decomposes it. In the same manner as Hydro-chloric acid unites with Ammonia, do Hydrobromic and Hydriodic acid unite with Hydrogen phosphide to H_4PBr and H_4PI. 11·19 Litres of Hydrogen Phosphide weigh 17 Grammes.]

When Hydrogen Phosphide is obtained by the action of Phosphorus on water in presence of certain metallic hydrates, a LIQUID

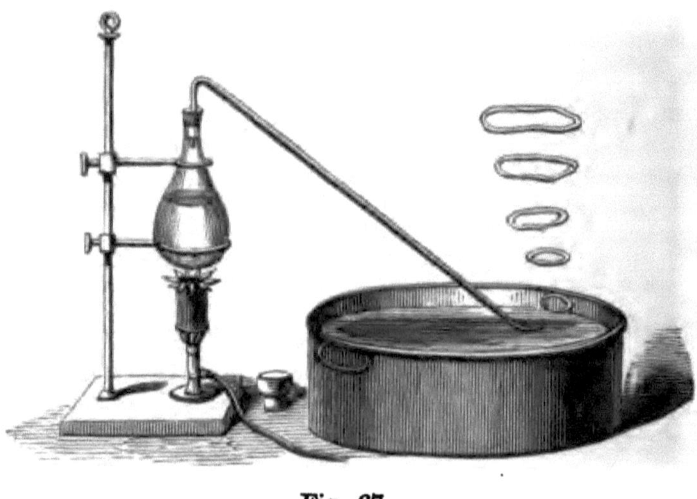

Fig. 67.

Hydrogen phosphide H_4P_2 is formed in minute quantities, which gives to the Hydrogen Phosphide H_3P the power of inflaming spontaneously in air, and burning to Water and Phosphoric anhydride. By heating Phosphorus with water and Potassium hydrate, Potassium hypophosphite is produced in the flask, Fig. 67, and Hydrogen

Phosphide, accompanied by the liquid Phosphide, is abundantly produced. $3KHO + P_4 + 3H_2O = 3(KPH_2O_2) + H_3P$.

[Finally, when liquid Hydrogen Phosphide is passed into Hydrochloric acid, a SOLID Phosphide H_2P_4 is produced. Thus: $5H_4P_2 = 6H_3P + H_2P_4$.]

[231. Phosphorus unites with Chlorine in two proportions. PHOSPHOROUS CHLORIDE $PCl_3 = 137{\cdot}5$. Colourless, volatile liquid. Sp. gr. 1·45. Boils at 78° C. It is made by passing Chlorine through melted Phosphorus: $P_4 + 6Cl_2 = 4PCl_3$. It fumes in air, and is decomposed by water into Hydrochloric and Phosphorous acids: $PCl_3 + 3H_2O = 3HCl + H_2PHO_3$. PHOSPHORIC CHLORIDE $PCl_5 = 208{\cdot}5$. A solid, crystalline compound, obtained by passing Chlorine through Phosphorous chloride. Water decomposes it into Hydrochloric and Phosphoric acids: $PCl_5 + 4H_2O = 5HCl + H_3PO_4$. But, with a small quantity of water, we obtain PHOSPHORIC OXY-CHLORIDE $POCl_3$. Thus: $PCl_5 + H_2O = 2HCl + POCl_3$.

232. Phosphorus unites also with BROMINE and IODINE. The compounds known, are: PBr_3 and PBr_5. PI_3 and P_2I_4.]

233. Phosphorus unites with the Dyads, Oxygen and Sulphur. The oxygen-compounds are very important.

[PHOSPHOROUS ANHYDRIDE $P_2O_3 = 110$. A snow-white powder, produced when Phosphorus is burnt in Oxygen or air, insufficient to produce the highest anhydride. It is more volatile than Phosphoric anhydride. By water, it is converted into PHOSPHOROUS ACID H_2PHO_3. In solution, it is best prepared by the decomposition of Phosphorous chloride with water, and expelling the Hydrochloric acid by evaporation: $PCl_3 + 3H_2O = 3HCl + H_2PHO_3$. It is di-basic. Its salts are hypo-phosphites. It reduces salts of Gold, Silver and Mercury. At a high temperature, it is decomposed into Hydrogen Phosphide and Ortho-phosphoric acid: $4(H_2PHO_3) = H_3P + 3(H_3PO_4)$.]

PHOSPHORIC ANHYDRIDE $P_2O_5 = 142$. A white powder, very deliquescent, with a greedy affinity for water. It is prepared by burning Phosphorus in excess of Oxygen. It is much used as a Desiccator or De-hydrant.

METAPHOSPHORIC ACID $HPO_3 = 80$. The common "Glacial phosphoric acid," produced by heating either Ortho-phosphoric or Pyro-phosphoric acids to redness. When Phosphoric anhydride is dissolved in water, it is formed: $P_2O_5+H_2O=2(HPO_3)$. Its solution coagulates albumen, and cannot be boiled without further assimilating the elements of water: $HPO_3+H_2O=H_3PO_4$. It is a mono-basic acid. Barium and Silver meta-phosphates are white. Sodium meta-phosphate $NaPO_3$ is made, by heating Sodium Ammonium Hydrogen Phosphate: $NaNH_4HPO_4=NH_3+H_2O+NaPO_3$.

PYRO-PHOSPHORIC ACID $H_4P_2O_7=178$. May be obtained in crystals, by evaporation in vacuo. It is tetra-basic. Barium and Silver pyro-phosphate are white. Its chief salt, SODIUM PYRO-PHOSPHATE $Na_4P_2O_7,10H_2O$ is prepared by heating Di-sodium Hydrogen Phosphate to 150° C., or more: $2(Na_2HPO_4)=H_2O+Na_4P_2O_7$. The acid is made from Lead pyro-phosphate $Pb_2P_2O_7$ by Hydrogen Sulphide H_2S.

ORTHO-PHOSPHORIC ACID $H_3PO_4=98$. Is known in the crystalline form. Salts are called ortho-phosphates or tri-basic phosphates. [They may be of three kinds: M_3PO_4; M_2HPO_4 and MH_2PO_4. Silver ortho-phosphate Ag_3PO_4 is canary-yellow, soluble in NH_3 and in HNO_3. DI-SODIUM HYDROGEN PHOSPHATE $Na_2HPO_4,12H_2O$ is the common "Rhombic phosphate." CALCIUM ORTHO-PHOSPHATE Ca_32PO_4 is the chief constituent of "bone-earth." Soluble orthophosphates give a crystalline precipitate of Magnesium Ammonium Phosphate $MgNH_4PO_4.6H_2O$ when Magnesium sulphate, to which Ammonium Chloride has been added, is rendered strongly alkaline by Ammonia: by ignition we obtain Magnesium pyro-phosphate $Mg_2P_2O_7$, containing 63·96 per cent. of Phosphoric anhydride.

When Ortho-phosphoric acid is heated to 213° C., Pyro-phosphoric acid is formed. Ortho-phosphoric acid is most easily prepared by boiling Phosphoric anhydride with water: $P_2O_5+3H_2O=2H_3PO_4$.] CALCIUM ORTHO-PHOSPHATE Ca_32PO_4 is employed in making PHOSPHORUS, as well as the common ortho-phosphates. A portion of the Calcium is removed by Sulphuric acid (as Calcium sulphate $CaSO_4$) from the Calcium Super-phosphate CaH_42PO_4. Thus: $Ca_32PO_4+2H_2SO_4=2CaSO_4+CaH_42PO_4$. The soluble phosphate is filtered off from the Calcium sulphate, boiled down with Charcoal, heated to bright redness in earthen retorts and the Phosphorus collected under water: $3(CaH_42PO_4)+16C=Ca_32PO_4$

$+16CO+6H_2+P_4$. If Di-sodium Hydrogen Phosphate is wanted, the solution of Calcium acid-phosphate is neutralized with Sodium carbonate, which precipitates the Calcium as Calcium carbonate: $CaH_42PO_4+2Na_2CO_3=CaCO_3+CO_2+H_2O+2(Na_2HPO_4)$.

[HYPO-PHOSPHOROUS ACID $HPH_2O_2=66$. An uncrystallizable syrup, feebly acid. Mono-basic. Changed by heat into Ortho-phosphoric acid and Hydrogen Phosphide: $2HPH_2O_2=PH_3+H_3PO_4$. If Cupric sulphate be added to Hypo-phosphorous acid and the liquid warmed, Cupric hydride CuH, is precipitated; on boiling, the latter breaks up into Copper and Hydrogen. The Hypo-phosphites reduce salts of Gold, Silver and Mercury. Barium hypophosphite $Ba2PH_2O_2$, soluble in water, is made by boiling Phosphorus in Baryta water: $3BaH_2O_2+2P_4+6H_2O=2H_3P+3(Ba2PH_2O_2)$: Barium is separated from this salt by means of Sulphuric acid. Thus: $Ba2PH_2O_2+H_2SO_4=BaSO_4+2HPH_2O_2$.]

[PHOSPHOROUS OXIDE P_4O. A red powder, perfectly neutral, obtained always when Phosphorus is burnt in air.]

[234. Phosphorus unites with Sulphur, to form Sulphides corresponding to the above-mentioned oxides. The three best known are: HYPO-PHOSPHOROUS SULPHIDE P_2S, PHOSPHOROUS SULPHIDE P_2S_3, and PHOSPHORIC SULPHIDE P_2S_5. They form double salts with alkaline sulphides. P_4S is also known.]

[235. PHOSPHAM N_2HP. A yellowish-white powder. Of no importance.]

236. The chief use of Phosphorus is in the manufacture of LUCIFER-MATCHES. Some forty years since, the tinder-box with flint and steel to obtain a spark, and the brimstone match to obtain a light, were in request in every household. The discovery that Phosphorus would inflame upon slight friction, and, so inflamed, would light a brimstone match, soon put an end to the use of tinder. In the common lucifer-matches, the wood is dusted over with Sulphur, and plunged, at the ends, into a mixture of Phosphorus and Potassium chlorate. The Phosphorus is generally made up into an emulsion with gum, and subsequently mixed with the Potassium chlorate; the Oxygen of the latter inflames the Phosphorus when the match is moved by friction over sand-paper. Other inflammable materials, such as Paraffine, may be substituted for the Sulphur,

and thus one disagreeable element removed in the construction of the match. But the most serious evil connected with the manufacture, is the injury to the health of those employed, arising from the vapours of the common Phosphorus, which produce painful caries of the jaw. This is entirely obviated by the use of Bryant and May's matches, in which red or amorphous Phosphorus is employed. Not until the match is rubbed upon the phosphorized surface, can it be kindled! In this way, too, many fires are prevented, which arise from the careless use of the common match.

CHAPTER XI.

ON THE TETRAD ELEMENTS SILICON AND TITANIUM, AND THEIR CHIEF COMBINATIONS WITH THE PRECEDING ELEMENTS.

237. SILICON in combination with Oxygen as Silicic anhydride. 238. A tetrad element, with three allotropic modifications like Carbon. Amorphous, Graphitoid and Adamantine Silicon. 239. Silicon hydride, SiH_4, the analogue of Methyl hydride CH_4. 240. Silicon chloride $SiCl_4$. Silicon bromide $SiBr_4$. 241. Silicon Fluoride SiF_4. Etching glass. Silico-fluoric acid H_2SiF_6. [242. Silicon sulphide SiS_2.] 243. Silicic anhydride SiO_2. As Rock-crystal in six-sided striated prisms, terminated in six-sided pyramids. Amethysts. Cairngorms. Properties of Silicic anhydride. Two hydrates. Ortho-silicic acid H_4SiO_4 is tetra-basic. Silicic acid H_2SiO_3 is di-basic. [224. Chryseon $Si_6H_6O_4$. Leukon $Si_3H_4O_5$.] XIII. TITANIUM Ti=50. [245. Titanium native as Rutile, Brookite, and Anatase, in combination with Oxygen as Titanic anhydride. Properties of the metal. 246. Titanic anhydride TiO_2. 247. Titanous oxide TiO. Titanous titanite TiO, TiO_2. 248. Titanic chloride $TiCl_4$. Titanic sulphide TiS_2.]

XIII. Silicon Si = 28·5.

237. THE element SILICON or SILICIUM is never found native; nevertheless, in combination with Oxygen, it is the most abundant constituent of the earth's crust. As Silicic anhydride, in the various forms of Rock-crystal, Flint, Sandstone, &c,, it is a well-recognized mineral, as well as in combination with Alumina as Clay.

238. Like the Tetrad element Carbon, Silicon may be obtained in three distinct allotropic modifications. Si*a*, AMORPHOUS SILICON is a dull brown powder, insoluble in water, which burns when strongly

heated in Oxygen, with great brilliancy to Silicic anhydride SiO_2. Hydro-fluoric acid alone dissolves it. Silicon is most easily prepared by heating Sodium in the vapour of Silicic fluoride. Thus: $SiF_4 + 2Na_2 = 4NaF + Si$. Inasmuch as Silicon has never been volatilized, nothing can be stated with regard to its molecular volume. Siβ or GRAPHITOID SILICON possesses almost metallic lustre, with much of the appearance of Graphite. Its sp. gr. is 2·49. In Oxygen it does not burn, but is slowly oxydized by a mixture of Nitric and Hydro-fluoric acids. Graphitoid Silicon is obtained by fusing Amorphous Silicon with Aluminum, by dissolving the Aluminum in Hydrochloric acid: Siβ remains in hexagonal plates. Siγ or ADAMANTINE SILICON is lustrous and steel-grey, frequently crystallized in six-sided prisms, terminated by three-sided pyramids derived from the octohedron. It is prepared by strongly heating Aluminum in the vapours of Silicon Chloride; thus: $3SiCl_4 + 4Al = 4AlCl_3 + 3Si$. After the volatilization of Aluminum chloride, Siγ remains in the crystalline form just described.

[239. Silicon unites with HYDROGEN to SILICON HYDRIDE $SiH_4 = 32{\cdot}5$. Is a colourless gas, not known in a pure state. In air, it inflames spontaneously, burning to water and Silicic anhydride: $SiH_4 + 2O_2 = 2H_2O + SiO_2$. The best mode of preparation is by decomposing Magnesium silicide with Hydrochloric acid: $Mg_2Si + 4HCl = 2MgCl_2 + SiH_4$.

240. The combination with Chlorine is known as SILICON CHLORIDE $SiCl_4 = 170{\cdot}5$. A colourless liquid, fuming in air, especially when the latter is moist. It is very volatile. By water, it is decomposed into Silicic anhydride and Hydrochloric acid. $SiCl_4 + 2H_2O = SiO_2 + 4HCl$. The liquid boils at 59° C., and has a sp. gr. of 1·52. 11·19 litres of the vapour weigh 85·25 Grammes. Although Silicon chloride may be prepared by burning heated Silicon in Chlorine, it is most easily made, by heating a mixture of lamp-black and Silicic anhydride, in a stream of Chlorine: the Carbon unites with Oxygen to Carbonic oxide, and the Chlorine volatilizes with the Silicon: $SiO_2 + 2C + 2Cl_2 = 2CO + SiCl_4$. SILICON BROMIDE $SiBr_4 = 348{\cdot}5$. A liquid of sp. gr. 2·813, boiling at 153° C.]

241. Silicon unites with Fluorine. SILICON FLUORIDE $SiF_4 = 104{\cdot}5$. A colourless gas, pungent and fuming in moist air. 11·19

Litres weigh 52·25 Grammes. By water, the gas is immediately but partially decomposed. Fluorine has an uncommon affinity for Silicon. The very principle of ETCHING glass depends upon this affinity. When Hydrofluoric acid and Silicic anhydride are brought together, or even with a Silicate, as in the case of glass, Water and Silicon fluoride are immediately formed. By heating together in a retort equal parts of sand and Calcium fluoride, with 12 parts of Sulphuric acid, Silicon fluoride is best obtained. $2CaF_2 + SiO_2 + 2H_2SO_4 = 2CaSO_4 + 2H_2O + SiF_4$. SILICO-FLUORIC ACID H_2SiF_6 or $2HF,SiF_4 = 144·5$. A sour, fuming liquid, forming salts called Silico-fluorides. It is prepared by passing Silicon fluoride into water. The tube from the retort *a*, in Fig. 68, must pass into a glass, *b*, the bottom of which is covered with mercury. If otherwise, the tube would soon be stopped up with Silicic acid. $3SiF_4 + 4H_2O = H_4SiO_4 + 2(H_2SiF_6)$. Silico-fluoric acid is a test for Potassium, as, with its hydrate, it immediately forms Water and an insoluble Potassium Silico-fluoride. $2KHO + H_2SiF_6 = 2H_2O + K_2SiF_6$. This salt is also employed for making amorphous Silicon: $K_2SiF_6 + 2K_2 = 6KF + Si$.

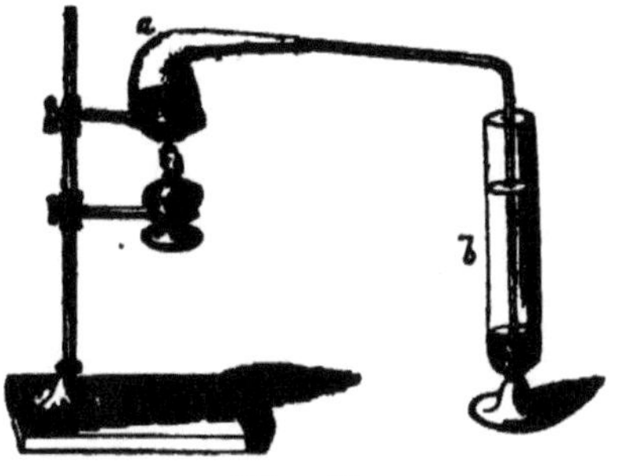

Fig. 68.

[242. SILICON SULPHIDE $SiS_2 = 92·5$. White, earthy-looking solid, soluble in water with decomposition into Hydro-sulphuric acid H_2S and Silicic acid. $SiS_2 + 4H_2O = 2H_2S + H_4SiO_4$. The sulphide is made, by passing Carbonic sulphide as vapour, over red-hot Silicic anhydride.]

243. The most important compound of Silicon, that with Oxygen, remains to be described.

SILICIC ANHYDRIDE $SiO_2 = 60·5$. Occurs in the form of Rock-crystal, in perfectly colourless and transparent, beautifully crystallized six-sided, transversely-striated prisms, terminated in six-sided pyramids. The finest crystals are cut into ornaments, or are employed as substitutes for glass, in spectacles and optical instruments. So used, they are termed pebbles, and they possess an advantage over glass in their extreme hardness, rendering them less liable to be scratched. The finest specimens of Rock-crystal are found in the mountains of Switzerland, Ceylon, Madagascar, and the Brazils. A purple variety of rock-crystal is called Amethyst, its colour being possibly due to Ferric acid. Brown and yellow Rock-crystals, of great beauty and value as stones, are found in the mountain of Cairngorm, in Scotland. The precious Opal is but a combination of Silica with water. Nor are these all, or nearly all, the varieties assumed by Silica, or Quartz. Chalcedony is but a mixture of crystalline and powdery Quartz; and Heliotrope, of Quartz with Chlorite. In fact, all the varieties of beautiful stones known as Agates, Bloodstones, Flint, Carnelians, Cats-eye, Onyx, Chrysoprase, &c., are but varying forms of Silica, coloured by metallic oxides.

Crystalline Silica, or Silicic anhydride, is a colourless, transparent, glass-like solid, of specific gravity 2·69. Being generally found in an insoluble form, it is, of course, tasteless. It is harder than the hardest glass, which it is therefore capable of scratching. Owing to the intense heat required for its fusion, it cannot be melted in any ordinary furnace, and can therefore only be applied as a substitute for glass, when found in such masses as to allow of its being mechanically wrought into the required form.

The amorphous variety of Silicic anhydride forms a white, porous powder. Under the ordinary tests, Silicic anhydride does not exhibit acid properties, for the obvious reason, that being insoluble, it cannot possess a sour taste, nor affect the colour of blue litmus. It is, nevertheless, one of the most powerful of acids, and is excelled by none in its power of uniting with metallic oxides to form salts, provided the temperature be sufficiently raised. This property, and the fact that many of its combinations are colourless and transparent as the most beautiful Rock-crystal, render Silex of the utmost value to mankind.

Silicic anhydride forms two acids.

ORTHO-SILICIC ACID H_4SiO_4 is tetra-basic. It is soluble in water at the moment of its liberation; but, when once evaporated to dryness, and heated to low redness, it is no longer soluble, and is in the condition of Silicic anhydride. Ortho-silicic acid is formed when a soluble Ortho-silicate is treated with Hydrochloric acid. $Na_4SiO_4 + 4HCl = 4NaCl + H_4SiO_4$.

SILICIC ACID H_2SiO_3 is di-basic. Evaporated to dryness, and heated to redness, Silicic anhydride SiO_2 is obtained.

Silicates are of two kinds; ortho-silicates and silicates. Glass is an acid silicate, or a mixture of such silicates. Most silicates are fusible, and all are insoluble in water, except certain alkaline silicates, containing an excess of alkali. When silicic anhydride is pure, it is infusible in the blow-pipe flame, fusible to a clear bead with Sodium carbonate, as well as with Borax, and capable of complete volatilization, when heated with Hydro-fluoric acid.

[244. Two other compounds can only be mentioned. CHRYSEON $Si_6H_6O_4$ is bright yellow. LEUKON $Si_3H_4O_5$ is white.]

[XIV. Titanium Ti = 50.]

[245. Titanium was discovered in 1791 by Gregor. It is not found native. Its chief ores are Rutile, Brookite and Anatase,—varieties of its chief compound, TITANIC ANHYDRIDE TiO_2.

Titanium may be obtained, as well in the amorphous state, as crystallized. It has a specific gravity of 5·3. From Titanic chloride $TiCl_4$, the metal may be obtained by heating Sodium in its vapour. $TiCl_4 + 2Na_2 = 4NaCl + Ti$.

246. TITANIC ANHYDRIDE $TiO_2 = 82$. When pure, is a yellowish-white infusible powder. TITANIC ACID H_2TiO_3 is di-basic. The great distinction between Silicic anhydride, which Titanic anhydride so closely resembles, is that Titanic anhydride is rendered soluble in water, when previously fused with Potassium Hydrogen Sulphate $KHSO_4$. The commonest ore of Titanium is Rutile, an impure Titanic anhydride. It is isomorphous with Tinstone.

247. A solution of Titanic acid in Hydrochloric acid, gives a purple, hydrated, TITANOUS TITANATE TiO_3TiO_2, xH_2O. TITANOUS OXIDE TiO is black, and results from a reduction of Titanic anhydride.

248. TITANIC CHLORIDE $TiCl_4 = 192$. A fuming volatile liquid. Obtained like Silicon chloride by substitution of Titanic anhydride.

249. TITANIC SULPHIDE TiS_2 is green.]

CHAPTER XII.

ON THE TRIAD ELEMENT BORON, WITH RECAPITULATIONS.

250. Boron never native. Three allotropic modifications: amorphous, graphitoid and adamantine Boron. [251. Boron chloride BCl_3. Boron bromide BBr_3.] 252. Boron fluoride BF_3. Boro-fluoric acid $6HF, B_2O_3$. Hydro-fluo-boric acid HF,BF_3. [Boron nitride BN.] 254. Boracic anhydride B_2O_3. Boracic acid HBO_2,H_2O. Borates. Borax $Na_2B_4O_7$, $10H_2O$. 255. Boron sulphide B_2S_3.] 256. The elements already described. 257. Molecular volume of Boron, Carbon, Silicon and Titanium unknown. 258. Classification of Hydrogen acids. Table of mono-di-tri-and tetra-basic acids. 259. The electro-positive, or "chlorous" element, which may be simple or compound, is separated at the Platinode of the battery. 260. Many compounds of preceding elements possess no acid properties. 261. Table of anhydrides. 262. Formulæ used in this book only express as far as we know the constituents of the compound. Examples. Problems.

XV. Boron B = 11.

250. BORON is always found in combination with Oxygen. The element itself is a Triad, capable of existing in three allotropic modifications, like Carbon and Silicon, although in its comportment it has distinctly the properties of a Triad. Amorphous Boron forms a brown powder, which decomposes both Nitric and Sulphuric acids. It burns with a reddish light, when strongly heated in Oxygen, to Boracic anhydride B_2O_3. Its molecular weight is unknown. Bα may be readily prepared by heating Boracic anhydride with Sodium, and removing the Sodium borate by means of acidulated water. $2B_2O_3+3Na_2=2Na_3BO_3+2B$. Graphitoid Boron Bβ is prepared by passing a current of Boron Chloride over melted Aluminum, and subsequent removal of Aluminum by Hydrochloric acid. $2BCl_3+2Al=2AlCl_3+2B$. Adamantine Boron Bγ, is nearly as hard as the

Diamond. In transparent, yellowish prisms, with square base. Sp. gr. 2·68. It is infusible, but oxydizable at high temperatures to Boracic anhydride. Aqua regia slowly dissolves it. Bγ is made by fusing Boracic anhydride with Aluminum. $B_2O_3+2Al=Al_2O_3+2B$. Alumina is then dissolved by Hydrochloric acid.

[251. No compound of Boron with Hydrogen, is known. BORON CHLORIDE $BCl_3=117·5$. A gas, fuming in air. Decomposed by water into Hydrochloric and Boracic acids. 11·19 Litres weigh 58·75 Grammes. It is made by passing Chlorine over a mixture of Boracic anhydride and lamp-black kept at red-heat in a tube. $B_2O_3+3C+3Cl_2=3CO+2BCl_3$. BORON BROMIDE $BBr_3=251$.]

[252. BORON FLUORIDE $BF_3=68$. A colourless gas, fuming in air. 11·19 Litres weigh 34 Grammes. Prepared by strong ignition of a mixture of 1 part of Boracic anhydride with 2 parts of Calcium fluoride. $4B_2O_3+3CaF_2=3(Ca2BO_2)+2BF_3$. Water dissolves 700 volumes of the gas, with formation of BOROFLUORIC ACID 6HF, B_2O_3. Thus: $2BF_3+3H_2O=6HF,B_2O_3$. HYDRO-FLUOBORIC ACID HF,BF_3. Is formed when Boro-fluoric acid is largely diluted with water. $4BF_3+2H_2O=HBO_2+3(HF,BF_3)$. Potassium boro-fluoride KF,BF_3 may be employed in making Boron, by the aid of Potassium and heat: $2(KF,BF_3)+3K_2=8KF+2B$.]

[253. Boron unites with Nitrogen at a red heat. BORON NITRIDE BN. White, infusible mass.]

254. The chief compound of Boron is with Oxygen. BORACIC ANHYDRIDE $B_2O_3=70$. Colourless glass-like substance, losing its transparency on exposure. BORACIC ACID HBO_2,H_2O. Crystallizes in scales of pearly lustre, soluble in 3 parts of boiling water and 25 parts of cold. Gentle heat removes the water, and at a red-heat we obtain a clear glass of Boracic anhydride. It feebly reddens litmus, and gives a brown stain to turmeric paper. In alcohol it is soluble, and communicates a green colour to flame. Boracic acid is a great solvent of metallic oxides. Its salts are called BORATES. SODIUM BIBORATE $Na_2B_4O_7,10H_2O$ is the well-known Borax. Alkaline borates are soluble in water: the insoluble borates are all soluble in Nitric acid. Boracic acid is mainly obtained from the steam which escapes from the Soffioni of Tuscany, which afford upwards of 2000 tons annually. Jets of steam, or fumerolles, issue from fissures in

the ground, and are conducted into artificial lagoons; the water, on evaporation, yields Boracic acid. Although Boracic acid is with difficulty volatilized, yet, especially in a current of steam, it may be partially.

Boracic acid is also obtained from commercial Borax (Tincal). This is found native in Thibet. It is decomposed by Hydrochloric acid. $Na_2B_4O_7+2HCl+5H_2O=2NaCl+4(HBO_2,H_2O)$.

[255. BORON SULPHIDE $B_2S_3=118$. By contact with water, it breaks up into Hydro-sulphuric acid and Boracic acid. $B_2S_3+6H_2O=3H_2S+2(HBO_2,H_2O)$.]

256. The elements and their compounds, as far as we have considered them, may be thus classified. Those printed in italics may be passed over.

Monads.		Dyads.		Triads.		Tetrads.	
Hydrogen	H= 1·	Oxygen	O = 16.	Nitrogen	N= 14·	Carbon	C= 12.
Chlorine	Cl= 35·5	Sulphur	S = 32·	Phosphorus	P= 31·	Silicon	Si= 28·5
Bromine	Br= 80·	[*Selenium*	Se= 79·5]	Boron	B= 11·	[*Titanium*	Ti= 50·]
Iodine	I=127·	[*Tellurium*	Te=129·				
Fluorine	F= 19·						

257. Of the molecular volumes of Boron, Carbon, Silicon and Titanium, nothing is known. The molecular volume of all the other elements is 2, except in the case of Phosphorus, the molecular volume of which is four times that of the atom.

258. The Hydrogen compounds of these elements may be thus classified. The Hydrogen acids are as follows: the less important ones are printed in italics.

Mono-basic acids.	
Hydro-chloric acid	$HCl=36·5$]
[*Hydro-bromic acid*	$HBr=81·$]
[*Hydriodic acid*	$HI=128·$]
Hydro-fluoric acid	$HF=20·$
[*Hydro-fluo-boric acid*	$HF,BF_3=88·$]
Nitric acid	$HNO_3=63·$
[*Nitrous acid*	$HNO_2=47·$]
Boracic acid	$HBO_2=44·$
Meta-phosphoric acid	$HPO_3=80·$
[*Hypophosphorous acid*	$HPH_2O_2=66·$]
[*Hypochlorous acid*	$HClO=52·5$]
[*Chlorous acid*	$HClO_2=68·5$]
Chloric acid	$HClO_3=84·5$
Perchloric acid	$HClO_4=100·5$

Di-basic acids.	
Hydro-sulphuric acid	$H_2S=34·$
[*Hydro-selenic acid*	$H_2Se=81·5$]
[*Hydro-telluric acid*	$H_2Te=131·$]
Carbonic acid	$H_2CO_3=62·$
[*Sulpho-carbonic acid*	$H_2CS_3=110·$
Silicic acid	$H_2SiO_3=78·5$
[*Fluo-silicic acid*	$H_2SiF_6=144·5$
Sulphurous acid	$H_2SO_3=82·$
[*Hyposulphurous acid*	$H_2S_2H_2O_4=132·$]
Sulphuric acid	$H_2SO_4=98·$
[*Dithionic acid*	$H_2S_2O_6=162·$]
[*Trithionic acid*	$H_2S_3O_6=194·$]
[*Tetrathionic acid*	$H_2S_4O_6=226·$]
[*Pentathionic acid*	$H_2S_5O_6=258·$]
[*Selenous acid*	$H_2SeO_3=129·5$]
[*Selenic acid*	$H_2SeO_4=145·5$]
[*Tellurous acid*	$H_2TeO_3=179·$]
[*Telluric acid*	$H_2TeO_4=195·$]
[*Phosphorous acid*	$H_2PHO_3=82·$]

TRI-BASIC ACIDS.		TETRA-BASIC ACIDS.	
Ortho-phosphoric acid	H_3PO_4= 98·	Pyro-phosphoric acid	$H_4P_2O_7$=178·
		Ortho-silicic acid	H_4SiO_4= 96·5

259. It will be observed, that with the exception of Oxygen, these elements unite to form acids with Hydrogen. Hydrogen is the ELECTRO-POSITIVE ELEMENT in the acid; the radical, whether simple, as in the case of Chlorine in Hydrochloric acid, or compound, as in the case of NO_3, in HNO_3, is the ELECTRO-NEGATIVE element. If Hydrogen be a metal, then the so-called acid is a SALT of Hydrogen. Hence, some works adopt the term Hydric chloride, for Hydro-chloric acid. A salt consists of an electro-positive element, either simple or compound, joined to an electro-negative. The electro-negative element separates at the Platinode of the battery.

260. Many of the most important compounds of these elements are not possessed of the properties of acids. Thus:—

Hydrogen Oxide	H_2O= 18·	Hydrogen di-carbide or Ethylene	H_4C_2= 28·
[*Hydrogen peroxide*	H_2O_2= 34·]	[*Hydrogen silicide*	H_4Si= 32·5]
[*Hydrogen persulphide*	H_2S_2= 66·]	Carbonic Sulphide	CS_2= 76·
Nitrous oxide	N_2O= 44·	Carbonic Chloride	CCl_4=154·0
Nitric oxide	NO= 30·	Silicic Chloride	$SiCl_4$=170·5
Nitrogen peroxide	NO_2= 46·	[*Titanium Chloride*	$TiCl_4$=192·]
[*Nitrogen chloride*	$NHCl_2$= 86·]	[Phosphorous chloride	PCl_3=137·5]
[*Nitrogen bromide*	$NHBr_2$=175·]	[*Phosphoric chloride*	PCl_5=208·5]
[*Nitrogen iodide*	NHI_2=269·]	Boron fluoride	BF_3= 68·
Hydrogen nitride, or ammonia	H_3N= 17·	[*Boron chloride*	BCl_3=117·5]
Hydrogen phosphide	H_3P= 34·		
Hydrogen carbide, or Methyl hydride	H_4C= 16·		

261. The Anhydrides, with the exception of Carbonic and Sulphurous, are also quite destitute of acid properties. The following list of Anhydrides will be found useful.

[*Hypochlorous anhydride*	Cl_2O= 87·]	Sulphurous anhydride	SO_2= 64·
[*Hypo-bromous anhydride*	Br_2O=176·]	Sulphuric anhydride	SO_3= 80·
[*Chlorous anhydride*	Cl_2O_3=119·]	[*Selenous anhydride*	SeO_2=111·]
[*Iodic anhydride*	I_2O_5=337·]	[*Tellurous anhydride*	TeO_2=161·]
[*Periodic anhydride*	I_2O_7=369·]	[*Telluric anhydride*	TeO_3=177·]
[*Nitrous anhydride*	N_2O_3= 76·]	Carbonic anhydride	CO_2= 44·
Nitric anhydride	N_2O_5=108·	Silicic anhydride	SiO_2= 60·5
Boracic anhydride	B_2O_3= 70·		

262. The formulæ adopted are to be considered simple, rather

than expressive of the whole truth. In the case of Sulphuric acid, the formula H_2SO_4, simply informs us of the fact that the acid is dibasic, as it contains two atoms of displaceable Hydrogen; but, although SO_4, represents the Sulphuric acid Radical, it is not pretended that any such compound exists. We know that when Sulphuric anhydride SO_3, is placed in water, it unites with one molecule, to form Sulphuric acid H_2SO_4; but how, exactly, that molecule of water is contained, we do not know for certain. This, however, we find: when Sulphuric acid is passed, in vapour, through a red-hot porcelain-tube, it breaks up into Water, Sulphurous anhydride and Oxygen. The Radical of Sulphuric acid is probably the Dyad SO_2, so that we might write SULPHURIC ACID $\left.\begin{matrix}SO_2\\H_2\end{matrix}\right\}O_2$. NITRIC ACID $\left.\begin{matrix}NO_2\\H\end{matrix}\right\}O$.

PROBLEMS: SOLVED.

A large room has 16 lbs. of Sulphurous anhydride passed into it. How much Sulphur does the atmosphere therein contain?

$$S = 32. \quad O = 16 \quad \therefore \; SO_2 = 64.$$

$$64 :: 16 = 32 : x$$

$$\therefore \; x = 8 \text{ lbs.}$$

It is found necessary to burn 100 Grammes of Sulphur. How much Oxygen is required in order to convert them into Sulphurous anhydride?

$$32 : 100 = 32 : x$$

$$\therefore \; x = 100.$$

A manufacturer buys a cargo of Brimstone. Supposing the Brimstone to be pure, how much Sulphuric acid H_2SO_4 can be obtained from 10 kilogrammes of it?

$$S = 32. \quad O = 16. \quad H = 1. \quad H_2SO_4 = 98.$$

$$32 : 10 = 98 : x$$

$$\therefore \; x = \frac{98 \times 10}{32} = 30{\cdot}625 \text{ kilogs.}$$

CHAPTER XIII.

ON THE TRIAD ELEMENT ARSENICUM, AND ITS COMBINATIONS WITH PRECEDING ELEMENTS.

263. ARSENICUM, like Phosphorus and Nitrogen, a triad element. 264. The molecule of Arsenicum contains 4 atoms. 265. Chief ores of Arsenicum. Properties. Preparation. 266. Arsenicum hydride AsH_3. 267. Marsh's test consists in the production of Arsenicum hydride. 268. Arsenicum chloride $AsCl_3$. Arsenicum bromide $AsBr_3$. Arsenicum iodide AsI_3. Arsenicum fluoride AsF_3. 269. Arsenious anhydride As_2O_3. Arsenious acid H_3AsO_3. Arsenites. 270. Arsenic anhydride As_2O_5. Forms 3 acids. Orth-arsenic acid H_3AsO_4. Pyr-arsenic acid $H_4As_2O_7$. Met-arsenic acid $HAsO_3$. 271. Analogies between Phosphoric and Arsenic acids. 272. Sulphur and Arsenic. Di-arsenicum di-sulphide As_2S_2. Sulph-arsenious anhydride As_2S_3. Sulph-arsenic anhydride As_2S_5. Not precipitated from solutions of Orth-arsenic acid. 273. Cases of suspected poisoning. 274. Reinsch's test.

XVI. Arsenicum As=75. As_4=300.

263. THE element ARSENICUM, is, like Nitrogen and Phosphorus, a Triad, possessing in its combinations the most striking analogies with the two latter elements.

264. Although the atomic weight of Arsenicum is 75, its molecular weight is 300, and its symbol As_4: the molecule contains four atoms and, like its analogue Phosphorus, furnishes two volumes of vapour.

265. Arsenicum is rarely found native; its chief ores are: MISPICKEL FeSAs; COBALT-GLANCE CoSAs; TIN-WHITE COBALT $CoAs_2$; ARSENICAL NICKEL $NiAs_2$; NICKEL-GLANCE NiSAs and

KUPFER-NICKEL NiAs. When free from tarnish, Arsenicum presents the appearance of a steel-grey solid, crystallizable in rhombohedra. Its sp. gr. is 5·75 to 5·9. At 180°C it volatilizes and sublimes. In water it is insoluble. It is so brittle, that it may be pounded in a mortar. When heated in air or Oxygen, Arsenicum burns with a bluish flame to Arsenious anhydride; at the same time a strong garlic odour is diffused, arising probably from a lower oxide. In Chlorine it takes fire; Bromine and Iodine also unite with it, when heated. Nitric acid converts Arsenicum into Arsenic acid. Arsenicum is obtained by heating Arsenious anhydride As_2O_3 with charcoal and collecting the sublimate. $2As_2O_3+3C=3CO_2+As_4$.

266. ARSENICUM HYDRIDE $AsH_3=78$. Also called ARSENIETTED HYDROGEN, is the analogue of AMMONIA NH_3 and of PHOSPHORETTED HYDROGEN PH_3. A colourless gas, of strong garlic odour, nearly insoluble in water. It is very poisonous. At a temperature of −40°C it is condensed into a colourless liquid. It is inflammable, and burns with bluish-white flame, to Water and Arsenious anhydride. 11·19 Litres weigh 39 Grammes; two volumes of this gas (AsH_3 occupies two volumes) require 3 vols. of Oxygen for complete combustion. Arsenicum hydride is prepared from Zinc arsenide, by decomposition with Sulphuric acid. Also, from any soluble arsenical compound, when brought together with nascent Hydrogen.

267. The well-known Marsh's test for Arsenicum, consists in the production of this gas. When the suspected liquid is introduced into a Hydrogen-generating apparatus furnished with pure Zinc, this gas accompanies the escaping Hydrogen; the Arsenicum may be recognized, by passing the mixed gases through a narrow glass tube, kept at a low red-heat, when a ring of steel-grey metal, more or less pronounced, will be found after a time deposited at each side of the source of the heat. Or the gas may be burnt, as it issues from a narrow jet; if then a plate, or the lid of a porcelain crucible, be depressed into the flame, brown spots of Arsenicum will be deposited as the plate or lid is shifted, and these spots will at once dissolve in solution of Chloride of Lime.

There exists also a solid Arsenicum hydride.

268. ARSENICUM CHLORIDE $AsCl_3=181·5$. A colourless liquid, fuming in air and immediately decomposed by water into Arsenious

and Hydrochloric acids. Boils at 132°C. It results when Arsenicum in powder is introduced into Chlorine, and may be made more readily by heating Arsenious anhydride in Chlorine. [ARSENICUM IODIDE AsI_3; a brick-red powder. ARSENICUM BROMIDE $AsBr_3$. ARSENICUM FLUORIDE AsF_3].

269. Arsenicum unites with Oxygen in two proportions. ARSENIOUS ANHYDRIDE $As_2O_3=198$. Exists in two modifications, vitreous and crystalline, the former passing into the latter. From a condition of transparency resembling glass, it passes into a dense white porcelain-like mass. From Hydrochloric acid solution, Arsenious anhydride may be obtained in octohedral crystals. Boiling water dissolves about 12 per cent. of the anhydride. When heated to 193°C, Arsenious anhydride sublimes in transparent octohedra. 11·19 litres of the vapour weigh 99 Grammes.

Arsenious anhydride is prepared by roasting any arsenical ores, especially the arsenio-sulphide of iron FeSAs; the Arsenicum oxydizes and volatilizes.

It is to be noted that it is impossible to evaporate even a dilute solution of Arsenious anhydride in Hydrochloric acid, without loss. For this reason, the process should always be conducted in a retort and the distillate collected in a receiver, when it is a matter for enquiry as to the presence of Arsenicum in a suspected liquid.

In solution, Arsenious anhydride behaves like an acid. It is tri-basic, and its hypothetical formula H_3AsO_3; its Salts are called ARSENITES. They are for the most part insoluble, hence, in poison cases, it is always sought to remove the Arsenious anhydride from solution, by giving an antidote like Magnesia or, better still, Ferric hydrate. Fowler's solution, so much used in medicine, is a POTASSIUM DI-HYDROGEN ARSENITE KH_2AsO_3. SILVER ARSENITE Ag_3AsO_3 is a canary-yellow powder, insoluble in water, but soluble in ammonia and in nitric acid. CUPRIC HYDROGEN ARSENITE $CuHAsO_3$ is Scheele's Green, also soluble in NH_3 and in HNO_3.

270. ARSENIC ANHYDRIDE $As_2O_5=230$. A white powder obtained by heating Arsenic acid to low redness. ORTHO-ARSENIC ACID $H_3AsO_4=142$ is common Arsenic acid. It may be prepared by boiling the "white arsenic" of the shops, or Arsenious anhydride, with Nitric acid. It crystallizes in needles, and forms salts, ORTH-

ARSENATES, isomorphous with ortho-phosphates. DI-SODIUM HYDROGEN ARSENATE $Na_2HAsO_4,12H_2O$ is isomorphous with Di-sodium hydrogen phosphate $Na_2HPO_4,12H_2O$. SILVER ORTH-ARSENATE Ag_3AsO_4 is liver-brown and soluble both in Nitric acid and in Ammonia. AMMONIUM MAGNESIUM ORTH-ARSENATE $NH_4MgAsO_4,6H_2O$ crystallizes in prisms, isomorphous with the corresponding ortho-phosphate. CUPRIC HYDROGEN ORTH-ARSENATE $CuHAsO_4$ is blue.

271. The complete analogy between arsenic and phosphoric acids may be seen from the following facts. Just as we have META-PHOSPHORIC ACID HPO_3, so we have MET-ARSENIC ACID $HAsO_3$; PYROPHOSPHORIC ACID $H_4P_2O_7$, PYRO-ARSENIC ACID $H_4As_2O_7$; ORTHOPHOSPHORIC ACID H_3PO_4, ORTH-ARSENIC ACID H_3AsO_4.

272. Arsenicum unites with Sulphur in three proportions. "Realgar" is DISULPHIDE $As_2S_2=214$. It is found native in ruby-red crystals and is a constituent of Indian White fire, composed of 2 parts of Realgar, with 24 parts of Nitre and 2 parts of Sulphur. SULPHARSENIOUS ANHYDRIDE $As_2S_3=246$. Is found native as Orpiment, in crystals of a brilliant yellow colour. It is insoluble in water and in dilute acids; with difficulty soluble in Hydrochloric acid; very soluble in Ammonia, Ammonium Hydrogen carbonate, and Ammonium Hydrogen Sulphide. It is easily prepared by passing Hydrogen Sulphide, through a solution of Arsenious acid in Hydrochloric acid. SULPHARSENIC ANHYDRIDE $As_2S_5=310$. A yellow powder. When a solution of Hydrogen Sulphide is passed through a solution of Arsenic acid, to which Hydrochloric acid has been previously added, a mixture of Sulphur and Sulpharsenious anhydride is slowly precipitated.

273. In cases of suspected poisoning, Potassium chlorate is employed together with Hydrochloric acid, for the purpose of oxydizing and destroying the organic matter present. In this process, the Arsenious becomes Arsenic acid: it is then better, first to bring back the Arsenic acid to the condition of Arsenious acid, before the precipitation with Hydrogen Sulphide is attempted. If it is desirable to test a wall-paper for Arsenic, it is best to boil the "green" paper with strong Hydrochloric acid in a retort, and to test the first few drops collected in the receiver with Hydrogen sulphide; an arsenical paper will at once give a bright yellow precipitate.

274. Besides the tests for Arsenicum already enumerated, *Reinsch's* test should be employed. A portion of the suspected liquid is acidulated with about $\frac{1}{10}$ of its volume of pure Hydrochloric acid and then boiled with pure copper. Cupric arsenide together with Arsenicum will separate upon the copper. The deposit should then be carefully dried, placed in a narrow glass tube and heated to redness; the Arsenicum in the form of octohedral crystals of Arsenious anhydride will be found to sublime.

CHAPTER XIV.

INTRODUCTION TO THE METALS.

275. The non-metallic elements. Classification. 276. The metals are good conductors of heat and electricity. 277. Solid, except Mercury. Fusion of metals. Temperature of the Oxy-hydrogen flame. 278. The metals may be volatilized by heat, if sufficient heat can be obtained. 279. Metallic lustre, a feature of the metals. 280. Opacity of metals. 281. Variations in colour. 282. The metals insoluble in water. 283. Vary in hardness. Wire-drawing. Tenacity. 284. Malleability. 285. Ductility. 286. Metals found native,—those with feeble affinities nearly always so. Often crystallized. 287. Specific gravity of metals varies greatly. 288. Alloys. 289. Amalgams. 290. Welding process. 291. Metallic salts. 292. Metallic oxides. Nomenclature of oxides and salts. Peroxides. Suboxides. 293. Anhydrides. 294. Sulphides, Selenides, Tellurides. 295. Table of Compounds. 296. Occurrence of metals as oxides, sulphides, chlorides, &c. Ores. Mining. 297. Ventilation of mines. Principles at pp. 16, 17. Double shaft. 298. Extraction of the Metal. Mechanical process. 299. Chemical process. Reverberatory furnace. Roasting of ores. Reduction. 300. Smelting. Fluxing. 301. Use of carbon as a reducing agent.

275. THE preceding Chapters have made us familiar with a class of elements, known as the NON-METALLIC. They afford excellent illustrations of each of the chief divisions among which the elements may be arranged and studied. The *Monads* embrace HYDROGEN, CHLORINE, BROMINE, IODINE, and FLUORINE; the *Dyads*, OXYGEN, SULPHUR, SELENIUM, and TELLURIUM; the *Triads* comprise NITROGEN, PHOSPHORUS, and ARSENICUM; the *Tetrads*, CARBON, SILICON, and TITANIUM. It is true that Titanium and Arsenicum partake more of the character of Metals, but they have these properties more especially in common with the other non-metallic elements.

viz., that their combinations with Oxygen and Hydrogen are acids, and they are closely allied in the one case to Silicon, in the other to Phosphorus.

276. The metals are, for the most part, GOOD CONDUCTORS of HEAT and ELECTRICITY. A good conductor of electricity, is also a good conductor of heat.

277. All of them are SOLID at ordinary temperatures, with the exception of Mercury, which melts above—39°·44C. Their physical condition is, however, simply a question of temperature. Some of them, as Tin, Lead, Cadmium, and Zinc melt below red-heat; others, as Silver, Copper, and Gold, melt somewhat above red-heat, yet at a temperature easily attainable in a furnace; while some, like Iron, Cobalt, and Nickel require a bright white-heat, before they will melt. A few, like Iridium and Platinum, are altogether intractable in ordinary furnaces. The temperature of the Oxy-hydrogen flame, to which all metals yield, is 2500°C.

278. Many metals may be VOLATILIZED by heat. Mercury, Cadmium, Zinc, Potassium, and Sodium, are obtained by distillation.

279. METALLIC LUSTRE, though a common, is not an essential feature of the metals. It is due to the reflection of light. All the metals are obtainable without this lustre, while several of the non-metallic elements constantly exhibit it.

280. The metals are perfectly OPAQUE, except when beaten out into very thin leaves. Gold-leaf transmits a green light.

281. The variations in COLOUR, are not so great as might be expected from so large a number of elements. Most of them present various shades of silvery whiteness; or the bluish colour of Zinc and Lead; the grey of Iron; the red of Copper; the pale-yellow of Barium and Calcium, and the bright yellow of Gold.

282. The metals are INSOLUBLE in water, unless chemical decomposition of the latter is accomplished.

283. They differ greatly in hardness. Steel may be rendered so hard, as to scratch glass; while Lead is so soft, as to be readily cut with a knife, and Potassium may be spread like butter. The terms soft and hard are but relative, the condition of metals in this respect being affected not only by temperature, but by the mode of manufac-

ture. A metal may be very hard and yet have but little TENACITY, by which we mean its power of resisting rupture by extension. Bismuth and Antimony are broken to pieces by a blow; Zinc can scarcely be bent, without its cohesion being overcome; while Iron, Copper, Palladium, Platinum, and Silver possess a very high degree of tenacity. Iron is 26 times more tenacious than Lead. The relative tenacity of metals is determined by testing the comparative strength of wires that have been drawn through the same draw-plate, Fig. 69, and are consequently of precisely the same diameter. Heat diminishes the tenacity of the metals.

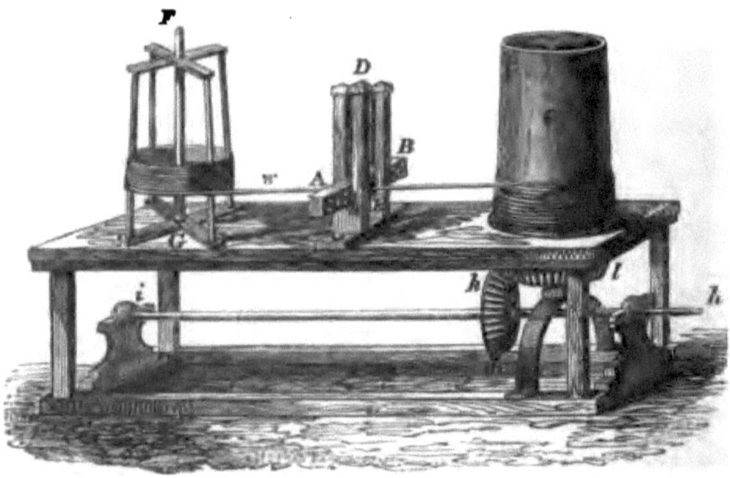

Fig. 69.

284. When a metal may be extended without rupture by hammering, it is said to be MALLEABLE. Gold, Silver, Copper, Platinum, Palladium, Iron, and Aluminum are the most malleable. Gold leaf is only $\frac{1}{280000}$ of an inch in thickness.

285. All malleable metals are DUCTILE, or capable of extension by drawing, but their ductility is not always in proportion to their malleability. A ductile metal is capable of being drawn into wire, but its value as a wire depends upon its tenacity. Gold, Silver, Platinum, Iron, Copper, and Palladium are the most ductile and are arranged in the order of their ductility.

286. The rarer metals, and especially those with feeble affinities, are nearly always found NATIVE, that is in the state in which we employ them. Gold, Platinum, and Bismuth are always so found; Silver and Copper frequently, but not mainly. Native metals often occur crystallized, the commonest forms being the cube, the octohedron, and the dodecahedron. Some metals may be obtained in crystals by melting several pounds weight in a crucible, breaking through the crust and pouring out the yet liquid metal in the way suggested at p. 89, par. 106.

287. The variations among metals, in SPECIFIC GRAVITY, are remarkably great. Lithium, the lightest of all, is of specific gravity 0·59, while Platinum is 21·53 times heavier than water.

288. Many of the metals are capable of combination with others to form ALLOYS; some of these being of great importance in the useful and fine arts, others being possessed of much beauty. Copper, for instance, is not, by itself, suited for castings; but, combined with Zinc, it forms the alloy Brass; with Tin, Bronze. Steel is a carbide of iron. Although in alloys the combination is perfect, yet they are held together by the weakest affinities. Heat alone is sufficient to separate much of the Zinc, from the Copper, in Brass, or Mercury from its alloy with Silver.

289. A combination of Mercury with another metal is not however called an Alloy, but an AMALGAM.

290. Some of the metals, as Iron and Platinum, possess the very curious and highly-valuable property of softening before fusion, and in this state separate pieces of them may be readily united by compression,—a mode of union which is known as WELDING.

291. The metals unite with the various non-metallic elements. With the chlorous elements, Chlorine, Bromine, Iodine, and Fluorine, they form SALTS, in which the metal is always the electro-positive element, and is separated by electrolysis (p. 59, par. 61.) at the Zincode or negative pole. With the compound radicals, the metals form the large class of Oxy-salts. Whether in the state of metal or as oxide, the term BASE applies to the electro-positive "element" in a salt.

292. Oxygen unites with the metals in various proportions, and forms OXIDES. Such an oxide is called a BASIC OXIDE, when it can

displace an equivalent quantity of HYDROGEN in an acid, and form a SALT. When only one basic oxide exists, we speak of salts of that particular metal. Thus :—ARGENTUM NITRATE or SILVER NITRATE $AgNO_3$, is the common nitrate of silver. LEAD NITRATE $Pb2NO_3$, is nitrate of lead. When a metal forms two basic oxides with Oxygen, we distinguish them by the affices OUS and IC. Thus, Iron unites with Oxygen to form two basic oxides. The higher oxide is called FERRIC OXIDE Fe_2O_3; the lower, FERROUS OXIDE FeO. So we have FERROUS CHLORIDE $FeCl_2$ and FERRIC CHLORIDE Fe_2Cl_6. When a metal forms an oxide, which only becomes basic by the loss of a portion of its oxygen, it is called a PEROXIDE: thus, MANGANESE PEROXIDE MnO_2 Such Peroxides evolve Chlorine from Hydrochloric acid, Bromine from Hydrobromic, and Iodine from Hydriodic acids; all of them are sources of Oxygen when heated. Sometimes a metal unites with Oxygen in such proportions, that the oxide cannot form a basic oxide until it has lost a portion of the metal; this is the case with lead. LEAD SUBOXIDE Pb_2O becomes basic, only when an atom of Lead is withdrawn.

293. When a metal by uniting with Oxygen, forms an ANHYDRIDE, the same nomenclature applies as in the case of the elements called Non-metallic.

294. SULPHUR-COMPOUNDS bear a remarkable resemblance in composition to the Oxy-compounds. And so also do those of SELENIUM and TELLURIUM.

295. The following combinations may serve as illustrations, and will afford the student an opportunity of contrasting the formulæ of a few representative instances of Monad, Dyad, Triad, and Tetrad metals in combination.

H_2O	H_2S	H_2Se	H_2Te	HCl	HBr	HI
K_2O	K_2S	K_2Se	K_2Te	KCl	KBr	KI
Na_2O	Na_2S	Na_2Se	Na_2Te	NaCl	NaBr	NaI
CaO	CaS	CaSe	CaTe	$CaCl_2$	$CaBr_2$	CaI_2
MgO	MgS	MgSe	MgTe	$MgCl_2$	$MgBr_2$	MgI_2
Bi_2O_3	Bi_2S_3	Bi_2Se_3	Bi_2Te_3	Bi_2Cl_6	Bi_2Br_6	Bi_2I_6
SnO_2	SnS_2	$SnSe_2$	$SnTe_2$	$SnCl_4$	$SnBr_4$	SnI_4

HF	HNO_3	H_2SO_4	$H_4P_2O_7$	H_3PO_4
KF	KNO_3	K_2SO_4	$K_4P_2O_7$	K_3PO_4.
NaF	$NaNO_3$	Na_2SO_4	$Na_4P_2O_7$	Na_3PO_4.
CaF_2	$Ca2NO_3$	$CaSO_4$	$Ca_2P_2O_7$	Ca_32PO_4.
MgF_2	$Mg2NO_3$	$MgSO_4$	$Mg_2P_2O_7$	Mg_32PO_4.
Bi_2F_6	$Bi3NO_3$	Bi_22SO_4	$BiHP_2O_7$	$BiPO_4$.
SnF_4	$Sn4NO_3$	$Sn2SO_4$	SnP_2O_7	Sn_34PO_4.

296. It is not surprising that most of the metals should occur in nature either as oxides, sulphides, chlorides, bromides, and iodides; as carbonates, silicates, phosphates, and sulphates. A metallic compound from which the metal is usually extracted, is called an ORE. Hæmatite is an ore of Iron, Ferric oxide. Malachite, an ore of Copper,—a Carbonate. The Silicates and Carbonates of Calcium, Aluminum, and Magnesium are the most abundant of the earth's constituents, and form the solid rocks and the known portions of the earth's crust. Metallic ores do not generally compose large beds or extensive strata in the crust of our globe, but are usually found in clefts, rents, or fissures, called VEINS. The process of obtaining the ores from these veins is called MINING—a term also applied to the getting of coal, salt, &c. The mode of proceeding varies. The mining operations are the simplest when the vein is in strata, hills, rocks, or mountains. If the vein be exposed at the surface of the ground, the mineral is simply dug out, and the excavation thus made, serves as a passage to the interior of the mountain in following the vein. When the vein does not appear externally, or when it takes a new direction, after being followed for some distance, access to it is obtained by adits or levels,—horizontal galleries dug from the sides of the hill, till the vein is arrived at. Similar galleries are also sometimes constructed to carry off the water, which drains through the higher part of the mountain, and which would otherwise impede operations. When the mineral lies in strata considerably below the surface of the earth, then a perpendicular pit or shaft is sunk to the requisite depth; and from its bottom or sides, horizontal galleries are carried to the beds, veins, or strata. The mode of supporting the superincumbent mass of earth, or rock, after the excavation, depends upon the nature of the mineral. Where it is valuable, the roof, or cavern overhead, left by the removal of the

ore, is propped up by timber or pieces of masonry; but in mines of coal or salt, the whole bed is not dug out, but masses of it are left like columns to support the roof of the mine. Of course the ventilation of mines is an important consideration. The mode usually adopted is to cause a current of fresh air, from the surface of the earth, to descend one shaft, or one-half of a shaft, to supply the place of the impure or noxious air, which is made to rise through another shaft, or through the other half of the same. The current is created, in most cases, by large fires at the bottom of the shaft, the impure air in which, being thereby heated, ascends, and to obviate the formation of a vacuum, fresh air must descend into the mine by another passage.

297. The VENTILATION of mines is dependent upon the principles described at pp. 16, 17, and illustrated in Figs. 11, 12, and 13. The lamentable accident at the Hartley Colliery, shows that the utility of a second shaft is not confined to its necessity for full ventilation; but it may also be of the last importance, as affording to the miners a second place of exit, in the event of one being closed by an accident.

298. The treatment of the ores for the EXTRACTION of the metal, resolves itself into two distinct operations; one MECHANICAL, the other CHEMICAL. The MECHANICAL process adopted, depends upon the marketable value of the ore, as the greater its worth, the more labour can be profitably expended on its working.

299. The CHEMICAL manipulation depends on the nature of the ore which must determine whether the extraneous matters, Oxygen, Sulphur, &c., can be removed at once, or whether their removal can only be effected after the addition of more Oxygen.

If the ore be found as a Sulphide, as is the case with Lead, Copper, Zinc, &c., the first process resorted to is that of washing. Subsequently the ores are roasted in heaps between layers of coal, or in the reverberatory furnace, as shown in section in Fig. 70. The fuel is seen in the illustration, to be burning on a hearth separate from the bed upon which the ores are laid, the flames being led over them by the bridge of the furnace. Air is admitted to the interior by means of side doors, by the opening or closing of which, Oxygen may be admitted or excluded at will. When the doors are wide open,

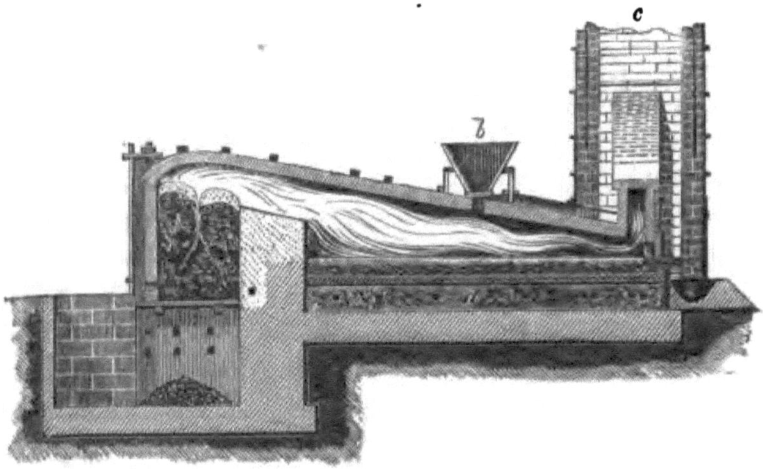

Fig. 70.

oxydation, more or less complete, takes place; while with closed doors, the ore is de-oxydized, or reduced.

The process of *roasting* consists in oxydation. The Sulphur of the compound unites with Oxygen and volatilizes as SULPHUROUS ANHYDRIDE SO_2; whilst the metal remains in the form of au Oxide. Reduction implies literally, on the contrary, the bringing back of the metal from its state of combination to an elementary condition. The great reducing agents are, Hydrogen and Carbon. When Hydrogen is passed over an oxide heated to redness, it reduces the oxide, with formation of Water and separation of the Metal. The reducing power of Carbon also depends upon the facility with which it unites with Oxygen; but the form in which it will pass off, whether as Carbonic oxide, or Carbonic anhydride, is determined altogether by the nature of the ore. If the metallic oxide is easily reduced, Carbonic anhydride will be given off, because the temperature required for its reduction is low; but if the temperature necessary for the reduction of the ore is high, the Carbon will pass off as Carbonic oxide.

300. The process of separating a metal from its ore, is called SMELTING. In carrying out this process, it becomes necessary to get rid of the Silica contained in many ores, by means of a Flux.

This, as the word itself indicates, is something that will liquefy. The Flux employed in smelting an ordinary Iron-ore, consists of Calcium oxide, which, uniting with the Silica of the ore, forms a liquid slag of Calcium silicate. In smelting Copper-ores, which contain Sulphur, Iron, and Copper, in chemical combination, the Iron oxydizes first, and is got rid of by means of a Flux: in this case, Silica or Sand being added, forms a fusible slag with the Ferric oxide, which can be skimmed off, and thus allow the heated air of the flame of the furnace to act upon the Cupric sulphide.

301. It is often necessary to employ Carbon in the reduction of the metallic oxides. The use of the reverberatory furnace alone, even with the variation produced by open and closed doors, does not always suffice to effect complete reduction; the special modifications required in the treatment of the metals will, however, be mentioned in the description of each.

CHAPTER XV.

ON THE METALS OF THE ALKALIES, ALL OF WHICH ARE MONADS.

302. POTASSIUM, SODIUM, LITHIUM, CAESIUM, RUBIDIUM and hypothetical AMMONIUM. 303. XVI. KALIUM or POTASSIUM, K=39. Properties. 304. How obtained. 305. Occurrence in sea-water, &c. 306. Potassium oxide K_2O, the only basic oxide. Potassium hydrate KHO. 307. Salts of Potassium. 308. Tests for Potassium. 309. XVII. SODIUM Na=23. Never native. Properties of Sodium. 310. Sodium oxide Na_2O, the only basic oxide. Sodium hydrate NaHO. 311. Salts of Sodium. 312. Tests for Sodium. [XVIII. LITHIUM L=7. 313. Never native. Properties. 314. One basic oxide. Lithium oxide L_2O. Hydrate LHO. Salts of Lithium. 315. Tests for Lithium. XIX. RUBIDIUM Rb=85·5. 316. Resembles Potassium. Properties. 317. One basic oxide, Rubidium oxide Rb_2O. Hydrate RbHO. XX. CAESIUM Cs=133. 318. Accompanies Rubidium. 319. One basic oxide. Caesium oxide Cs_2O. Hydrate CsHO.] XXI. AMMONIUM (hypothetical) NH_4=18. 320. Reasons for believing in ammonium as a quasi-metal. 321. Ammonia NH_3 described at p. 70, par. 77. 322. Ammonium hydrogen sulphide NH_4HS, the common Laboratory test. Ammonium chloride NH_4Cl and other salts. 323. Tests for Ammonium as Ammonia, or as Ammonium Platinic Chloride, $2NH_4Cl,PtCl_4$.

XVII. Kalium or Potassium. K = 39. K_2 = 78.

302. NONE of the metals of this group are ever found native. They comprise **Potassium, Sodium, Lithium, Caesium, Rubidium,** and the compound metal **Ammonium.**

303. KALIUM or POTASSIUM was discovered by Davy in 1807. When kept from atmospheric air in hermetically-sealed tubes, it is a silver-white metal, of specific gravity 0·865. It melts at 58°C, boils at a low-red heat and distils; the vapour being of a bright green colour. It is so soft that it may be cut with a knife, and it is

malleable at ordinary temperatures. Owing to its affinity for Oxygen, it must be preserved under naphtha. When thrown upon water, Potassium unites with the Oxygen with such energy, that the liberated Hydrogen burns with flame, tinted rose-red from the presence of Potassium in vapour; Potassium hydrate is found in solution. $K_2+2H_2O=2KHO+H_2$. Potassium displaces Hydrogen atom for atom; it is, therefore, a Monad Metal.

304. Potassium was obtained by Davy through the electrolytic decomposition of Potassium hydrate; the metal, together with Hydrogen, separates at the Zincode. It is, however, generally prepared by strongly igniting a mixture of Potassium carbonate with charcoal. The Carbon unites with the Oxygen, and passes off as Carbonic oxide. $K_2CO_3+2C=3CO+K_2$.

305. Potassium is found in Sea- and other Waters, in the form of Potassium Chloride KCl and Potassium sulphate K_2SO_4; it is contained in the ashes of plants in the form of Potassium carbonate K_2CO_3; but its chief supplies are treasured up in certain Silicates and Double-Silicates, which will be considered under the head of the metal Aluminum.

306. Potassium unites with Oxygen in three proportions. There is but one basic oxide: POTASSIUM OXIDE K_2O.

[POTASSIUM TETROXIDE K_2O_4. Results from the burning of Potassium in Oxygen. It is of a yellow colour. On addition of water, it is decomposed; Oxygen is evolved and a solution is obtained of POTASSIUM DI-OXIDE K_2O_2.]

POTASSIUM OXIDE $K_2O=94$. A white, caustic, deliquescent solid, fusible at a red-heat and volatile at high temperatures. It is formed when Potassium spontaneously oxydizes in dry air; but, most readily, by heating one molecule of Potassium with two molecules of Potassium hydrate. $K_2+2KHO=2K_2O+H_2$.

POTASSIUM HYDRATE KHO=56. A greyish-white solid, which unites with more water, with heat, to form a crystalline hydrate $KHO,2H_2O$. It is almost as soluble in ethylic alcohol (spirit of wine) as in water, and this affords a ready means of separating Potassium carbonate from Potassium hydrate. It fuses at a red-heat, and is then commonly cast in moulds, as in Figs. 71 and 72, and sold as Caustic potash.

Fig. 71.

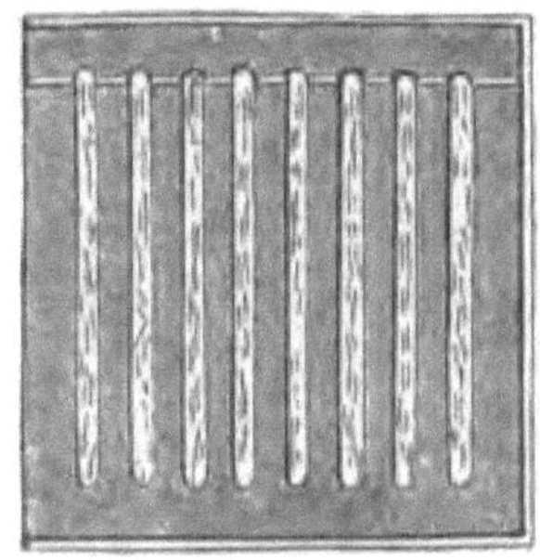

Fig. 72.

Potassium hydrate is prepared from Potassium carbonate. The latter is dissolved in ten parts of water, and milk of lime, equal in weight to one-half the quantity of Potassium carbonate, is stirred into the boiling liquid. Calcium carbonate separates, and, on cooling, is removed by decantation. The solution is evaporated in an iron or silver basin and the residue fused, if required. In symbols, the action is thus represented :—$K_2CO_3+CaH_2O_2=CaCO_3+2KHO$.

Potassium hydrate is best obtained pure, from its alcoholic solution. It should not effervesce on addition of Nitric acid to its solution in distilled water, and the solution thus acidulated, should not be precipitated by Silver nitrate, or by Barium chloride. These tests would prove the absence of Carbonates, Chlorides and Sulphates, which, together with Silicates of Aluminum, Iron and Calcium are always more or less contained in common caustic potash.

A solution of Potassium hydrate, containing 6 per cent. of the latter, is used in medicine. It destroys animal and vegetable matters, dissolves Silica from glass and porcelain, and is indeed one of the most powerful chemical agents. It is intensely alkaline (p. 70, *note*), and blues red litmus. Next to Caesium- and Rubidium hydrates, Potassium hydrate is the most powerful base. It displaces Ammonia from all its salts. Thus :—$NH_4Cl+KHO=NH_3+H_2O+KCl$. And it separates the chief metallic oxides from their solutions. Thus :—$CuSO_4+2KHO=K_2SO_4+CuH_2O_2$. It absorbs many gases, as CO_2, SO_2, N_2O_3, Cl_2, HCl, H_2S, and similar ones. Indeed, the affinity of Potassium hydrate for Carbonic anhydride, renders it indispensable in the laboratory of the chemist.

307. The salts of Potassium are most numerous. They are for the most part soluble in water. POTASSIUM-PLATINIC CHLORIDE $2KCl,PtCl_4$ is the salt, which, on account of its insolubility, is employed for estimating the quantity of Potassium in a solution. POTASSIUM CHLORIDE KCl, crystallizes in cubes, soluble in 3 parts of water. It is largely contained in Kelp, and is consequently a constituent of sea-water. POTASSIUM BROMIDE KBr. Also in cubes. Is much used in medicine. Prepared by dissolving Bromine in Potassium hydrate, and heating to low redness, to decompose the Potassium bromate formed at the same time. $3Br_2+6KHO=5KBr+KBrO_3+3H_2O$. POTASSIUM IODIDE KI, in cubes, very soluble. Also much used in medicine. When free from iodate, the solution does not turn brown on the addition of tartaric acid. Prepared, like the Bromide, by substituting Iodine, but better still from Ferrous iodide by Potassium carbonate: $K_2CO_3+FeI_2=FeCO_3+2KI$. POTASSIUM FLUORIDE KF. POTASSIUM SILICO-FLUORIDE K_2SiF_6 or $2KF,SiF_4$. Gelatinous, almost insoluble in water. Therefore, Silico-fluoric acid $2HF,SiF_4$ is a test for Potassium, and the Silico-fluoride a source of Silicon by heating with the metal Potassium. $2KF,SiF_4+2K_2=6KF+Si$. POTASSIUM CYANIDE KCN, also in cubes, very soluble in water. Much used in Photography and in Electro-plating and Gilding. It is very poisonous. POTASSIUM FERRO-CYANIDE $K_4FeC_6N_6,3H_2O$, crystallizes from water in tables of lemon-colour. It is formed by heating animal matters containing Nitrogen with Potassium carbonate and Iron-filings in an iron pot. In the Laboratory, it is employed as a test for Ferric oxide. When fused with Potassium carbonate and Charcoal, Potassium cyanide results. Thus :—$K_4FeC_6N_6+K_2CO_3+2C=Fe+3CO+6KCN$. POTASSIUM FERRI-CYANIDE $K_3FeC_6N_6$, in right rhombic prisms of rich red colour, soluble in 2½ parts of cold water. A test for Ferrous salts. It is produced, when Chlorine is passed through a solution of Potassium ferro-cyanide. $2K_4FeC_6N_6+Cl_2=2KCl+2K_3FeC_6N_6$. POTASSIUM SULPHO-CYANIDE KCNS resembles

nitre, but deliquesces. Strikes a deep-red colour with Ferric salts. POTASSIUM SULPHIDE K_2S. Not well known. POTASSIUM HYDROGEN SULPHIDE KHS results from the absorption of Hydrogen sulphide, by Potassium hydrate. Also a BI-SULPHIDE K_2S_2 obtained by the oxydation of an alcoholic solution of KHS. K_2S_3. K_2S_4. PENTA-SULPHIDE K_2S_5 is liver-brown. When Sulphur is boiled with solution of Potassium hydrate, this compound is formed in addition to Potassium hyposulphite and Water. $6KHO+6S_2=K_2S_2H_2O_4+2H_2O+2K_2S_5$. POTASSIUM CARBONATE K_2CO_3. A white, granular salt, deliquescent, very soluble in water. Also called Pearl-ash. Insoluble in alcohol. Its taste is acrid, and it is strongly alkaline to test paper. It fuses at a red-heat. POTASSIUM HYDROGEN CARBONATE $KHCO_3$, in right rhombic prisms, requiring 4 parts of water for solution. Much more free from impurities than the carbonate, which is therefore often prepared from it, by heating to redness. POTASSIUM ORTHO-SILICATE K_4SiO_4. POTASSIUM META-SILICATE K_2SiO_3. When Silicic anhydride is heated to bright redness with Potassium carbonate, Carbonic anhydride is expelled and a silicate formed. See Chapter XVI., ON GLASS. POTASSIUM NITRATE KNO_3, in 6-sided, striated prisms, with dihedral summits, soluble in $3\frac{1}{2}$ parts of cold and $\frac{1}{3}$ of boiling water. Melts at 339°C, and forms *Sal Prunella* of the shops. Commonly called Nitre or Saltpetre. Used in making gunpowder, a mixture of 2 atoms of Nitre with 3 atoms of charcoal, and 1 atom of sulphur; when detonated, it resolves itself into Potassium sulphide, Carbonic anhydride and Nitrogen. $4KNO_3+6C+S_2=2K_2S+6CO_2+2N_2$. The expansion of the gases is 1500 times the bulk of the powder. Nitre may be employed in making Nitric acid. $KNO_3+H_2SO_4=KHSO_4+HNO_3$. 101 parts yield 63 parts of acid. When Potassium nitrate is heated to bright redness, POTASSIUM NITRITE KNO_2, results. Gives a green cupric nitrite with Cupric sulphate. POTASSIUM CHLORATE $KClO_3$, in crystalline scales. It is made from Calcium chlorate by the addition of Potassium carbonate: $Ca2ClO_3+K_2CO_3=CaCO_3+2KClO_3$. Chief source of Oxygen. POTASSIUM PERCHLORATE $KClO_4$, in prisms, with difficulty soluble in water. When Potassium chlorate is carefully heated, it changes into a mixture of Potassium perchlorate and chloride, before it loses the whole of its Oxygen: $2KClO_3=KCl+$

$O_2 + KClO_4$. POTASSIUM SULPHATE K_2SO_4, in 6-sided prisms, terminated in 6-sided pyramids. Soluble in 16 parts of water. Owing to the presence of mechanically enclosed moisture in the crystals, they *decrepitate* (crepitas, noise) when heated, from the expansive force of the steam. POTASSIUM HYDROGEN SULPHATE $KHSO_4$, in rhomboids, very soluble in water. It forms the residue from the Nitric acid manufacture, when Nitre is employed. [POTASSIUM ORTHO-PHOSPHATE K_3PO_4. DI-POTASSIUM HYDROGEN PHOSPHATE K_2HPO_4. POTASSIUM DI-HYDROGEN PHOSPHATE KH_2PO_4. Unimportant.]

308. TESTS FOR POTASSIUM. The salts of Potassium communicate a violet colour to flame. They are nearly all soluble in water. They are not precipitated by Hydrogen sulphide, nor by Ammonium hydrogen sulphide, nor by an Alkaline carbonate. Platinic chloride precipitates a yellow salt. POTASSIUM PLATINIC CHLORIDE $2KCl, PtCl_4$, almost insoluble in water, and quite insoluble in a mixture of alcohol and ether; this salt contains 16 per cent. of Potassium. See also tests for Sodium, p. 175, par. 312.

XVIII. Natrium or Sodium Na=23. Na_2=46.

309. The metal Natrium is never found native; its chief and commonest salt is SODIUM CHLORIDE, NaCl, but it is also met with as SODIUM SULPHATE $Na_2SO_4,10H_2O$; SODIUM BI-BORATE $Na_2B_4O_7,10H_2O$, and, less abundantly, as SODIUM BROMIDE NaBr and SODIUM IODIDE NaI.

Sodium is a bluish-white metal, of specific gravity 0·97. It melts at 98° C, and burns, when more strongly heated, with bright-yellow flame to Sodium oxide Na_2O. When introduced into water, it rapidly oxydizes and dissolves as SODIUM HYDRATE, with evolution of Hydrogen, and thus affords a means of preparing Hydrogen (p. 55, Fig. 28). $Na_2 + 2H_2O = H_2 + 2NaHO$. Owing to its affinity for Oxygen, Sodium must be preserved under naphtha. Sodium greatly resembles Potassium, but is less energetic; on account of its cheapness, it is much employed as a reducing agent. It is prepared by strongly heating Sodium carbonate with charcoal; some powdered chalk is introduced to prevent the Sodium carbonate, when it fuses, from withdrawing from the action of the charcoal. $Na_2CO_3+2C=3CO+Na_2$.

310. Sodium forms TWO oxides. SODIUM OXIDE $Na_2O=62$, is the only basic oxide. It is yellowish-white, with a greedy affinity for water, with which it unites to form SODIUM HYDRATE $NaHO$, so much used in the making of hard soaps. On a small scale, it may be prepared from Sodium carbonate, a solution of which is treated with Calcium hydrate in the same manner as in the preparation of Potassium hydrate. $Na_2CO_3+CaH_2O_2=CaCO_3+2NaHO$. Heat alone will not drive off the water. [SODIUM PEROXIDE Na_2O_2. Quite unimportant.]

311. The salts of Sodium are most numerous. The metal being a Monad, the salts have a composition analogous to those of Potassium. All but one salt, are soluble in water. Containing, as they do, for the most part, water of crystallization, many of the Sodium salts "effloresce" by loss of water, whilst those of Potassium "deliquesce" by attracting moisture. Many compounds are quite unimportant and need not be mentioned. SODIUM CHLORIDE $NaCl$ crystallizes in cubes. It constitutes the "common salt" of our households, and is the source of many salts of sodium. Water dissolves 35·5 per cent. at 0° C., and 41·2 per cent. at 109°5 C., the boiling-point of a saturated solution. Sea-water about our coasts contains 2·7 per cent., which is equal to rather more than four ounces per gallon. The crystals decrepitate when suddenly heated, and fuse at a bright red-heat. In absolute alcohol, this salt is insoluble.

The chief deposits of salt are met with near Northwich, at Belfast, Cardona, Wielitzka in Poland, and in Canada. From Sodium chloride, Chlorine is made. $2NaCl+MnO_2+2H_4SO_4=Na_2SO_4+MnSO_4+2H_2O+Cl_2$. SODIUM PLATINIC CHLORIDE $2NaCl,PtCl_4,6H_2O$, is very soluble in water. SODIUM BROMIDE $NaBr$, and SODIUM IODIDE NaI, are both contained in sea-water. From the latter, contained in Kelp, Iodine is made: $2NaI+MnO_2+2H_2SO_4=Na_2SO_4+MnSO_4+2H_2O+I_2$. SODIUM SULPHIDE Na_2S. SODIUM SULPHATE $Na_2SO_4,10H_2O$. Crystallizes in four-sided prisms, with di-hedral summits. It is very soluble and has a saline taste. It is commonly known as *Glauber's Salts*. By exposure, it may lose all its water of crystallization; indeed, no salt is more efflorescent. On a very large scale, anhydrous Sodium sulphate, technically called "Salt-cake," is prepared by heating Sodium chloride with Sulphuric acid; Sodium sulphate and Hydrochloric acid result: ($2NaCl + H_2SO_4 = 2HCl + Na_2SO_4$). [SODIUM HYDROGEN SULPHATE $NaHSO_4$ is first formed when Sulphuric acid is added to Sodium chloride. $NaCl + H_2SO_4 = HCl + NaHSO_4$. By heat, it loses water: $2NaHSO_4 = H_2O + Na_2SO_4,SO_3$. And this salt, by a stronger heat, is a ready source of Sulphuric anhydride. $Na_2SO_4,SO_3 = Na_2SO_4 + SO_3$.] SODIUM SULPHITE $Na_2SO_3,10H_2O$, in oblique prisms, soluble in 4 parts of water. SODIUM HYPOSULPHITE $Na_2S_2H_2O_4,4H_2O$, in striated, oblique rhombic prisms (p. 95, par. 115). It dissolves Silver chloride AgCl, forming SODIUM SILVER HYPOSULPHITE $NaAgS_2H_2O_4$. SODIUM CARBONATE $Na_2CO_3,10H_2O$. Crystallizes in colourless rhomboids of large size, very soluble in water. It effloresces, and readily parts with its water of crystallization; at a red-heat it melts. It used to be made from *kelp*, the ash of sea-weed, which affords, in addition, Potassium chloride and large quantities of "Plate-Sulphate of Sodium" K_3Na2SO_4, together with small quantities of Sodium iodide. The manufacture is now conducted by the process of Leblanc. Sodium sulphate or salt-cake is reduced to the condition of SODIUM SULPHIDE Na_2S, through the instrumentality of small coal. $Na_2SO_4 + 4C = 4CO + Na_2S$. Chalk and small coal are, indeed, added at the same time, and the operation is conducted in a reverberatory furnace. Calcium now combines with the Sulphur, and Sodium, in part at least, with Carbonic acid radical; an excess of Calcium carbonate, supplies the necessary Calcium oxide, to convert the Sulphide into an insoluble CALCIUM OXY-SULPHIDE $2CaO,5CaS$. In symbols, the change may be thus further expressed. $5Na_2S + 7CaCO_3 = 2CO_2 + 2CaO,5CaS + 5Na_2CO_3$. This constitutes "Ball soda," or "Black-ash," which contains in fact much of the Sodium as Sodium oxide. By solution in water, Sodium carbonate and Sodium hydrate are separated from unburnt coal, Lime, and Calcium-oxysulphide: on evaporation and cooling, crystals of Sodium carbonate are obtained, and the Sodium hydrate is converted into Sodium carbonate, by evaporation to dryness, gentle roasting with sawdust, and crystallization from its solution n water. SODIUM HYDROGEN CARBONATE $NaHCO_3$, is the common

Bi-carbonate of soda. It requires 10 parts of water for solution, and has a less unpleasant taste than the carbonate. It has generally the appearance of a white powder, but crystallizes in rectangular, four-sided prisms. SODIUM SESQUI-CARBONATE $2Na_2CO_3,H_2CO_3,3H_2O$, is the "Natron" or "Trona salt" of commerce, found native on the borders of certain lakes in Thibet, the Caspian Sea, and Egypt. DI-SODIUM HYDROGEN PHOSPHATE $Na_2HPO_4,12H_2O$, in transparent, rhombic prisms, soluble in 4 parts of cold water. The "common phosphate" or "rhombic phosphate." SODIUM PYRO-PHOSPHATE $Na_4P_2O_7,10H_2O$, in prisms. Obtained by heating the former salt to low redness, and crystallization from water. SODIUM META-PHOSPHATE $NaPO_3$, is deliquescent. Fuses to a clear glass. Easily obtained by heating Microcosmic salt (p. 177, par. 322). SODIUM NITRATE $NaNO_3$. Crystallizes in oblique rhomboids, soluble in two parts of water. Found native near Iquique in Peru: known in commerce as "cubic nitre," and much employed in making Nitric acid (p. 73, par. 81), and as a manure. SODIUM BI-BORATE $Na_2B_4O_7,10H_2O$. Well known as Borax; found native in Thibet: called "Tincal," in commerce. Crystallizes in flattened, 6-sided prisms, soluble in 12 parts of cold and one-half their weight of boiling water. When heated, borax loses water and melts to a clear glass. It is much used as a flux. SODIUM BORATE $NaBO_2,4H_2O$, in oblique rhombic prisms. SODIUM ORTHO-SILICATE Na_4SiO_4 and SODIUM META-SILICATE Na_2SiO_3, are more or less soluble in water. The Sodium silicate in window-glass may be represented as $Na_2O,3SiO_2$, and is quite insoluble in water.

312. TESTS FOR SODIUM. The salts of Sodium communicate an intense yellow colour to flame. They are so soluble in water, that Sodium can only be precipitated as a bi-metantimonate. Like Potassium, Sodium salts cannot be precipitated by an alkaline carbonate. It is usual when Potassium and Sodium occur together, to convert them into chlorides, to weigh them as such, and to separate Potassium chloride by Platinic chloride; after deducting the Potassium chloride from the total weight, the residue is calculated as Sodium chloride.

[XIX. Lithium L=7; L_2=14.

313. Lithium derives its name from λίθος, lithos, a stone. It is never found native, but is more widely-diffused than was first supposed. It is found in minute quantities in many springs and in the ash of tobacco: it is, however, chiefly obtained from certain Silicates, as Petalite, Lepidolite, and Spodumene. Lithium is a white metal of specific gravity 0·59. It melts at about 80° C., and burns when strongly heated with brilliant white light. In properties it resembles Potassium and Sodium: it is an alkaline metal and a Monad.

314. Lithium has but one oxide. LITHIUM OXIDE L_2O. Its HYDRATE LHO, fuses below redness. LITHIUM CHLORIDE $LCl,2H_2O$, is one of the most deliquescent of salts. From the fused, anhydrous chloride, Lithium is prepared by electrolysis (p. 59, par. 61). Lithium chloride is very soluble in mixed alcohol and ether. LITHIUM SULPHATE L_2SO_4,H_2O, in flat tables. LITHIUM PHOSPHATE L_3PO_4, is insoluble in alkaline phosphates, but very soluble in dilute acids. LITHIUM CARBONATE L_2CO_3, is little soluble in water.

315. TESTS FOR LITHIUM. Salts very fusible and corrode Platinum. The soluble salts communicate a purple-red colour to flame. Potassium carbonate precipitates Lithium carbonate, soluble in hot water. The precipitation by Di-sodium hydrogen phosphate, the most characteristic test. Lithium chloride can be separated from Potassium and Sodium chlorides, by means of mixed Alcohol and Ether.]

[XX. Rubidium Rb=85·5; Rb_2=171.

316. Rubidium is a Monad Metal, resembling Potassium and Sodium. Like Lithium, it is found in many springs and in the ashes of many plants. It owes its name to *rubidus*, dark red, from the dark red lines in the spectra of its salts. Rubidium is a silver-white metal, soft, and fusible at 38·°5 C.

317. RUBIDIUM OXIDE Rb_2O, is a more powerful base than Potas-

sium oxide. Its HYDRATE RbHO, is very deliquescent. RUBIDIUM CHLORIDE RbCl, in cubes. With Platinic chloride, it forms a double salt $2RbCl,PtCl_4$. RUBIDIUM CARBONATE Rb_2CO_3, is deliquescent.]

[XXI. Caesium Cs=133; Cs_2=266.

318. Caesium usually accompanies Rubidium. It derives its name from caesius, lavender-blue, from the bright blue bands in the spectrum of its salts. It is the most electro-positive of the metals.

319. CAESIUM OXIDE Cs_2O; HYDRATE CsHO. CAESIUM CHLORIDE CsCl, in cubes. CAESIUM PLATINIC CHLORIDE $2CsCl, PtCl_4$, is the most insoluble of the corresponding alkaline salts. CAESIUM CARBONATE Cs_2CO_3, is deliquescent.]

Ammonium NH_4=18.

320. Ammonium is a hypothetical metal, behaving like a Monad, and in many of its combinations forming salts, which are isomorphous (p. 97, par. 131) with those of Potassium. If an amalgam of Sodium with Mercury be moistened with a saturated solution of Ammonium chloride, the Amalgam swells up considerably, becomes puffy, pasty, and lustrous, and almost instantaneously evolves Ammonia. This amalgam is supposed to consist of Ammonium with Mercury: at —17·°7 C. it crystallizes in cubes. All attempts to obtain Ammonium have failed: it breaks up into Ammonia NH_3 and Hydrogen H.

321. AMMONIA NH_3, has been already described at p. 70, par. 77. The dry gas unites with the anhydrides to form AMMONIDES. Thus we have: SULPHURIC AMMONIDE $(NH_3)_2SO_3$, and SULPHUROUS AMMONIDE $(NH_3)_2SO_2$. When these are boiled with water, they appropriate each a molecule, H_2O and become Ammonium salts.

322. AMMONIUM HYDROGEN SULPHIDE NH_4HS=51, is the common Laboratory-test for certain metals. It is readily prepared by

passing Hydrogen Sulphide through Ammonia. $NH_3+H_2S=NH_4HS$. Although colourless when first made, it becomes deep yellow from the formation of AMMONIUM BI-SULPHIDE $(NH_4)_2S_2$; Oxygen brings about the change by the simultaneous production of AMMONIUM HYPOSULPHITE $(NH_4)_2S_2H_2O_4$ and WATER H_2O. Thus: $8NH_4HS+5O_2=2H_2O+2[(NH_4)_2S_2H_2O_4]+2[(NH_4)_2S_2]$. Ammonium Hydrogen Sulphide precipitates Cobalt, Nickel, and Ferrous salts as black Sulphides; Manganous salts as Manganous sulphide, of a flesh colour; Zinc, as white Zinc sulphide; Aluminum and Chromium respectively as white and greenish hydrates. AMMONIUM BISULPHIDE $(NH_4)_2S_2$, in yellow crystals may be obtained by passing Ammonia and the vapour of Sulphur through a red-hot tube. All these compounds possess a fœtid odour. AMMONIUM CHLORIDE NH_4Cl. Commonly called Sal Ammoniac, is the chief salt from which Ammonia is made. $2NH_4Cl+CaO=CaCl_2+H_2O+2NH_3$. It forms semi-transparent, tough, fibrous masses, or small fern-shaped crystals, requiring 3 parts of water for solution. It is volatile and sublimes with readiness: the vapour occupies 4 volumes, instead of 2 volumes. AMMONIUM PLATINIC CHLORIDE $2NH_4Cl,PtCl_4$, a yellow, insoluble salt, containing 7·65 per cent. of Ammonia. When heated to redness, spongy Platinum remains. AMMONIUM SESQUI-CARBONATE $2[(NH_4)_2CO_3]CO_2$, has been already described at p. 71, par. 79. It is the common "Smelling Salts" of the Chemist. AMMONIUM HYDROGEN CARBONATE NH_4HCO_3, requires 8 parts of water for solution. AMMONIUM NITRATE NH_4NO_3, crystallizes in long striated prisms, very soluble in water. It melts at 107·°7 C., and decomposes completely at 249° C. into Nitrous oxide and water. $NH_4NO_3=2H_2O+N_2O$. Nitrous oxide is prepared from this salt, which is made by neutralizing Nitric acid with diluted Ammonia. $HNO_3+NH_3=NH_4NO_3$. AMMONIUM PHOSPHATE $(NH_4)_2HPO_4$ SODIUM AMMONIUM HYDROGEN PHOSPHATE $NaNH_4HPO_4,4H_2O$—Commonly called "Microcosmic salt." Crystallizes in large transparent prisms. When heated, it loses water and ammonia, and fuses to a clear glass. SODIUM METAPHOSPHATE $NaPO_3$. Thus:—$NaNH_4HPO_4=H_2O+NH_3+NaPO_3$. The salt is prepared by mixing saturated solutions of Ammonium chloride (1 part), with Di-sodium Hydrogen Phosphate (6 parts). $Na_2HPO_4+NH_4Cl=NaCl+NaNH_4HPO_4$; the phosphate separates on cooling. AMMONIUM SULPHATE $(NH_4)_2SO_4$, in flattened prisms, isomorphous with Potassium sulphate. It is soluble in 2 parts of cold water, and melts at 140° C. On a large scale it is made from the Ammonia in gas-liquor, by passing the gas into Sulphuric acid.

323. TESTS FOR AMMONIUM. The salts of ammonium are volatile, either with, or without decomposition; they are colourless, except those with a coloured acid. Potassium- and Sodium-hydrates expel Ammonia NH_3 from all its combinations, and this may be detected by the smell, the blue colour given to red litmus paper, and the white fumes of Ammonium chloride which are formed, when a glass rod dipped into Hydrochloric acid is brought near. The best test is *Nessler's* test (see Mercury). It should be remembered that Potassium and Ammonium so resemble each other, that Ammonia must be sought for, and, if present, removed by heat, before Platinic chloride is employed as a test for Potassium. Tartaric acid precipitates Ammonium Hydrogen tartrate, if added in excess to Ammonia, or Ammonium carbonate.

CHAPTER XVI.

THE CHEMISTRY OF GLASS.

324. Dr. Johnson's remarks. 325. The discovery of the art of Glass-making. 326. Derivation of the word Glass. 327. Egypt, the birth-place of the discovery. The Portland Vase. 328. The art of staining Glass. 329. Introduction of glass-windows. 330. Rock-crystal is Nature's glass. 331. Silica or Silicic anhydride. 332. Most widely diffused. As Quartz, Mica, and Felspar in Granite. Silicates. 333. Glass is a mixture of different Silicates. 334. Potassium and Sodium carbonates used in Glass. 335. Kelp. Now re-placed by Soda-ash. 336. Calcium carbonate or Limestone, and Calcium oxide or Lime. 337. Lead oxide or Litharge and Red lead. Lead silicate. 338. Broken glass or Cullet. Decolourizing agents. Manganese peroxide and Red Lead. 339. The mixture of materials called Frit. 340. Transparency of glass long in coming. 341. Air-bubbles. 342. Crown glass. Composition. 343. Sheet-glass. 344. Sheet-glass less brilliant and more wavy than Crown-glass. 345. Plate-glass. 346. Annealing glass. Bohemian glass. Composition. 348. Flint-glass. 349. Cutting and roughing glass. 350. Bottles. 351. Reaumur's Porcelain. 352. Etching glass. 353. Painted and Stained glass. 354. Enamel.

324. Dr. Johnson has wisely remarked, that it might contribute to dispose one to a kinder regard for the labours of his fellows, if he were to consider from what unpromising beginnings the most useful productions of art have arisen. Who, when he first saw the sand or ashes, by casual intenseness of heat melted into a metalline form, rugged with excrescences and clouded with impurities, would have imagined, that in this shapeless lump lay concealed so many conveniences of life, as would in time constitute a great part of the happiness of the world? "Yet by some such fortuitous liquefaction,

was mankind taught to procure a body at once in a high degree solid and transparent, which might admit the light of the sun and exclude the violence of the wind; which might extend the sight of the philosopher to new ranges of existence, and charm him, at one time with the unbounded extent of the material creation, and at another with the endless subordination of animal life; and, what is yet of more importance, might supply the decay of nature and succour old age with subsidiary sight. Thus was the first artificer of glass employed, though without his own knowledge, or expectation. He was facilitating and prolonging the enjoyments of light, enlarging the avenues of science, and conferring the highest and most lasting pleasures; he was enabling the student to contemplate nature: and the beauty to behold herself."

It is certain, that of the many discoveries effected by the ingenuity of man, none is so beautiful, so perfect, and in fact so nearly approaching to a production of nature, as Glass.

325. When, or by whom, the ART of glass-making was discovered, is unknown; but it is quite certain that it was practised at an early period in the history of mankind. A very slight knowledge of the matter will satisfy us, that scarcely a kiln of bricks could be burned, or a batch of pottery baked, without the conversion of some of the bricks or of a portion of the ware, at least superficially, into Glass.

326. The derivation of the name is much disputed. It has been supposed to be derived from the German, gleissen, to shine; and the older writers call by the name of Glass, almost everything that glitters, whether transparent or not.

327. Egypt has generally been considered as the birth-place of the invention. Thence, in all probability, it passed to Greece; from Greece, to Rome and Modern Europe. Tyre, Sidon, and Alexandria were long celebrated for the production of glass, and supplied the chief wants of imperial Rome. Nero is said to have paid a sum equal to £50,000 for two glass cups with handles, of no great size, but valued chiefly on account of their transparency.

One of the most celebrated specimens of ancient art, is the Portland vase in the British Museum. It was found about three centuries since, enclosed in a marble sarcophagus, in the tomb of Alexander Severus, who died A.D. 235. The body of the vase is of deep blue glass, and is ornamented with delicate white opaque figures, in low relief, in the style of cameos.

328. The art of STAINING GLASS must have been nearly co-eval with its discovery; or rather, it may be asserted, that it was at all times easier to obtain a coloured, than a colourless glass. The imitation of precious stones so commonly found with Egyptian mummies, shows that the knowledge of the various colours obtainable in glass, must have been even then very complete. The so-called painted glass, used in our ancient churches, is only superficially coloured; the method of colouring the glass throughout, now generally adopted, was acquired in the fifteenth century.

329. GLASS WINDOWS were first introduced into England in the seventh century. The venerable Bede mentions the glazing of the church and monastery of Wearmouth, A.D. 674. Their use, however, was for centuries confined to ecclesiastical buildings, the windows of dwelling-houses even among the wealthy being filled with oiled paper, or wooden lattice-work, until about the thirteenth century. The finest sort of window-glass is said to have been made in 1557, at Crutched Friars; and shortly afterwards the first flint-glass was made at the Savoy-house in the Strand. The first plate-glass for looking-glasses, coach-windows, and the like purposes, manufactured in England, was made in Lambeth, in 1670, by workmen brought from Venice, by the second Duke of Buckingham. From that time English glass began to rival the best manufacture of France and Venice, while of late years the removal by Sir R. Peel of all fiscal restrictions on its manufacture, has made it possible for England to become the source of supply of the largest and finest descriptions of glass, to a considerable portion of the globe.

330. Glass, as is well known, is entirely an artificial production; but in QUARTZ or ROCK-CRYSTAL, we recognize a natural substance resembling Glass in many of its most valuable properties. It occurs often perfectly colourless and transparent, beautifully crystallized in six-sided striated prisms terminated in six-sided pyramids. The finest crystals are cut into ornaments or are used as substitutes for glass, in spectacles and optical instruments; they are called Pebbles, and they possess an advantage over glass in their extreme hardness, rendering them less liable to be scratched. The finest specimens of

Rock-crystal are found in the mountains of Switzerland, Ceylon, Madagascar, and the Brazils. A purple variety of Rock-crystal is called Amethyst. Brown and yellow Rock-crystals, of great beauty and value, are found in the mountain of Cairngorm, in Scotland. The precious Opal is but a combination of Rock-crystal with water. Nor are these all, or nearly all, the varieties assumed by SILICA, as Quartz or Rock-crystal is scientifically termed. Chalcedony is but a mixture of crystalline and powdery quartz; and Heliotrope, of Quartz with Chlorite. In fact, all the varieties of beautiful stones, known as Agates, Bloodstones, Flint, Carnelians, Cats-eye, Onyx, Chrysoprase, &c., are but varying forms of Silica coloured by metallic oxides,—Glass, in fact, of nature's manufacture.

331. Pure SILICA, SILEX or SILICIC ANHYDRIDE (p. 145, par. 242), is a colourless, transparent, glass-like solid. Being generally found in an insoluble form, it is, of course, tasteless. It is harder than the hardest glass, which it is therefore capable of scratching. Owing to the intense heat required for its fusion, it cannot be melted in any ordinary furnace, and can therefore only be applied as a substitute for glass, when found in such masses as to allow of its being mechanically wrought into the required form.

Under the ordinary tests, Silex does not exhibit acid properties, for the obvious reason, that being insoluble, it cannot possess a sour taste, nor affect the colour of blue litmus. It is, nevertheless, one of the most powerful of acids, and is excelled by none in its power of uniting with metallic oxides to form Salts, provided the temperature be sufficiently raised. This property, and the fact that many of its combinations are colourless and transparent as the most beautiful Rock-crystal, render Silex of the utmost value.

332. Silica is by far the most abundant and widely distributed constituent of the earth, being met with not only in a separate state as Sandstone, Flint, Quartz, &c., but as the substance giving the character to most rocks. Granite consists of Quartz, and of combinations of Silica with Potash, Alumina, and Magnesia, in the forms of Felspar and Mica. Syenite, Basalt, Serpentine, and Lava, are all Silicates, while common Clay, a constituent of every fertile soil, is a silicate of alumina.

333. Glass is not composed alone of Silica, but is a mixture, in

varying proportions, of SILICATES of POTASH, SODA, LIME, LEAD OXIDE, BARYTA, MAGNESIA, and ALUMINA. In its most familiar form, Glass is a transparent, brittle substance, very ductile just before the point of fusion, and therefore very easily wrought into any desired form. The extent of its fusibility depends altogether upon the nature of the Silicates employed. In the manufacture of the superior qualities of Glass, everything depends upon the judicious selection of materials.

334. POTASSIUM CARBONATE (p. 170, par. 307), must always be employed, when glass is required perfectly colourless. SODA, although it furnishes a glass of greater lustre than POTASH, communicates to it a greenish tinge.

335 Not more than fifty years since, KELP, or the ash of seaweed was used in admixture with Sand, for making Glass; but now, our manufacturers rely for the supply of the valuable alkali Soda, upon Sodium sulphate, or the Sodium carbonate made from Sodium chloride (p. 173, par. 311).

336. LIME may be employed as quick-lime, or in the form of Calcium carbonate. It is chiefly used in the manufacture of Flint-glass, for which purpose it is essential that it should be free from Ferric oxide.

337. Next to the three OXIDES of POTASSIUM, SODIUM and CALCIUM, LEAD-OXIDE ranks in importance as an ingredient in glass-making, its presence being the distinguishing characteristic of Flint-glass. It may be prepared by exposing the grey powder which forms on melted Lead, to the action of heat and air, until it acquires an uniform yellow colour, when it is commonly called Litharge. By further heating this oxide for many hours at a temperature of about 315°C, under full exposure to air, it attracts a larger quantity of Oxygen, and becomes converted into Red Lead. The great use of the Lead-oxide lies in its power of forming very fusible Silicates, possessing a high metallic lustre; but unfortunately LEAD SILICATE is very soft and easily scratched.

338. Among the essentials to glass-making, must be mentioned BROKEN GLASS, and DECOLOURIZING AGENTS.

Broken glass, waste glass, or "Cullet," as it is called, being more fusible than any of the raw materials, is employed for the purpose of facilitating their fusion. The object of the decolourizing agents is to remove accidental impurities, such as carbonaceous matters from the fuel, and the Ferric oxide which colours all common glass. The most useful material for this purpose is MANGANESE PEROXIDE, which has already been alluded to, at p. 57, par. 61, as a source from whence Oxygen may be obtained for experiment. The facility with which this substance gives off Oxygen, imparting it to the iron, renders it a most fitting agent in the removal of the chief discolouring matter in glass. Another agent for this purpose, RED LEAD, has been already mentioned: this oxide is preferred in glass-making, on account of its finer state of division, and because its excess of oxygen over that of Litharge, is given off at a red-heat, oxydizing the charcoal and any other organic impurities that may have become mixed with the ingredients.

339. The mixture of the materials for glass-making is technically called "Frit." Their fusion is effected in large conical crucibles, Fig. 73, made of the most infusible Fire-clay, which have been pre-

Fig. 73.

viously heated nearly to whiteness. The whole of the materials is not introduced at once, as the crucible will contain double the quantity of melted glass that it can of the frit. The chemical action is simple, and one explanation will suffice for all kinds of material. If we suppose, for example, the Silica to be mixed with Potassium Carbonate, Carbonic anhydride is expelled, and a Silicate of Potash produced. As long as the evolution of Carbonic anhydride lasts, the whole mass is kept in agitation by the escape of the gas, and thus the mixture of the materials is promoted.

340. The Glass, however, does not for a long time become transparent, owing partly to the unwillingness of the last gas-bubbles to make their escape, and partly to the excess of Lime and of other earthy impurities that will not fuse. For the purpose of allowing these to settle, and the gas to escape more freely, the temperature of the furnace is raised so as to render the glass as fluid as possible, the process occupying, in all, about 48 hours. This being accomplished, the temperature is gradually lowered, by regulating the draught, so as to allow the Glass to assume the pasty consistency in which it may be readily shaped at pleasure into the required form.

341. As nothing so much detracts from its beauty as air-bubbles, the liquid Glass is sometimes poured into water, by which means it is reduced to a fine powder, which, when dried and re-heated, loses every trace of air long before it reaches the fusing-point.

342. And now a few words with reference to the manufacture of the chief varieties of Glass. PLATE-GLASS, and common WINDOW-GLASS consist chiefly of Silicates of Soda and of Lime, and are, therefore made from the raw materials Sand, Soda-ash, Calcium carbonate and Cullet. Great care is required that the Lime be not in excess, as the Glass would in that case appear milky on cooling. When the melted mass is sufficiently cooled, a quantity is taken out by the workmen on the end of an iron tube. This, by gentle rotation of the rod in the air, is allowed to cool, and a fresh layer is then added to it by re-dipping into the molten Glass. When a quantity of about nine pounds weight has been thus gathered, it is rolled on a metallic bed until it assumes a pear-like shape, while a boy blowing down the tube, expands it into the form of a flask similar to those in which Lucca oil is imported. By repeated heating and blowing, the Glass is made to assume the form of an enormous decanter, with a very flat bottom and short neck. The tube being supported horizontally on an iron rest, a workman approaches, holding an iron rod tipped with a lump of molten glass, which, by pressure against an iron point, he shapes into the form of a little cup, which he then attaches to the bottom of the decanter-shaped mass of Glass; it immediately adheres, forming the point of attachment from which the Glass has now to be worked into shape. By means of a piece of cold iron, the iron tube is now detached, and the mass of Glass held before

the "nose-hole of the furnace, where it soon begins to melt; it is then made to revolve before this fierce heat, when, by centrifugal force, it, in a minute, is transformed into a flat, circular sheet, of what is known as CROWN-GLASS.

The composition of Crown-glass for optical purposes may be expressed by the formula: $K_2O,2SiO_2.CaO,2SiO_2$.

343. This mode of manufacture has long been abandoned on the Continent, as the Glass so produced can only form panes of very limited size, in consequence of the central lump or bull's-eye, and of the circular form of the plate. In their manufactories, and, indeed, in many of ours, Glass is first blown into the form of a spheroid, which, when its ends are cut off, leaves a cylinder, which is divided by means of shears, or by a straight line, traced with a drop of water: it is then taken to the furnace, to be spread out or flattened by means of an iron rule into a sheet, Fig. 74.

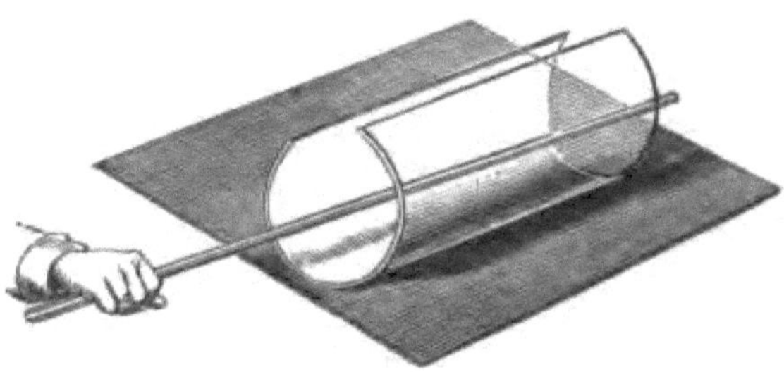

Fig. 74.

344. Sheet-glass is far less brilliant, and more wavy than Crown-glass, and it is much improved by grinding and polishing.

345. In the manufacture of PLATE GLASS, the judicious selection of materials is a most essential element of success. The ingredients employed at St. Gobain consist of 300 parts of pure Sand, 100 parts of Sodium carbonate, 15 parts of Lime, 300 parts of Cullet and decolourizing matters in the shape of Arsenious anhydride and Manganese peroxide. The glass, when melted, is poured on to a table previously heated by hot coals, and furnished with a ledge, varying in height according to the intended thickness of the Glass. It is then spread out by means of a very heavy roller, and afterwards removed to the annealing-oven, where it remains for a week or a fortnight.

This oven, being first heated to a dull red-heat, has all its air-openings closed, and the Plate-glass is left in it, until it has cooled down to the temperature of the surrounding air. The Glass, when polished, is fit for use.

346. The process of Annealing is of great importance, all glass being inclined to brittleness, and liable to fly, as it is termed. The perfection of the process depends entirely on the temperature of the furnace; if it is too high, the glass would (partially) melt and lose its shape; if too low, the plates would be badly annealed, and would be likely to fly when taken out. Tumblers and similar articles, although supposed to be properly annealed when bought, are treated by careful housewives somewhat in the following manner.

They are put into cold water, upon which is placed a little hay, and the water is carefully raised to the boiling-point. If the water is then allowed to cool slowly and the glasses are not removed from it till quite cold, they will be far more likely than before to withstand, without fracture, sudden changes of temperature. The use of the hay, or of similar material, is simply to prevent any sudden contact of cold: in fact to permit the cooling of the water to proceed at an even rate throughout.

347. In BOHEMIAN GLASS, Potash and Lime Silicates are the chief ingredients. The ornamental varieties contain in addition much Silicate of Alumina, for the purpose of promoting their fusibility.

The composition of Bohemian glass may be represented by the formula:—

$$K_2O,3SiO_2,CaO,3SiO_2.$$

That used for vases, or the like ornamental purposes, by the formula:—

$$K_2O,2SiO_2.2(CaO,2SiO_2).Al_2O_3,6SiO_2.$$

348. FLINT-GLASS is chiefly employed for the numerous articles of domestic use, for which England is so greatly celebrated. No one who has had an opportunity of comparing the glass on English and French dining-tables, can hesitate to acknowledge the great superiority of the former. Flint-glass is so designated, because the Silex of which it is composed was formerly exclusively obtained from Flints calcined and ground. It is distinguished from other glass by its containing Lead-oxide, in the place of Lime. It is more transparent,

more lustrous, more fusible, more free from colour, and heavier than other descriptions of glass. Its one defect is its comparative softness, rendering it liable to be scratched.

According to Faraday, the composition of Flint-glass is as follows :—

Silex	51·93
Potash	13·77
Alumina	0·47
Oxide of lead	33·28
Oxides of iron and manganese	0·27
Impurities and loss . .	0·28
	100·00

This composition would correspond, approximately, with the formula :—

$$K_2O,3SiO_2,PbO,3SiO_2.$$

349. The CUTTING of glass is effected by a grinding process ; the roughing, by the use of sand.

350. BOTTLES are made of the cheapest materials. The sand always contains Ferric oxide, which not only gives colour, but greater fusibility. A heated pipe having been dipped into the melted glass and a certain quantity thus collected, is withdrawn by a continuous rotary motion ; when the glass has become sufficiently consistent not to bend on itself, the blower blows through the pipe, Fig. 75, and gives the glass somewhat of the form of an egg. After having introduced it into a mould of the proper form, as soon as the bottle is formed, the blower withdraws it from the mould, and by a see-saw motion raises it on high, as in Fig. 75, and indents the bottom of the bottle. Then, taking a drop of water, he applies it to the neck of the bottle, which is immediately carried to a small cavity in the side of the furnace, and separated from the pipe by a dexterous jerk.

Fig. 75.

The Bottle being thus prepared, the blower turns it, and fastening the pipe to its base, extracts from the pot with another pipe a small quantity of melted glass, which elongates like a thread. The end of this he brings to the neck of the bottle, and by a rotary motion surrounds the mouth with a small glass cord; he then introduces the neck into the working-hole, and finishes the mouth with pincers. The Bottle being completed, an assistant takes it from the hands of the master-workman, carries it to the annealing-furnace, and detaches the pipe by a dexterous blow. (*Regnault*)

351. There is a peculiar accident, which not unfrequently befalls Glass, especially Bottle-glass, during its manufacture, by which it is converted into what is called REAUMUR'S PORCELAIN. If heated to the point of softening without fusion, and after being exposed to this heat for some time, allowed to cool, it will frequently be found to have the appearance of Porcelain. It is harder than Glass, and less liable to crack. This change is due to a process of De-vitrification, consisting in the crystallization of one or other of the Silicates, but especially those of Lime and Alumina. A mass of Glass is often found to be filled with white, worm-shaped, deposits of this peculiar Porcelain.

352. ETCHING, or engraving on glass, is effected by Hydrofluoric acid, p. 145, par. 241. The art was first practised at Nuremburg, and is now in general use. After careful cleaning, the Glass is covered with a varnish made from wax and turpentine. The graver or burin is then passed over the varnish, cutting through it so as to expose the glass, and sketching whatever the artist may desire. The fumes of Hydrofluoric acid, when directed against this surface, attack the glass only where it is uncovered, scratching, as it were, the design upon it. The scales on thermometers are best engraved by this process.

353. The materials used in COLOURING Glass are very numerous. In PAINTED GLASS, the colours are simply laid on, and the glass is then re-heated until the colouring matter is fused into its surface. In STAINED GLASS, the colours are mixed with the ingredients during the manufacture.

Blue is produced by Oxide of Cobalt, the colour of which is so intense, that the addition of one-thousandth part of the oxide renders colourless glass visibly blue. The most brilliant green is produced by Oxide of Chromium; a less brilliant one, by Cupric oxide. Ruby-red is obtained by the use of the Cupreous oxide; rose-red and scarlet, by the Oxide of Gold. The brightest yellows are derived from Oxide of Uranium; an inferior yellow, from the Oxide of Antimony. All these metallic oxides depend, for their successful application, on their power of withstanding heat and of forming fusible Silicates.

354. ENAMEL consists of an easily-fusible glass, rendered white and opaque through the dissemination of Stannic oxide.

CHAPTER XVII.

ON THE DYAD METALS, BARIUM, STRONTIUM AND CALCIUM.

355. Barium, Strontium and Calcium, never native. Oxides called Alkaline Earths. XXII. BARIUM Ba=137. 356. Properties of Barium. 357. Barium oxide or Baryta BaO. Prepared by heating Barium nitrate $Ba2NO_3$ to redness. 358. Barium peroxide BaO_2, a source of Hydrogen peroxide H_2O_2. 359. Also a source of Oxygen. 360. Barium sulphide BaS, a source of oxide. 361. Salts of Barium. Barium chloride $BaCl_2$, $2H_2O$. Barium silico-fluoride BaF_2, SiF_4. Barium nitrate $Ba2NO_3$, used in making Green-fire. Barium carbonate $BaCO_3$, native as Witherite. Barium sulphate $BaSO_4$, native as Heavy-Spar. 362. Tests for Barium. XXIII. STRONTIUM Sr=87·5. 363. Resembles Barium. 364. One basic oxide, Strontia SrO. Strontium hydrate SrH_2O_2. 365. Salts of Strontium. Strontium chloride $SrCl_2,6H_2O$. Soluble in absolute alcohol. Strontium nitrate $Sr2NO_3$, $5H_2O$. Strontium carbonate $SrCO_3$, native as Strontianite. Strontium sulphate $SrSO_4$, native as Celestine. 366. Tests for Strontium. XXIV. CALCIUM Ca=40. 367. Calcium, the base of Lime. A dyad metal, like Barium and Strontium. 368. One basic oxide, Calcium oxide, CaO. Calcium hydrate CaH_2O_2. Uses as Quick-lime and Slaked lime; in mortars and cements; as a manure; as an absorbent of Hydrogen sulphide and Carbonic anhydride in the gas-purification. Calcium sulphide CaS. Calcium di-silicide $CaSi_2$. Calcium phosphide. 369. Salts of Calcium. Calcium chloride $CaCl_2,6H_2O$. Calcium oxy-chloride $CaOCl_2$, is Chloride of lime. Calcium fluoride CaF_2 is Fluor-spar. Calcium carbonate $CaCO_3$. Calcium bi-carbonate $CaCO_3,CO_2$ in temporarily hard waters. Calcium di-silicate $CaO,2SiO_2$. Calcium sulphate $CaSO_4,2H_2O$ is Gypsum or Selenite. Calcium phosphate Ca_32PO_4, is bone-earth. Calcium acid phosphate, CaH_42PO_4 is Superphosphate of lime. 370. Tests for Calcium.

355. The Metals **Barium, Strontium** and **Calcium** are never found native; they have the greatest possible affinity for Oxygen, with which they each combine to form one basic oxide. Barium

oxide, Strontium oxide and Calcium oxide, are soluble in water. Their solutions turn red litmus paper blue, and they are precipitated by Carbonic anhydride as Carbonates. Their oxides are commonly called ALKALINE EARTHS.

XXII. Barium Ba = 137.

356. Barium is but little known. It is of a yellowish colour, lustrous, and of specific gravity about 4. It tarnishes immediately on exposure to air, and decomposes water at common temperatures. It is best prepared from the fused Barium chloride by electrolysis, and separates at the Zincode of the battery.

Barium derives its name from the Greek βαρὺς, *barys*, heavy, on account of the weight of its compounds.

357. It has two oxides, only one of which is basic. BARIUM OXIDE, or BARYTA $BaO = 153$. Grey, porous substance, fusible in Oxy-hydrogen flame. When mixed with water, it unites with the latter, and forms a HYDRATE of BARIUM BaH_2O_2. This hydrate is white, very alkaline, and soluble in 20 parts of cold, and three parts of boiling water; from a saturated, boiling solution, much of it crystallizes out on cooling. These crystals contain eight molecules of water: $BaH_2O_2,8H_2O$. Barium hydrate is a fearful poison, as well as all its soluble salts: the best antidote is a solution of Sodium sulphate or Magnesium sulphate, with the acid of which the Barium unites to an insoluble, an inert, Barium sulphate.

Barium oxide is best prepared by heating Barium nitrate to redness.

358. If heated strongly in a stream of Oxygen, or of air from which the Carbonic anhydride has been removed, it unites with another atom to form BARIUM PEROXIDE BaO_2. A yet better mode of preparing the peroxide, is by heating a mixture of Barium oxide with Potassium chlorate to redness: $KClO_3 + 3BaO = KCl + 3BaO_2$. By water the Potassium chloride may be dissolved, leaving the peroxide as a white, bulky, HYDRATE $BaO_2,6H_2O$, insoluble in water. Barium peroxide is an interesting oxide, on account of the facility with which it imparts the Oxygen which constitutes the Peroxide, to Hydrogen oxide, thus converting the latter into Hydrogen peroxide

H_2O_2, p. 62, par. 65. The latter may be prepared directly from hydrated Barium peroxide and Hydro-chloric acid; thus, $BaO_2, 6H_2O + 2HCl = BaCl_2 + 6H_2O + H_2O_2$.

359. Barium peroxide is also a source of Oxygen: when it is heated to bright redness, it parts with an atom of Oxygen, and is re-converted into Barium oxide. At a red-heat: $2BaO + O_2 = 2BaO_2$; then, at a bright-red heat $2BaO_2 = 2BaO + O_2$.

360. BARIUM SULPHIDE BaS. Is prepared by heating Barium sulphate, with Charcoal, to bright redness: $BaSO_4 + 4C = 4CO + BaS$. When dissolved in a little boiling water, a yellow solution is formed, being a mixture of Barium Hydrogen sulphide, and Barium hydrate: $2BaS + 2H_2O = BaS,H_2S + BaH_2O_2$. Oxygen, from the air, changes the Barium Hydrogen sulphide into PERSULPHIDE: $2(BaS,H_2S) + O_2 = 2H_2O + 2BaS_2$. Barium sulphide is an admirable source of Barium salts and of Barium hydrate; for, when its solution in water is boiled with Cupric oxide, insoluble Cupreous sulphide is produced, together with solutions of Barium hydrate and Barium hypo-sulphite: $6BaS + 6H_2O + 8CuO = 4Cu_2S + BaS_2H_2O_4 + 5BaH_2O_2$.

361. SALTS OF BARIUM. BARIUM CHLORIDE $BaCl_2,2H_2O$. In flat, 4-sided tables, soluble in 2 parts of water. It may be prepared either from Barium carbonate, or from Barium sulphide, by dilute hydrochloric acid. It is an important test for sulphuric acid and soluble sulphates, but it must be remembered that Barium chloride is itself precipitated by strong nitric and hydrochloric acids. BARIUM SILICO-FLUORIDE BaF_2,SiF_4. In minute crystals, nearly white and insoluble. BARIUM NITRATE $Ba2NO_3$ crystallizes in octohedra, requiring 8 parts of cold and 3 of boiling water, for solution. It is employed not only as a test, but in the production of *Green-fire*, a mixture of 450 parts of dry Barium nitrate, 150 parts of Sulphur, 100 parts of Potassium chlorate, and 25 parts of Lampblack. The materials must be separately pounded, and not mixed in a mortar by rubbing. BARIUM CARBONATE $BaCO_3$, is found native as *Witherite*, a mineral abundant in the Lead-mines of Derbyshire, both massive and in 6-sided prisms, terminated in 6-sided pyramids. When artificially prepared, it is perfectly white, and insoluble in water. It is used in the making of Flint-glass. BARIUM SULPHATE $BaSO_4$:

constitutes the mineral *Heavy Spar*. It is insoluble in water, and in Nitric and Hydrochloric acids: in order to render it available as a source of soluble Barium salts, it must be strongly heated with $\frac{1}{10}$ of its weight of Charcoal. 100 parts of Barium sulphate contain 34·34 parts of SO_3.

362. Tests for Barium. The soluble Barium salts are precipitated by a solution of Sodium carbonate or Ammonium carbonate. Barium sulphate is precipitated by Sulphuric acid quantitatively. Those which are soluble communicate a yellowish-green colour to flame. Most of the Barium salts are with difficulty soluble in water, hence is Barium chloride much used as a test for many acids; all of them, except the sulphate, are soluble in dilute Hydrochloric or Nitric acid. Calcium sulphate immediately occasions a precipitate in solution of Barium salts.

XXIII. Strontium Sr = 87·5.

363. Strontium resembles Barium, and is also a Dyad. It is pale-yellow, malleable and of specific gravity 2·54. It tarnishes readily, and decomposes water with liberation of Hydrogen and formation of Strontium hydrate. Like Barium, it can be prepared from the fused chloride by electrolysis.

364. Strontium unites with Oxygen to Strontium oxide or Strontia $SrO = 103{\cdot}5$. It is the only basic oxide. From water, it crystallizes as Strontium hydrate $SrH_2O_2,8H_2O$, requiring 50 parts of water for solution. The solution is strongly alkaline, and is precipitated by Carbonic anhydride. Strontium oxide is made by heating the nitrate to bright redness.

365. Strontium so closely resembles Barium that the description of its salts may be more brief. Strontium chloride $SrCl_2,6H_2O$ in deliquescent needles, soluble in alcohol (Barium chloride is not). Strontium nitrate $Sr2NO_3$, in octohedra requiring 5 parts of water for solution. Crystallizes also with 5 molecules of water. The anhydrous salt is much used in *Red-fire*, a mixture of 40 parts of the Nitrate, with 13 parts of Sulphur, 10 parts of Potassium chlorate, and 4 parts of Antimonous sulphide. Requires the greatest care in mixing. Strontium carbonate $SrCO_3$. Native as "Strontianite," a

mineral found near Strontian in Scotland. It is white, often crystalline, and insoluble in water. STRONTIUM SULPHATE $SrSO_4$, is the "Celestine" of Mineralogists. It is white and insoluble in water and acids. In Sicily, it is commonly associated with crystals of native Sulphur. When heated with charcoal, a Sulphide may be obtained, which is employed for making the salts of Strontium.

366. TESTS FOR STRONTIUM. The soluble salts communicate a bright-red colour to flame. Ammonium carbonate precipitates Strontium as Carbonate from its solutions; thus both Barium and Strontium can be removed from a solution containing alkalies. Barium chloride being insoluble in alcohol, Strontium chloride can be removed from a mixture of Barium and Strontium chlorides by means of alcohol. Strontium silico-fluoride is soluble in water. Calcium sulphate only precipitates Strontium after a time, or immediately on heating.

XXIV. Calcium Ca = 40.

367. Calcium is the basis of LIME OR CALCIUM OXIDE. Like Barium and Strontium, it is a dyad metal, never found native. It is a light, yellowish metal, of specific gravity 1·578. It is malleable, and fusible at a red-heat. Calcium decomposes water and forms a hydrate: it does not, however, tarnish as readily as Barium and Strontium. Calcium is made by decomposing a fused mixture of Calcium and Strontium chlorides; the Calcium chloride is alone decomposed. It may, however, be prepared by fusing Sodium with Calcium iodide: $CaI_2+Na_2=2NaI+Ca$.

368. Calcium forms but one basic oxide, CALCIUM OXIDE $CaO=56$. It is more commonly called *Lime* and *Quick-lime*. When pure, it forms a white, porous substance, infusible in the Oxy-hydrogen flame. When ignited, it glows with an intense white light, known as the Lime-light, and the Oxy-calcium light. When moistened with water, it enters into combination with one molecule, evolves great heat, and falls to powder; CALCIUM HYDRATE CaH_2O_2 requires 778 parts of cold, and about 1200 parts of boiling water for solution. This water it loses at a red-heat, and from the condition of *Slaked*, returns to that of *Quick-lime*. Suspended in water, the hydrate forms *Milk of lime*.

Calcium oxide is much employed in making Mortars and Cements. Although Quick-lime sets when mixed with water, it cracks and shrinks too much to be of any use; but, when mixed with fine river sand, the shrinking is prevented. By slow degrees it hardens more and more, some of the Lime becomes changed into Calcium carbonate, with loss of Water, and some unites with the Silica to form Calcium silicate. Calcium oxide is also much employed to absorb both Carbonic anhydride and Hydrogen Sulphide in the process of Gas-purification: $CaO + CO_2 = CaCO_3$. And $CaO + H_2S = H_2O + CaS$. As *lime-wash*, or Milk of lime, Calcium hydrate is also much used. The reason why a newly-whited room is damp, is owing to the absorption of Carbonic anhydride and the consequent liberation of Moisture: $CaH_2O_2 + CO_2 = H_2O + CaCO_3$ Lime is also much used as a manure.

Calcium oxide is a powerful base. Its solution blues red litmus paper, and is an important test for Carbonic anhydride. It expels Ammonia from its salts and is employed for this purpose. Thus: $2NH_4Cl + CaO = CaCl_2 + H_2O + 2NH_3$ (p. 70, par. 77). Its combinations and salts are most numerous. CALCIUM SULPHIDE CaS is insoluble in cold water, and phosphorescent. It may be prepared by heating Calcium sulphate to redness with Charcoal: $CaSO_4 + 4C = 4CO + CaS$. CALCIUM OXY-SULPHIDE, $2CaO,5CaS$ is the "Soda waste" in the Sodium carbonate manufacture. When Calcium hydrate is boiled in water with excess of Sulphur, a mixture of CALCIUM PENTA-SULPHIDE CaS_5, and of CALCIUM HYPO-SULPHITE $CaS_2H_2O_4$, is obtained: $3CaH_2O_2 + 6S_2 = 2CaS_5 + CaS_2H_2O_4 + 2H_2O$. CALCIUM PHOSPHIDE CaP, is obtained in lumps of a red-brown colour, when the vapours of Phosphorus are made to pass through Calcium oxide heated to redness: it is a mixture of the Phosphide with Calcium pyro-phosphate. $7P_4 + 28CaO = 20CaP + 4Ca_2P_2O_7$. It is a good source of Hydrogen Phosphide, when thrown into water. CALCIUM DI-SILICIDE $CaSi_2$.

369. SALTS OF CALCIUM. CALCIUM CHLORIDE $CaCl_2$, white, porous, deliquescent mass, much employed for desiccation. With water, it forms striated prisms $CaCl_2,6H_2O$, very deliquescent, and easily obtained by dissolving Calcium carbonate in Hydrogen chloride. $CaCO_3 + 2HCl = H_2O + CO_2 + CaCl_2$. A saturated solution of this

salt boils at 179°·5 C. CHLORIDE OF LIME $CaOCl_2$. A white or yellowish powder, not deliquescent, emitting the odour of Chlorine. Practically it is a mixture of Calcium Oxy-chloride with Calcium hydrate. It is prepared by exposing Calcium hydrate, in thin layers, to the action of Chlorine. The Calcium oxy-chloride is alone soluble in water. See also p. 129, par. 195. When Chlorine is transmitted rapidly and in excess into milk of lime, a mixture of Calcium chlorate and chloride results, which is employed in making Potassium chlorate. Thus: $6CaH_2O_2 + 6Cl_2 = 6H_2O + 5CaCl_2 + Ca2ClO_3$. Then: $Ca2ClO_3 + 2KCl = CaCl_2 + 2KClO_3$. The use of Chloride of lime as a bleaching and disinfecting agent is mentioned in Chapter XX. CALCIUM FLUORIDE CaF_2, is found native as fluorspar, as well massive as in cubes and octohedra. It is always contained in small quantities in teeth and bones, in certain waters, and in various seeds, especially Cereals. It is sometimes as transparent as glass, but more generally blue, or amethystine, and on account of its beauty is much employed for ornamental objects. When heated, it is very phosphorescent. As a source of Hydrofluoric acid, it has already been mentioned: $CaF_2 + H_2SO_4 = CaSO_4 + 2HF$. (Also it is employed in making Boron fluoride BF_3. by heating Calcium fluoride with Boracic anhydride: $3CaF_2 + 4B_2O_3 = 3(Ca2BO_2) + 2BF_3$). CALCIUM CARBONATE $CaCO_3 = 100$. Crystallyzes in rhomboids as "Iceland Spar" and "Calcspar," and in 6-sided prisms as "Aragonite;" as it crystallizes in these two irreconcileable geometrical forms, Calcium carbonate is said to be *dimorphous*. Limestone, Chalk, and Marble are varieties of Calcium carbonate, and so also is Coral. Shells of birds and fishes also consist for the most part of Calcium carbonate. In water it is nearly insoluble, as one litre only dissolves three milligrammes; or one gallon, nearly two grains. When, however, water contains Carbonic acid, it is much more soluble therein, and gives rise to CALCAREOUS waters. Calcium carbonate, when first precipitated by Carbonic anhydride from Lime-water or Aqua Calcis, is re-dissolved by more Carbonic anhydride as CALCIUM BI-CARBONATE $CaCO_3,CO_2$; on boiling such a water, CO_2 is evolved and Calcium carbonate is precipitated. Such waters are commonly met with, and give rise to "Stalactites" and "Stalagmites" (see Chapter XIX. on Water).

These calcareous waters may then be softened, either by boiling, or by the introduction of Calcium hydrate in quantity sufficient to precipitate the Carbonic anhydride. A solution of Soap may also be employed as a test of hardness (p. 235, par. 449). CALCIUM SILICATE $CaO2SiO_2$, is contained in ordinary window-glass. CALCIUM NITRATE $Ca2NO_3,4H_2O$, in prisms. CALCIUM SULPHATE $CaSO_4,2H_2O = 172$. Is found native as Selenite, Gypsum, and Alabaster. It crystallizes in transparent, flattened prisms, requiring 400 parts of water for solution. This salt is contained in many waters, and confers the property of "permanent hardness," as distinguished from the "temporary hardness" of calcareous waters containing Calcium bi-carbonate; such a water deposits in boilers and kettles, a "crust" or "fur," difficult to remove. When heated to 260° C., Calcium sulphate loses its water, and acquires the property of *setting* when mixed with water; "Plaster of Paris" is anhydrous Calcium Sulphate $CaSO_4$, found also native as "Anhydrite." CALCIUM PHOSPHATE Ca_32PO_4, is bone-phosphate, the chief mineral constituent of bones. It is white, and almost insoluble in water. When this salt is acted upon by Sulphuric acid, it gives rise to the ACID ORTHO-PHOSPHATE OF CALCIUM SUPERPHOSPHATE CaH_42PO_4, so much used as a manure: $Ca_32PO_4 + 2H_2SO_4 = 2CaSO_4 + CaH_42PO_4$. From the Superphosphate, Phosphorus is made: $3CaH_42PO_4 + 16C = Ca_32PO_4 + 16CO + 6H_2 + P_4$. The Superphosphate is also the source of DI-SODIUM HYDROGEN PHOSPHATE $Na_2HPO_4,12H_2O$. Thus: $CaH_42PO_4 + 2Na_2CO_3 = CaCO_3 + CO_2 + H_2O + 2Na_2HPO_4$.

370. TESTS FOR CALCIUM. The salts are colourless, and give a reddish-yellow tint to flame. Sodium carbonate precipitates Calcium carbonate, white. From the salts of Barium and Strontium, Calcium is distinguished by producing no precipitate with a solution of Calcium sulphate. Ammonium oxalate precipitates CALCIUM OXALATE, as a white salt, insoluble in water and in Acetic acid, but soluble in Hydrochloric and Nitric acids. By heating Calcium oxalate to dull redness, we obtain Calcium carbonate, 100 parts of which contain 56 parts of Calcium oxide; and, as Calcium carbonate, Calcium is alwyas weighed.

CHAPTER XVIII.

THE CHEMISTRY OF THE DYAD METALS, MAGNESIUM, ZINC, AND CADMIUM.

371. The metals Magnesium, Zinc, and Cadmium are volatile and capable of distillation. They have but one basic oxide. Although carbonates of these metals are found, they cannot be precipitated from solutions of their salts except as basic carbonates.

XXV. Magnesium Mg=24.

372. Magnesium is a silvery-white, malleable and ductile metal. Specific gravity 1·743. It is nearly as volatile as Zinc. In Oxygen it burns with dazzling whiteness to Magnesia MgO. Magnesium

also unites directly, when heated, with Chlorine, Bromine, Iodine, Sulphur, and Nitrogen. It slowly tarnishes in damp air, and dissolves with the greatest rapidity in Hydrogen chloride. $Mg+2HCl=MgCl_2+H_2$. Magnesium is never found native. Its chief salts are the Carbonate, Chloride, Sulphate, and Silicate.

Magnesium is prepared by heating Sodium with Magnesium chloride. $MgCl_2+Na_2=2NaCl+Mg$. Also from the fused chloride, by electrolysis.

373. MAGNESIUM OXIDE $MgO=40$, is the well-known "Magnesia" and *Magnesia usta* of Pharmacy. It forms a white, bulky powder, almost tasteless and infusible. In water it is nearly insoluble, but yet distinctly blues red litmus paper. With water, Magnesia forms a HYDRATE MgH_2O_2, which slowly absorbs Carbonic anhydride from the air.

374. MAGNESIUM SULPHIDE MgS, is very little soluble in water. A soluble salt of Magnesium is precipitated by Potassium sulphide, but not by Ammonium hydrogen sulphide.

375. The SALTS of Magnesium are numerous and important. MAGNESIUM CHLORIDE $MgCl_2,6H_2O$, is contained in sea-water. It crystallizes in silken, deliquescent prisms. AMMONIUM MAGNESIUM CHLORIDE $NH_4Cl,MgCl_2$, may be converted into Magnesium chloride by a red-heat, and thus fitted for making Magnesium. MAGNESIUM BROMIDE $MgBr_2$, is also contained in sea-water. MAGNESIUM CARBONATE $MgCO_3=84$. Occurs native as *Magnesite*, a white mineral of considerable hardness. *Magnesia alba* $MgH_2O_2,2(MgCO_3.H_2O)$ is a basic carbonate. It is obtained by adding a hot solution of Sodium carbonate to a boiling solution of Magnesium sulphate. The basic carbonate dissolves in Carbonic acid water, and the solutions deposit a hydrated carbonate $MgCO_3,3H_2O$. *Dolomite* is a double carbonate of Calcium and Magnesium $CaCO_3,MgCO_3$: it constitutes a superb Building Stone. [MAGNESIUM SILICATES are abundant minerals. *Olivine* or *Chrysolite* is MAGNESIUM ORTHO-SILICATE Mg_2SiO_4. *Talc* $4MgO,5SiO_2$. *Steatite* or French Chalk $3MgO,4SiO_2$, is also called "Soap-stone." *Meerschaum* $2MgO,3SiO_2,4H_2O$, is Magnesium sesqui-silicate. *Serpentine* $(2[MgO,SiO_2,]MgO,2H_2O)$, often contains a portion of the Magnesia displaced by Ferrous oxide. *Augite* $(CaMg)O,SiO_2$, a mineral in which both Ferrous and Man-

ganous oxides replace Lime and Magnesia. *Hornblende* is yet more complicated in structure. See Aluminum silicates.] MAGNESIUM SULPHATE $MgSO_4,7H_2O$, is "Epsom salts." It crystallizes in right-rhombic prisms, soluble in 3 parts of water at 15° C and 1⅓ parts at 100° C. The taste is too well known to need description. On a large scale it is made by the action of Sulphuric acid upon Magnesian-Lime; Magnesium sulphate is removed by washing, from Calcium sulphate. MAGNESIUM NITRATE $Mg2NO_3,6H_2O$, in deliquescent prisms. MAGNESIUM HYDROGEN PHOSPHATE $MgHPO_4,7H_2O$. With difficulty soluble. In prisms, or as a fine white powder. MAGNESIUM AMMONIUM PHOSPHATE $MgNH_4PO_4,6H_2O$, is the well-known "triple phosphate." When a salt of Magnesium, to which an ammoniacal salt has been added, is mixed with Di-sodium hydrogen phosphate, this compound is precipitated in minute crystals. Heated to redness, it becomes MAGNESIUM PYRO-PHOSPHATE $Mg_2P_2O_7$, containing 36·04 per cent of MAGNESIA MgO.

376. TESTS FOR MAGNESIUM. The salts are colourless, and are not precipitated by Sodium Hydrogen carbonate until boiled. Ammonium salts hinder the precipitation. Lime-water precipitates Magnesium hydrate, and so also does Ammonia. In presence of ammoniacal salts, Ammonium oxalate does not precipitate Magnesia; so that, after the precipitation of Calcium as Calcium oxalate, and separation by filtration, Magnesium may be thrown down as Magnesium Ammonium phosphate, on addition of Di-sodium Hydrogen phosphate mixed with much Ammonia. It is weighed as Magnesium pyro-phosphate, $Mg_2P_2O_7$.

XXVI. Zinc Zn=65.

377. ZINC Zn=65, is a metal never found native. It is bluish-white, hard, lustrous, and crystalline. At common temperatures, Zinc is brittle, but somewhat below 150° C., it is both malleable and ductile. It melts at 412° C., and volatilizes at a red heat. Specific gravity 6·8 to 7·1. In the air, Zinc tarnishes slowly and very slightly, but when strongly heated in air or Oxygen, it burns with bluish-white flame to ZINC OXIDE ZnO. Owing to its lightness and

durability, it is much employed as a substitute for lead. In the Galvanic battery, Zinc forms the electro-positive element or generating metal. Sheet-iron coated with Zinc constitutes galvanized iron,

The acids dissolve Zinc with greediness, and the corresponding salts are obtained. Potassium Hydrate also dissolves Zinc. ($Zn + H_2SO_4 + 2H_2O = H_2 + 2H_2O + ZnSO_4$. $Zn + 2HCl = H_2 + ZnCl_2$. $3Zn + 8HNO_3 = 2NO + 4H_2O + 3Zn2NO_3$ $4Zn + 10HNO_3 + nH_2O = N_2O + 5H_2O + nH_2O + 4(Zn2NO_3)$. $Zn + 2KHO = H_2 + K_2O,ZnO$).

378. Zinc oxide $ZnO = 81$. Like Magnesia, the only basic oxide. It forms a white powder, insoluble in water, and becomes canary-yellow whenever it is heated to redness. It dissolves readily in acids, and unites with water to form a hydrate ZnH_2O_2. *Zinc white*, sometimes used as a substitute for White lead, is mainly Zinc oxide.

379. Zinc sulphide $ZnS = 97$. Constitutes the mineral *Zinc blende*, from which Zinc is generally prepared. Pure Zinc sulphide is white, and insoluble in Potassium hydrate. It does not fuse, when heated. Ordinary Zinc blende is brown or black from the presence of Sulphide of iron; sometimes massive, but often crystallized in rhombic dodecahedra. When roasted in air, both the Zinc and the Sulphur oxydize; Zinc oxide remains, and Sulphurous anhydride is evolved. $2ZnS + 3O_2 = 2SO_2 + 2ZnO$. The oxide is reduced by Charcoal.

380. Salts of zinc. Zinc chloride $ZnCl_2$, is a white, deliquescent substance, fusible at 250° C., and volatile at a red-heat. It is very corrosive, and is used by Surgeons as an Escharotic. A solution of it has been long and effectively employed for sanitary purposes and as an antiseptic, under the name of *Burnett's Disinfecting Fluid*. It is poisonous: Magnesia is the best antidote. Zinc carbonate $ZnCO_3 = 125$. Found native as *Calamine*, both massive and crystallized in rhomboids. When pure it is white, and insoluble in water; it is, however, more generally grey or yellowish from the presence of Ferric oxide. Basic oxy-carbonate of zinc ($8ZnO,3CO_2,6H_2O$) is precipitated from soluble Zinc salts by Sodium carbonate. By roasting, Calamine loses Carbonic anhydride,

and the Zinc oxide is subsequently reduced by Charcoal. ZINC ORTHOSILICATE $Zn_2SiO_4.H_2O$, is also known as *Electric Calamine*. ZINC SULPHATE $ZnSO_4.7H_2O$, in colourless four-sided prisms, soluble in $2\frac{1}{2}$ parts of water. It is used both in medicine and in the arts. On a large scale, it may be prepared by roasting Zinc sulphide at a low temperature: $ZnS+2O_2=ZnSO_4$.

381. ZINC AND ITS METALLURGY. The two ores are Calamine and Blende; the former is the most valuable. The metal seems to have been unknown to the Ancients, who were, however, acquainted with oxydized Zinc under the name of Cadmia. Pliny makes mention of it as necessary for making Brass, and adds that Cadmia not only improved the colour of Copper, but also increased its weight. Albertus Magnus, too, in the thirteenth century, was well acquainted with the use of furnace calamine (ore of zinc) in making Brass. The first Author who makes use of the name Zinc is Paracelsus Theophrastus, who died in 1546. Yet even as late as the middle of the sixteenth century, the metal must have been scarce. In China and India, however, Zinc has been long known; and utensils of Zinc, inlaid with various metals, are not uncommon in those countries. In the extraction of Zinc, the ore is first crushed and then roasted. When the Blende is mixed with Galena or Lead sulphide, as at Laxey in the Isle of Man, and other places, the latter has to be carefully separated first, or the crucibles in which the smelting is accomplished would soon be destroyed. The roasting of the Blende is very tedious; the Zinc becomes oxydized and the Sulphur is volatilized as Sulphurous anhydride. In the case of Calamine, Carbonic anhydride is expelled by heat, and Zinc oxide remains.

Fig. 76.

The oxide is then mixed with half its weight of coal-dust, and introduced into a retort of clay. These retorts, *b*, are arranged in a furnace one above the other as represented in Fig. 76. The large ends of the retorts rest on ledges in the wall at a level somewhat higher than the other ends, which are supported in front on iron plates. Each retort is connected with a conical receiver of cast-iron, which receives the metal, and a second one, *a*, with an opening for the escape of gas.

In England, a somewhat different method prevails. The reduction is carried on in crucibles of peculiar construction, as represented

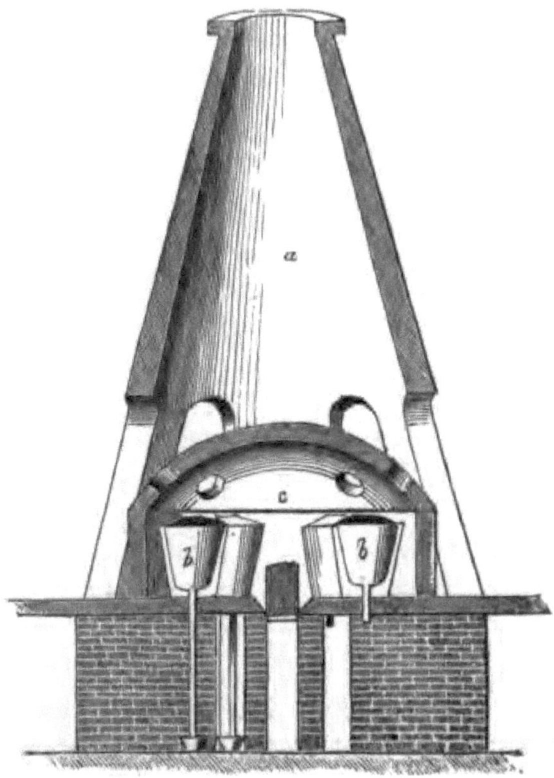

Fig. 77.

in Fig. 77. A circular furnace is employed, in which are ranged large crucibles, *b b*, through the bottoms of each of which an iron tube enters which terminates in vessels fitted to receive the Zinc. A mixture of the calcined ore and Coke is then introduced into each crucible, and the cover very firmly fastened down. As the heat increases, carbonic oxide is evolved, the Zinc distils and is received in the vessels placed for the purpose.

382. TESTS FOR ZINC. The salts of Zinc are colourless, and of an astringent, metallic taste. Sodium carbonate precipitates a basic carbonate. Zinc is not precipitated from its solutions by H_2S, but as white Zinc sulphide, on the addition of Ammonium hydrogen sulphide, and this precipitate is insoluble in Potassium hydrate. Zinc hydrate, white, gelatinous, and soluble in excess, is precipitated by Potassium, Sodium and Ammonium hydrates. Zinc is weighed as Zinc oxide, which contains 80·24 per cent. of Zinc.

[XXVII. Cadmium Cd = 112.

383. CADMIUM Cd=112. Is a tin-white metal, of specific gravity 8·65, never found native. It fuses at 227°·8 C. In air, Cadmium is unchangeable. It is soluble in the acids. Cadmium is used in Dentistry.

Cadmium was discovered by Stromeyer in 1818; it always accompanies ores of Zinc. It is, however, so much more volatile than Zinc, that it sublimes first. In order to purify it from Zinc, advantage is taken of the fact that Cadmium is precipitated in acid solutions by Hydrogen sulphide, in the form of Cadmium sulphide; the latter is dissolved in Hydrochloric acid, precipitated by Ammonium carbonate, and reduced with charcoal. The metal sublimes.

384. CADMIUM SULPHIDE CdS, is bright yellow and insoluble in Ammonium hydrogen sulphide. It is found native, crystallized in 6-sided prisms, and is known as *Greenockite*.

385. CADMIUM OXIDE CdO=128. Is brown; its HYDRATE CdH_2O_2, is white, and soluble in Ammonia. CADMIUM CHLORIDE $CdCl_2,2H_2O$. CADMIUM IODIDE CdI_2, is much used in Photography. CADMIUM SULPHATE $CdSO_4,4H_2O$.

386. TESTS FOR CADMIUM. These salts resemble those of Zinc, but Hydrogen sulphide precipitates Cadmium from acid solutions, as Cadmium sulphide. Cadmium hydrate is insoluble in Potassium hydrate. Cadmium is weighed as oxide; 100 parts contain 87·5 parts of the metal.]

CHAPTER XIX.

PHYSICAL AND CHEMICAL PROPERTIES OF WATERS.

387. Water, one of the elements of the Ancients. 388. The importance of water. 389. Blue, when viewed in mass. 390. Water is HYDROGEN OXIDE H_2O. 391. It is 770 times heavier than air. One Litre weighs one Kilogramme. 392. Water as ice at 0° C. Increases in bulk below 4° C. 394. Expansive force of ice. 395. Breaks water-pipes in winter. 396. Vapour at 100° C. Boils at 100° C., when Barometer at 760mm. Influence of Barometer. 397. Increase of pressure raises the boiling-point. Papin's digester. 398. Distilled water. 399. Moisture in the air. 400. Deliquescence and Efflorescence. 401. Specific heat of water. 402. Latent heat of water. 403. A bad conductor, but good conveyer of heat. 404. Slightly compressible. 405. Mobility of water. 406. Equal pressure. 407. Upward pressure. 408. Capillarity. 409. Level surfaces. Water-level. Artesian Wells. Water-supply. 410. Pressure on the bottom of a vessel is independent of the shape. Bramah's press. 411. Displacement of fluids. 412. Specific gravity. 413. Hydrometers. 414. Pneumatics and Hydrostatics. 415. The Syphon. 416. Distilled water alone chemically pure. 417. Rain-water contains certain gases and salts. 418. The solvent powers of water. Collecting waters for the estimation of Carbonic acid. 419. CALCIUM BICARBONATE $CaCO_3,CO_2$ in waters. Stalactites. Stalagmites. Travertine. 420. CALCIUM SULPHATE $CaSO_4,2H_2O$, in water. 421. SODIUM CHLORIDE $NaCl$, in water. 422. Description of the Dead Sea. 423. The Sea, the great reservoir of salts. River-water polluted. 424. Self-purification of rivers. 425. Hardness of Spring-waters. Crystal Palace Fountains. 426. Temperature of waters. Calcareous, Chalybeate, Hepatic, and Saline waters. 427. Geysers of Iceland. 428. Sea-water gives indication of the earth's structure. 429. Epsom salts or MAGNESIUM SULPHATE $MgSO_4,7H_2O$, and MAGNESIUM CHLORIDE $MgCl_2,6H_2O$, in sea-water. Analysis of sea-water. 430. SODIUM IODIDE NaI. 431. Traces of salts of Silver, Copper, and Lead in sea-water.

387. WATER, one of the so-called elements of the old philosophers, is met with chiefly in its liquid form. A very superficial acquaint-

ance with the physical condition of the earth, will enable us to judge of the mighty effects produced in times past upon its surface, by the agency of vast masses of water, at rest or in motion. The deposit of all the so-called sedimentary rocks, as well as their arrangement in successive Strata, is due entirely to its action; while the accumulation of Drift, and many other geological phenomena, may be traced to the same agency.

388. Even now, water plays such an important part in the economy of nature, and is employed in our households, as well as in the useful arts, for such an infinite variety of purposes, that a knowledge of its nature and properties cannot fail to be both interesting and useful.

389. In small quantities pure water presents the appearance of a colourless, transparent fluid, without taste or smell: but, seen in masses, it appears blue.

390. It is a compound of Hydrogen with Oxygen. Two atoms of Hydrogen are united with one atom of Oxygen. The chemical symbol for HYDROGEN OXIDE or WATER is therefore H_2O (p. 60, par. 64).

391. It is 770 times heavier than air: one cubic inch of water at 15°·5 C., with a Barometer at 30″, weighs 252·458 grains. A Litre of water at 4° C., weighs 1000 Grammes or 1 Kilogramme.

392. Among the many interesting properties of water, not the least is its transformation into ice at a temperature of 0°·C. If a vessel filled with broken ice be placed in a warm room, the ice will gradually melt: but a thermometer placed in it will continue to indicate the same temperature (0° C.), as long as any ice remains unmelted. This constant temperature of freezing water has been adopted as one of the fixed points of all thermometers (p. 8, par. 13).

393. Water increases in volume by congelation; but, in its mode of increase, it forms a nearly singular exception to the very general and regular contraction of bodies by cold. Above 4° C., water expands by heating; below that point, it expands by cooling. It therefore attains its maximum of density at 4° C. This singular fact has the most important consequences in nature; on it depends, in some degree, the very life of aquatic plants and animals, in the temperate and colder regions of the earth. For, if the contraction of water by cold continued down to its freezing-point, there would be a

continual fall of cold water to the bottom and rise of warm water to the top, till the whole was frozen and every living thing in it must die. Were ice to obey the ordinary laws of contraction, it would sink in water almost beyond the reach of the sun's heat: as it is, it floats above the warmer water, and, being a bad conductor of heat, actually protects it, so that whatever may be the temperature of the air, the deep water is never below 4° C. Accordingly, a severe winter may kill all the fish in a shallow lake, but deep lakes such as those of North America, are not liable to such an accident; they contain such a mass of warm water below, that they are never even frozen over.

394. Among the beneficial results of the Expansion of cold water and of ice, must be enumerated its effects upon soils during a severe winter. The water, always contained in the pores of soils, freezes, and, in expanding, causes a breaking-up or comminution of the hard clods. Freezing water is also, by its wedge-like expansion, one of nature's chief agents in breaking up or disintegrating rocks for the formation of soil: it is thus capable of producing results equal to those of blasting.

395. This expansion of freezing water is sometimes unpleasantly brought home to us by certain domestic inconveniences. Of these, the breaking of water-jugs and the bursting of water-pipes are instances. The thaw has usually the credit for the mischief. Nevertheless, the fracture takes place during the frost; the thaw, by liquefying the ice, only brings it to light. Very slight precautions to protect the pipes from the frost will entirely prevent such accidents.

396. Water readily assumes a gaseous state, evaporating at all temperatures: but it is said to BOIL, that is, to become converted into a vapour of the same elastic force as the air, at 100° C. But this is not universally true, or, rather, it is true only under certain conditions; for the temperature at which water boils, varies with the pressure of the air. The lighter the air, the more readily does it boil at a lower temperature. The temperature at which it boils, or, as it is briefly called, the "boiling-point," is 100° C. when the Barometer stands at 760mm. Under the receiver of an Air-pump, when the air is exhausted, it may be made to boil at 21° C.; and as the

pressure of the air diminishes with the increasing height from the earth's surface, the boiling-point falls 1° C. for 327 metres of ascent. So that the boiling-point of water may be, and is, commonly used for determining the heights of mountains. On the other hand, when the pressure is increased, as in deep mines, the boiling-point is higher than 100° C. Under a pressure of two atmospheres, it is 120°·8 C.; under that of twenty atmospheres, 213° C.: so that the barometric pressure must always be observed, before any deduction can be drawn from the temperature of boiling water.

397. In the open air, the Boiling-point cannot be raised, because the pressure cannot be raised. But if water be boiled in a close vessel, as in the boiler of a steam-engine, the steam which is generated, having no escape and continually increasing in quantity, gradually raises the pressure, more and more, to any point that may be required; and, with the pressure, the temperature. Now, the steam which rises from the water under pressure, is called High-pressure steam. Temperature and pressure are, indeed, in such close relationship, that the engineer judges of the pressure by the thermometer. The temperature of the Steam is always the same as that of the Water from which it rises.

Advantage is frequently taken of this fact, when a steady temperature above 100° C. is desired, it being found that the solvent powers of water are greatly increased at these higher temperatures. It is turned to account in "Papin's Digester," so much used for the extraction of Gelatine from bones. This consists of a strong iron vessel, supplied with a safety-valve by which the pressure of Steam can be regulated, and the vessel prevented from blowing up. Some such apparatus becomes absolutely essential to persons living on high mountains, as the temperature at which the water there boils in the open air, is too low for ordinary purposes. Thus, on the summit of Mont Blanc, water boils at 85°·4 C.

398. The formation of Steam, and its subsequent condensation, is the only mode of obtaining perfectly pure water. No water, except that specially prepared, is absolutely pure in a chemical sense. The purest natural water is rain; but, even that contains not only Gases in solution, which it has washed from the air in its descent, but also many organic and other impurities derived from the same source:

and if it has once touched the soil, its character is, of course, further altered by the solution of certain Salts. All these impurities may be removed by distillation, the only process by which it is possible to separate a liquid from the solids which it has dissolved (p. 61, par. 64).

399. The MOISTURE in the atmosphere varies in quantity according to its temperature (p. 40, par. 32). In rainy weather, the air is generally near its point of Saturation; in other words, it contains nearly as much moisture as it can absorb.

400. Certain substances possess the power of absorbing, or attracting to themselves, this vapour from the air: they are then said to DELIQUESCE. A good illustration of Deliquescence is afforded by Potassium carbonate, which becomes liquid on exposure to air. Some substances, on the other hand, which contain water, give off a portion of it to the surrounding air, when the latter is not saturated with moisture: they EFFLORESCE. Of Efflorescence, Sodium carbonate is a good illustration: it becomes dry and powdery in a dry atmosphere.

401. The high SPECIFIC HEAT of water, as compared with all other liquids and solids, has been already alluded to (p. 23. par. 22). By virtue of this property, the waters of the earth contribute in no small degree, towards moderating the transitions of the atmosphere from heat to cold, or from cold to heat, by reason of the large quantities of heat which they absorb; at the same time, their own temperature varies but little.

402. The LATENT HEAT of water is equally remarkable. In the liquefaction of a single pound of ice, no less than 79° C. of heat become latent; consequently, when water freezes, for each pound of ice formed, 79° C. of heat are set free (see p. 25, par. 23). The opposite action is of equal importance, in preventing ice from thawing too rapidly.

403. Water is a bad CONDUCTOR of heat; it may be boiled at its surface without communicating heat downwards. But, although a bad conductor, it is a good CONVEYOR; and hence its value in conveying heat, by its circulation through pipes, to different parts of a large building. This property is also operative in the case of the Gulf-Stream (p. 12, par. 19).

404. Experiments have shown that water is but very slightly COMPRESSIBLE. For each additional pressure of 1·033 Kilogrammes per square Centimetre (or of 15lbs. upon a square inch of its surface), Water is only compressed 47 millionths of its bulk. In this respect it contrasts strongly with air, which, under the same amount of pressure, is reduced to one-half of its bulk

405. No attentive observer of the characters of water, can fail to be struck with the fact of the extreme MOBILITY of its particles. How slight is the effect of the weight of a body of water falling from a height, compared to that which would be produced by it, if converted into ice! Owing to the opposition of the air to its fall, the water becomes divided into innumerable drops, and falls as an irregular shower instead of a solid mass. What an important effect has this property on a fall of rain, which, if not broken up by the resistance of the air, would descend in a mass and with great velocity, carrying destruction and desolation before it! How much, too, does this property add to the beauties of Nature, in providing the spray of the waterfall or the rivulet!

406. In consequence of the complete Mobility of its particles, water, like air, exerts an EQUAL PRESSURE in all directions, each layer pressing with equal force downwards and upwards. This may be easily demonstrated by taking a straight glass lamp-chimney, Fig. 78, *a*, one end of which is covered by a circular disc of tin, *c*, held in its place, against the bottom of the tube, by the piece of string, *b*. On forcible immersion of this end into water, the tin will be kept in its place by the upward pressure of the water, until the latter rises to the same height within, as without the tube, when, the pressure on both sides of it being equalized, the tin will fall, by its own gravity, to the bottom. On account of this upward pressure of fluids, when a hole is made in a ship's bottom, the water rushes in, to effectually oppose which, a force must be applied equal to the weight of a column of water, of which the base is of the same area

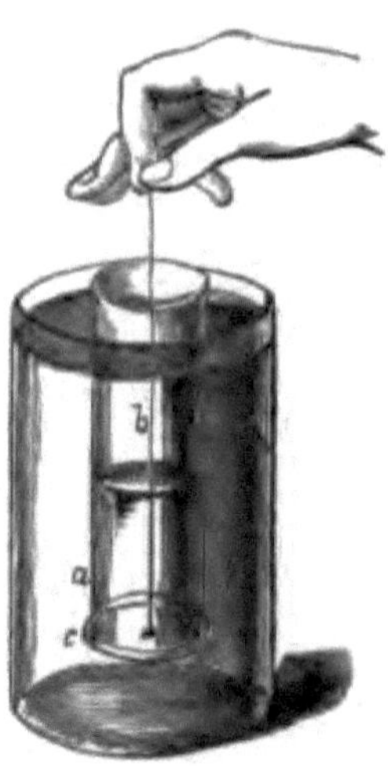

Fig. 78.

as that of the aperture in the vessel, and the length equal to the depth of the hole from the surface of the water. Hence, in vessels of large draught, the under-surfaces should possess considerable strength, to enable them to oppose the upward pressure exerted by the water in which they float.

407. It is in consequence of this equal pressure of water in all directions, that, when at rest, it assumes a perfectly HORIZONTAL surface. Not that this property is peculiar to Water: it belongs equally to all fluids and gases, that is, to all bodies whose particles are free to move on each other. It is well illustrated and turned to account, in that useful instrument, the SPIRIT-LEVEL. This is a glass tube, closed at each end, and so nearly filled with spirit of wine as to leave space only for a single bubble of air. When the instrument rests on a perfectly horizontal surface, this bubble remains stationary at its centre; but if it be inclined ever so little, the bubble immediately rises to the higher end.

408. There is one apparent exception to this law of the level surfaces of liquids. If a tube be plunged into water, the liquid will be seen to stand at a higher level inside the tube, than outside; and it rises towards this tube, as in Fig. 79 (*a* and *b*), so as to give the liquid a curved surface.

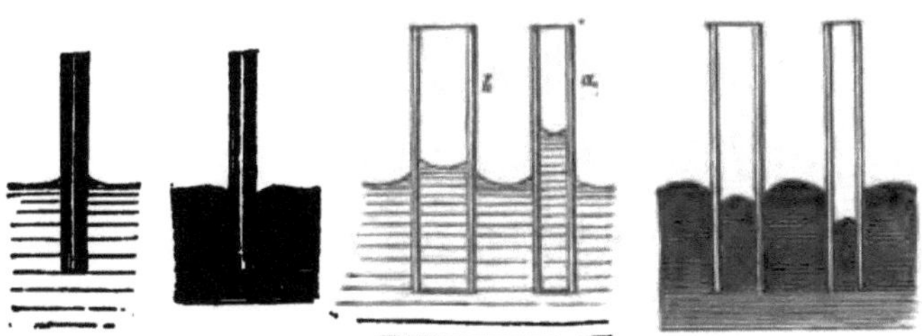

Fig. 79.

The phenomenon of CAPILLARY ATTRACTION (p. 121, par. 181) is due to the adjustment of the forces of Adhesion and Cohesion between Liquids and Solids, under the simultaneous influence of gravity. The smaller the tube, the higher the water will rise in it. The cause of the rise of the water is the adhesion between its particles and those of the glass; the limits to that rise are the action of

Gravity and the force of Cohesion among the liquid particles. The height of the column is inversely as the diameter of the tube. Capillary attraction takes place wherever a liquid touches the side of the vessel containing it. It may be readily seen in a tea-cup; the bubbles we make in stirring our tea, always make their way to the side of the cup, because, as soon as they happen to come within reach of its capillary attraction, they are drawn towards it and held there.

But if, instead of water, the tube be plunged into some liquid which cannot wet it, such as mercury, instead of attracting, it appears to repel it, and the liquid will not rise so high within the tube as without it. This is called Capillary Depression. The narrower the bore of the tube, the greater is the Depression. This Depression requires a certain correction in reading off the height of the Mercury in the Barometer. But, by employing a tube of 16-20 millimetres in the bore, this correction is sufficiently trifling to be ignored.

409. However large a mass of water may be, whether a lake or the ocean, it must obey this Law of level surfaces. It is not even necessary that the surface should be everywhere continuous. If two bodies of water, in different vessels or reservoirs, are connected by a tube, or underground channel, so long as they are at rest, their surfaces must be at the same level. No matter what the size or shape of the different containing vessels may be, the one law which the liquid obeys, is, that its surface in all must be at the same level. Thus, if several glass tubes, (*a*, *b*, and *c*) of different shapes and sizes (as in Fig. 80), are placed in communication (through *f*, with the receiver *d*), water will not attain a condition of rest in any of them, until it stands at the same level in all.

Fig. 80.

An application of this condition of even level, in vessels communicating with one another, is seen in the WATER-LEVEL. The instrument is much used in levelling, to ascertain how much one point is higher than another. It consists of a Metallic Tube, bent at both

Fig. 81.

ends, and fitted with glass tubes, *a* and *b*, Fig. 81. It is placed on a Tripod, and Water introduced until it stands at the same level in both tubes. The levelling staff being held vertically, the Observer looks at it through the Water-level, along the line of *a*, *b*, and *c*. The height of the staff is then measured and, by subtracting it from the height of the level, the Difference between the Height of the spot at which the boy is placed and that upon which the Tripod stands, is obtained.

The supply of water to houses from various Reservoirs depends on

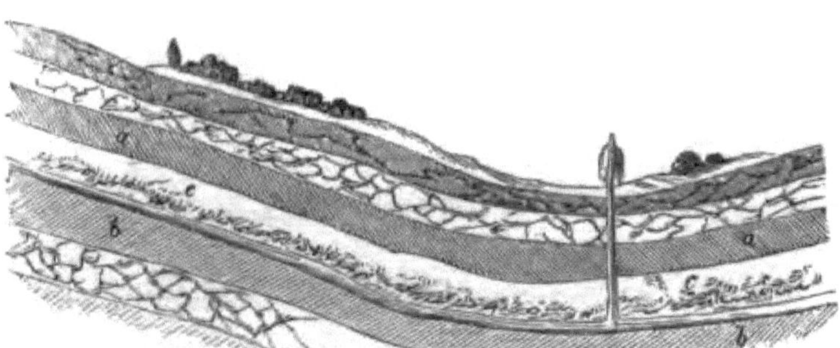

Fig. 82.

this law. No matter how long, or how circuitous, may be the pipes which the water has to traverse, it is only necessary that the cistern to be supplied should not be at a higher level, than the reservoir from which the water is to flow.

The same is the case with Artesian wells. Let us suppose a basin, Fig. 82, to enclose a layer of sand (*c*) between two layers (*a* and *b*) of impermeable clay. Following the natural fall of the ground, water will collect in the hollow of the basin, and will be prevented from escaping by the very nature of the clay. If a boring be now made into the stratum which contains the water, the latter will rise, fountain-like, and to a height nearly corresponding to the difference between the two levels.

410. A little consideration of the last paragraph, will render it clear to any-one, that the pressure exerted by Water, or by any liquid, on the bottom of the vessel which contains it, is entirely independent of the shape of the vessel, or of the actual quantity of water in it. It depends on the Height of the water, and is equal to the Weight of a column of water having a Diameter equal to the

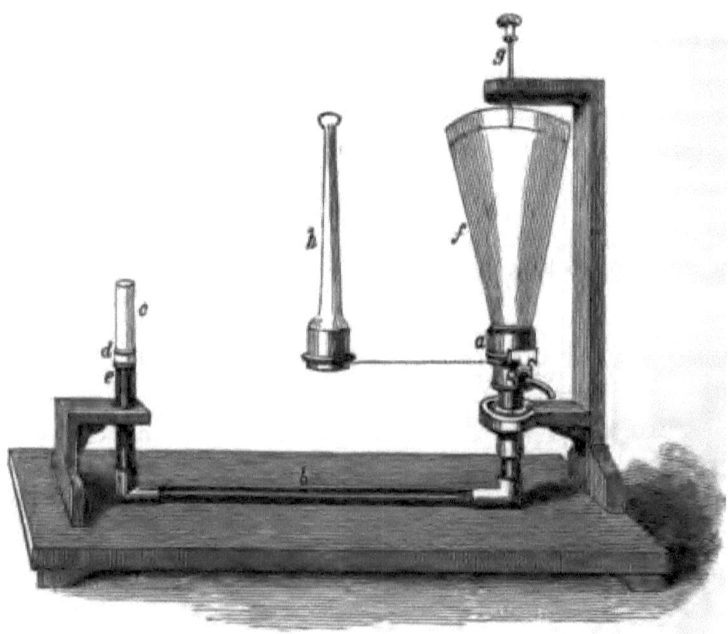

Fig. 83.

Base of the vessel, and a Height equal to the extreme Depth of the Water. This may be rendered more clear by the apparatus shown in Fig. 83. The bent glass tube (*a*, *b*, *c*) is provided with a screw at *a*, so that vessels of various shapes, as *f* or *h*, may be attached to it. Mercury is poured into it, till it stands at the same level as *e*. If water be now poured into *f*, up to the level marked by the index *g*, the mercury in the other limb of the bent tube will be forced up to the level marked *d*. Now let *f* be removed and the vessel *h* be put into its place; and let *h* be filled with water up to the same level as marked by the index *g*: the mercury in *c* will again be pressed up to the point *d*, the same point it reached before. In other words, the pressure on the mercury is just the same in both cases, although the vessel *f* takes so much more water than *h*; showing that it is only the weight of the column immediately above the mercury, which is felt by it. The downward pressure of the remainder is supported by the sloping sides of *f*, so that the pressure at *a*, is regulated by the height of the water, and not by its bulk.

Now, it is manifest that, if this be the case, any amount of Pressure may be obtained, simply by increasing the Weight of the column of Water, no matter what may be its Width. Suppose, for instance, a common cask be taken, and a very long tube inserted into its top, and the whole filled with water. The pressure upon every point of the inside of the cask, will be just as great, as though the cask itself was as high as the tube. So that if the tube had an area of one square inch, and it was long enough to contain a hundred-weight of water, every square inch of the interior of the cask, would be pressed upon with the force of a hundred-weight. By this simple expedient, then, enormous power may be obtained.

The principle has been taken advantage of, in the construction of Bramah's Hydraulic Press, Fig. 84. Only, as a very long tube is practically inconvenient, the effect of a very high column of water is obtained by the mechanical pressure of a force-pump, *c*, on a shorter pipe. One end of the "cask" or cylinder with which this pipe communicates is a movable piston, *a*, and, of course, the force bearing upon this piston is in proportion to its size, without any alteration in the size of the tube. The power of this machine exceeds, perhaps, that which could be obtained by any other known means. Such enormous weights as the tubes of the Britannia Bridge are easily lifted by it; while it affords the readiest and most satisfactory means of testing the strength of beams, and performing numerous important engineering operations.

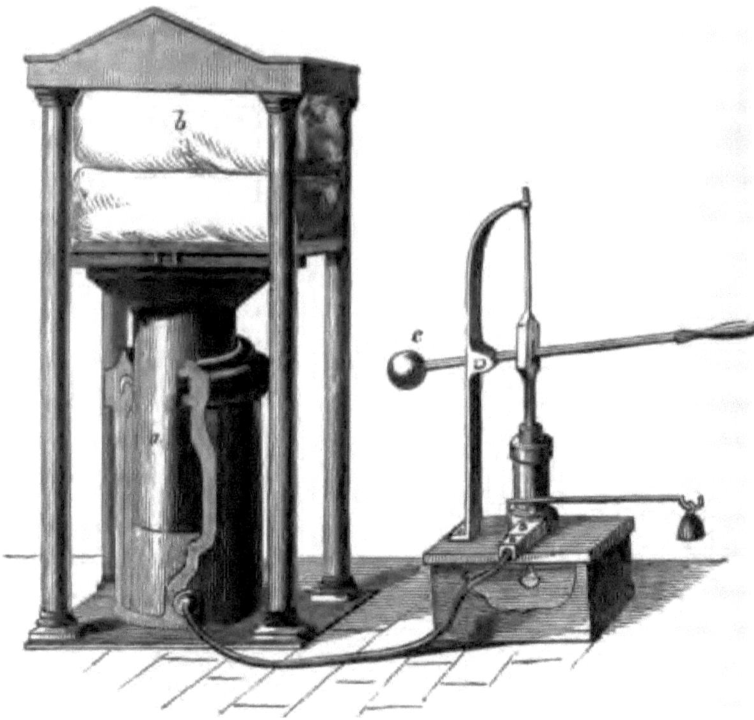

Fig. 84.

411. When a solid body is immersed in water, it displaces a quantity of the latter exactly equal to its own bulk. If the solid be of less weight than the quantity of fluid displaced, in other words, if it be of less specific gravity than the fluid, it will swim; if of greater, it will sink. But if it be of equal weight with it, it will remain at rest in the water in any position in which it may be placed. Fish appear to be in this latter condition. To enable them to ascend or descend in the water, they are furnished with an air-bladder which they compress or expand at will, and so regulate their own weight.

The amusing toy of children, called the Cartesian devil, affords a good illustration of this effect. A hollow glass figure, with a small hole at the lower end, nearly filled with water, but containing just enough air to cause the figure to float, is placed in a tall glass jar nearly filled with water, and closed on the top with caoutchouc. When this caoutchouc is pressed upon by the hand, the pressure is transmitted, through the water, to the bubble of air in the figure. A part of the space which this bubble occupied, being now filled with the heavier water, the figure sinks, but rises again, as soon as the removal of the hand enables the air to expand to its former bulk.

412. Upon the principle, that a body immersed in water loses as much of its Weight as is equal to that of the water which it displaces, depends the mode of determining the SPECIFIC GRAVITY of such substances as are heavier than water. If, therefore, we have ascertained the weight of a body in air, the difference of the two is the weight of an equal bulk of pure distilled water, with which the weight, in air, of the substance, is to be compared. The mode of weighing is represented in Fig. 85. The operation consists first in weighing the solid in air; then, having suspended it by means of horse-hair (*c*) from the bottom of the scale-pan, *b*, immersing it in distilled water at 15°·5 C., and again weighing. The weight of an equal bulk of water, is to the specific gravity of water (taken as 1·000), as the weight of the body in air is to the specific gravity required. The rule for obtaining the specific gravity of a solid may be thus expressed: divide the weight of the body in air by the loss which it experiences when weighed in water. The quotient is the Specific gravity.

Fig. 85.

If a solid is lighter than water, it is made heavier by fastening to it another substance, the weight of which, in air and in water, has been previously ascertained. The simplest mode of estimating the specific gravity of liquids, is by comparison of equal bulks of distilled water, and of the fluid in question, in a small bottle. By dividing the weight of the fluid, by the weight of the water, the specific gravity of the latter is determined.

413. Sometimes Hydrometers (from the Greek ὕδωρ, *hydor*, water, and *μέτρον*, *metron*, a measure) are employed for taking specific gravities, their action likewise depending ou the fact, that a floating body displaces an amount of the liquid in which it floats, equal to its own weight. The Hydrometer, Fig. 86, consists of a graduated scale, so constructed as to float vertically in the fluid, by means of a hollow ball of glass weighted by mercury. The measure of its sinking, will be the measure of the liquid displaced, and, therefore, of its specific gravity. If the liquid is light, the hydrometer will sink deep: and the reverse, if heavy. Some hydrometers are so graduated that their degrees express the specific gravity directly. With us in England, that of Twaddell is most in use. The degree of gravity marked on Twaddell's scale must be multiplied by 5 and the product added to 1000, to give the actual specific gravity.

Fig. 86.

414. As air, at a fixed temperature, is employed as the standard of comparison for the weight of gases, so is water for that of all other substances—and in both instances, on account of their worldwide diffusion and great importance. In France, water at 4° C. is accepted as the standard; with ourselves, water at $15^{\circ}{\cdot}5$ C. For the same reason, the name of PNEUMATICS has been given to the study of the Laws of Gases, and HYDROSTATICS to those of Fluids—air being the gas, and water the fluid—the laws of which we have the best opportunity of observing.

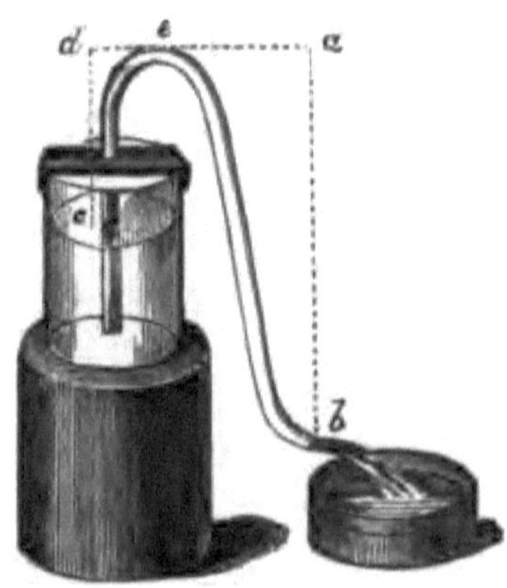

Fig. 87.

415. An apparatus founded on the properties of the air, and much employed for transferring liquids

from one vessel to another, may not inaptly be introduced into the present Chapter on Water. The Syphon, Fig. 87, consists of a bent tube, *e*, open at both ends. Being filled with some fluid, and closed by the fingers at both ends, the shorter limb is immersed in the fluid at *c*; the latter continues to flow out at *b*, as long as its end dips in the liquid. The flow will be more rapid in proportion as the difference of level between *b*, and the surface of the liquid at *c*, is greater.

416. It has been already stated that scarcely any Water, in a natural state, is CHEMICALLY PURE. Now, although distilled, or chemically pure, water is of great value to the chemist, in studying the varied properties of this important liquid, it would, in this puro state, answer comparatively but few of the purposes for which it was so plentifully provided by Nature. Its great power of dissolving Solids, Liquids and Gases, a power which it possesses in a higher degree than any other liquid, adapts it to its varied uses and gives distinctive characters to rivers and seas and waters of every description.

417. Rain-water recently collected in the open country, and not stored in a cistern from which it can derive anything soluble, is as free from solid constituents as water can ever be in a natural state. But as it has power to dissolve gases, rain always contains varying quantities of Oxygen, Carbonic anhydride, and Nitrogen, with minute quantities of Ammonium Hydrogen Carbonate, and, during storms, of Ammonium nitrate and of Nitric acid. By the help of these gases and other substances which it washes down, the rain assists materially in fertilizing the soil. The rain-water of towns is always rendered impure by the various exhalations, by the products of combustion and decay, and by the chemical processes there carried on.

418. The Solvent-power possessed by rain, is exerted as strongly upon Solids as upon Gases. The character of the water of a spring, brook, or river, will, therefore, entirely depend on the nature and solubility of the soils, through which its waters have passed. Carbonic acid, which is always present in rain, though its acid properties are feeble, is possessed of wonderful solvent power when diffused through water. The great importance of this power will be shown,

by a brief consideration of the characteristics of certain of the salts, which are found, more or less, universally in soils.

The correct estimation of the quantity of Carbonic anhydride in a water, is a matter of extreme importance. The method devised by Dr. Mohr for collecting the water, is illustrated by Fig. 88. The apparatus is intended to be used, when the water to be analysed is at a depth beyond the reach of the arm. A bottle, made of sufficiently stout glass to withstand the pressure of the water, is provided with a doubly-perforated cork, and fitted with two glass tubes. The one, of smaller bore, just enters the bottle, and is connected with an india-rubber tube, *c*, the object of which is to allow of the escape of the air at the will of the operator: the second tube, of larger size, reaches nearly to the bottom of the bottle, and is closed by a cork, fastened to a string held in the right hand. By means of the knotted string, held between the thumb and finger of the left hand, the whole apparatus is suspended, and retained in a vertical position by the metallic cover and weight (*a*, *b*).

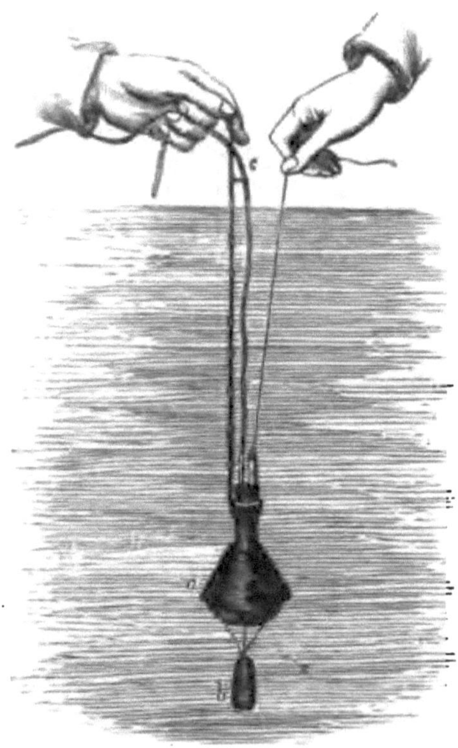

Fig. 88.

Such an apparatus can be filled at any depth in water, as complete command is had over the filling of the bottle by the exercise of pressure on the india-rubber tube, and by the retention of the cork in its place, until the requisite depth is attained.

419. One of the most abundant constituents of rocks and minerals, is Calcium carbonate (p. 198, par. 369). It is not only diffused through every fertile soil as Chalk, but is deposited under the several rms of Chalk, Limestone, Oolite, and the varieties of Marbles. It

is found also beautifully crystallized in Iceland spar in rhombohedral crystals, Fig. 89. It is a constituent of the skeletons of all animals, and forms the great bulk of Shells and Corals. All, or nearly all, of this Calcium carbonate has been, or is, derived from water, and yet half-a-gallon is required to dissolve a single grain of it! No wonder, then, that it should be found in all natural waters. How else could the inhabitants of the sea obtain their regular supply? The sea has it from the rivers, and the rivers from the soil, out of which it is dissolved in large quantities by the Carbonic acid carried there by the rain, and produced by the respiration of fish, and the decay of plants and animals. This acid solution of chalk, or CALCIUM BI-CARBONATE, gives to Springs and Sea-water one of their great characteristics,—Hardness. Carbonic acid thus breaks down some of the hardest limestone rocks, by a purely chemical process, and becomes the occasion of one of the commonest natural phenomena of lime-stone districts, viz: the enormous natural caverns peculiar to them. The

Fig. 89.

Fig. 90.

roofs of these are often found clothed with pendent masses of Calcium carbonate, like icicles (Fig. 90), called by the name of "Stalactites," while their floor is covered with a thick layer of the same nature, formed by the droppings from the stalactites, and which are known as "Stalagmites." Their formation is due to the facility with which the lime-laden water, when exposed to the air, parts with the excess of Carbonic anhydride; as it is, then, no longer able to hold in solution all the Calcium carbonate, a portion is deposited in the above beautiful forms.

In Italy, a building-stone is in common use, known as "Travertine," which is altogether composed of Calcium carbonate, and is a deposit from volcanic waters. Some springs in Tuscany, deposit as much as a foot in depth of carbonate of lime, in a single year. Such springs are also frequent in England, where they are known as petrifying wells. In fact, so common a constituent of waters is Calcium bi-carbonate, that scarcely any are found without it. When such a water is boiled, Carbonic anhydride is expelled, and Calcium carbonate is deposited in the vessel. The crust found in boilers is generally composed of this substance.

420. Another salt of very common occurrence in water, is CALCIUM SULPHATE, (p. 197, par. 369). It is generally met with in masses, as Gypsum, and is rarely found in a crystalline form. Calcium Sulphate being contained in most soils and being soluble in water, it is not surprising that all waters, whether Rivers, Springs, or the Ocean, should contain it. By its means, plants are supplied both with Lime and Sulphuric acid, and its presence in water is another cause of their hardness. When water containing much Calcium sulphate is boiled, it deposits the salt just in proportion to the quantity of water evaporated, a pint of water being able to contain in solution only about twenty-two grains.

421. In addition to Calcium sulphate and bi-carbonate, SODIUM CHLORIDE (p. 173, par. 311) is universally present in water. Nor, on reflection, will this fact create any surprise. That "common salt" is contained in every vegetable and animal, implies not only its solubility, but also its universal diffusion. Not only is it contained in every soil, but it exists in gigantic deposits in various parts of the earth. In the chalk-formation at Wielitzka, in Poland, there is a bed of Salt no less than 500 miles in length, 20 miles in breadth, and 1,200 feet in thickness. In Northwich, in Cheshire, and in

other places in England, are beds of Salt which have supplied this country with it for ages. To account for the saltness of the sea, is it too much to suppose that it has washed out immense beds of salt? Or, if this were only one of the sources of its saltness, it is not difficult to give another. The wide diffusion of Salt, which is washed by the descent of rains into the rivers, has supplied the sea with it; and as the sea presents a vast surface for evaporation, from which fresh water only arises, it is easy to conceive, that in the lapse of ages, it must have sensibly increased in saltness from this cause.

The sea is very far from being uniformly saline in all places. Wherever its evaporation is greatest, there is its saltness increased. Thus, at the Equator it is more salt than in the temperate zones. Of course, the salter it is, the heavier it is; and differences of density so caused, are quite as powerful a cause, if not more so, of currents in the sea, as that formerly mentioned, viz., its unequal heating. For example, there is a constant surface current from the ocean into the Mediterranean. In this inland sea, water evaporates; what remains, being rendered heavier, sinks and runs out as an under-current, which was found by Dr. Wollaston to contain actually four times as much salt as the surface-water. Where the sea obtains large supplies of fresh water, as at the mouths of rivers, its saltness is diminished. Off the coast of Patagonia, it receives such an enormous amount of rain from the slopes of the Andes (p. 47, par. 42), that it is actually fresh water. The same is the case, in a minor degree, in the Baltic and Black Sea. On the other hand, the Dead Sea, which is constantly receiving salt from the very salt waters of the Jordan, and which, having no outlet, cannot get rid of it, is saturated with it: and as the water evaporates, the salt is deposited at the bottom of the sea. This, therefore, is the heaviest known natural water.

422. The following note on the Dead Sea is taken from *Fraser's Magazine*:—"The April sun was shining down broad and bright on the clear, rippling waters of the splendid lake, which shone with metallic lustre, closed in between the high cliffs of the Judæan hills to the west, and the grand chain of Moab, like a heaven-high wall, upon the east. Over the distance, and concealing from us the

further half of the sea, hung a soft, sunny haze. There was nothing in all this of the Accursed Lake, nothing of gloom and desolation. Even the shore was richly studded with bright, golden Chrysanthemums growing to the edge of the rippling waters. There was but one feature of the scene to convey a different impression—it was the skeletons of the trees once washed down from the woody banks of Jordan by the floods into the lake, and then at last cast up again by the south wind on the shore, and gradually half-buried in the sands. They stood up almost like a blasted grove, with their bare, withered boughs in all fantastic shapes, whitened and charred as if they had passed through the fire. It had been my intention, of course, to bathe in the sea, so I was provided for the attempt, with the exception, unfortunately, of sandals; and the stones being of the sharpest, I was unable to follow the long shallow water, barefooted, far enough out, to test its well-known buoyancy for swimming. As few ladies, our dragoman told us (indeed, he absurdly supposed none) had bathed in the Dead Sea, I may as well warn any so disposed that the water nearly burnt the skin from my face, and occasioned quite excruciating pain for a few moments in the nostrils and eyes, and even on the arms and throat. The taste of it is like salts and quinine mixed together—an odious compound of the saline and the acridly bitter. No great wonder, since its analysis shows a variety of pleasing chlorides, bromides, iodides, and sulphates, of all manner of nice things; magnesia and ammonia among those more familiar to the gustatory nerves. The Dead Sea is 1,300 feet lower than the Mediterranean, and the evaporation from it (without any outlet) fully makes up for the supply poured in by the Jordan, so that the sea sinks as time goes on."

423. The Sea is obviously the great reservoir of all matters soluble in water, while Rivers are, as regards the Salts they contain, but very dilute Sea-water. The water of Rivers is often less pure than Spring-water, as it always contains more or less of animal or vegetable matter washed from the land; while in England, at least, it is generally polluted by the drainage of towns.

424. Rivers, and especially those whose current is strong, are, however, provided with a natural means of self-purification, in the oxydation of vegetable and animal matters, through the agency of the

air with which they are constantly being brought into contact; and also by the action of the various aquatic plants, which give off a perpetual supply of Oxygen at the very time, viz., in summer, when it is most needed.

425. Spring-waters are generally harder than River-waters, owing to the presence in them of a greater quantity of saline constituents —a difference due to the fact, that Rivers always contain a large quantity of Rain, which has passed direct to them, without washing through the soil. The rocks, constituting the earth's crust, are everywhere rent and torn into crevices and hollows, which form natural basins for the reception of water, which accumulates in them, till, at some point or other, it finds vent as Springs, very often at immense distances from the places where the water entered the ground. These underground basins and reservoirs are supplied with water from land at higher levels, and vent may often be given to the waters, artificially, by boring or well-sinking, as is illustrated in Fig. 82, p. 215. Our modern water-works are arranged on precisely similar principles, and the supply of water from an Artesian well, is exactly of the same character as that of the fountains at the Crystal Palace, substituting for the pipes, in the latter case, the natural crevices in the strata perforated, and for the artificial reservoirs, the natural hollows or basin-like formations in the earth, in the drainage-area above the level of the well.

426. Springs vary exceedingly in their properties. From what has been said in a former chapter, of the increasing temperature of the earth, as its crust is penetrated, it will be readily conceived that spring-waters may vary considerably in TEMPERATURE, according to the depth from which they rise. The temperature of the hot wells at Ems is 56°·1 C.; at Wiesbaden, 70° C. The water of some hot springs is nearly pure, that of others, abounds in salts. The latter are termed Mineral springs, and their constituents vary very greatly. Some are called carbonated or effervescing springs, from the Carbonic acid they contain. In such springs, the gas has been forced into the water under great pressure, and, as water, when relieved from this pressure, can only retain in solution its own bulk of this gas, the excess is given off with effervescence. Some springs flow through Common Salt, and are called "brine-springs." Others which con-

tain much Carbonic acid, and flow through beds of Ferrous bi-carbonate, contain a large quantity of Iron, and are characterized as "Chalybeates" (a word derived from the Greek, χάλυψ, *chalybs*, steel). Other springs are impregnated with salts of Magnesium. Springs flowing over Chalk districts contain much Lime, and hence are called "Calcareous"; those in the Chalk-district round London contain about 20 grains of Chalk, in the form of "Calcium bi-carbonate," per gallon. "Hepatic" or "Sulphurous" springs, like those of Harrowgate, contain Sulphuretted Hydrogen (p. 91, par. 109,) and give off the offensive smell of putrid eggs. A spring at Kreuznach is celebrated for the presence in it, of MAGNESIUM BROMIDE. This salt is also contained in sea-water, in quantity amounting to about 1 grain per gallon.

427. Among the most remarkable springs, are the hot springs of Iceland. These geysers, as they are called, contain large quantities of SILICA (p. 145, par. 243) with which they encrust everything that is placed in contact with them. This Silica they derive from a rock, common in Iceland, which contains a very soluble modification of it.

428. Most springs give indications of the character of the earth through which they pass; but even more information as to the earth's chemical constituents, is derived from a knowledge of the components of the Sea. As has already been indicated, the sea contains the washings of the whole globe for unnumbered ages, and in it may be found the rarest mineral treasures.

429. Sea-water is possessed of a peculiar bitter taste, due to the presence of two salts, MAGNESIUM SULPHATE (p. 202, par. 375) and CHLORIDE. Both of these salts are present in considerable quantities, as will be seen from the following analysis of sea-water:—

	Water from the English Channel.
Water	963·74372
Sodium chloride	28·05948
Potassium chloride	0·76552
Magnesium chloride	3·66658
Magnesium bromide	0·02929
Calcium sulphate	1·40662
Magnesium sulphate	2·29578
Calcium carbonate	0·03301
Sodium bromide	distinct trace
Sodium iodide	trace
	1000·00000

Specific gravity 1027·4 at 15°·5 C.

430. Of the above list of the constituents of sea-water, one, SODIUM IODIDE (p. 173, par. 311) though existing only in minute quantities, as in analysis to be found only in "traces," is of vast importance, owing to its great value as a medicine.

431. Among the "traces" in sea-water, are salts of Silver, Copper, and Lead. The ship, as it sails through the ocean, actually receives on its copper-bottom deposits of Silver and Lead, and attempts have even been made to make use of the sea as a source of silver, but the quantity derived from it has never proved sufficient to make the experiment remunerative.

CHAPTER XX.

ON SOAP, WITH SOME ACCOUNT OF BLEACHING AND DISINFECTING AGENTS.

432. Soap, as an article of domestic use. 433. Mentioned by Pliny. Derivation of Sapo from Sepe. 434. Modes of cleansing in olden times. 435. Use of lye of ashes. 436. Cleansing power of wood-ashes. 437. No agents equal to POTASSIUM HYDRATE KHO and SODIUM HYDRATE NaHO. 438. Process of saponification. Fixed and volatile oils. 439. Drying and non-drying oils. 440. Insoluble in water, soluble in ether. 441. Consistency of fats and oils, various. 442. OLEINE $C_{57}H_{104}O_6$, a liquid. MARGARINE $C_{54}H_{104}O_6$, a soft solid. STEARINE $C_{57}H_{110}O_6$, a hard solid. GLYCERINE $C_3H_5H_3O_3$, soluble in water. MARGARINE, by saponification into a margarate, and glycerine; OLEINE, into an oleate, and glycerine; STEARINE, into a stearate and glycerine; PALMITINE $C_{51}H_{98}O_6$, into a palmitate, and glycerine. PALMITIC ACID $H,C_{16}H_{31}O_2$; MARGARIC ACID $H,C_{17}H_{33}O_2$; STEARIC ACID $H,C_{18}H_{35}O_2$; OLEIC ACID $HC_{18}H_{33}O_2$. 443. Hard and soft soaps. 444. The soap-lye. Manufacture of curd soap. 445. Yellow and Mottled soap. 446. Cocoa-nut oil soap. 447. Marseilles and Naples soap. 448. Soap, a true salt. 449. Hard waters decompose soap. 450. Sea-water upon soap. 451. Employed in cleansing linen. 452. Description of the skin. Action of soap upon it. 453. Fuller's earth. 454. Bleaching. 455. Bleaching by Chlorine. CALCIUM OXY-CHLORIDE $CaOCl_2$; is "Chloride of Lime." 456. Mode of bleaching by Chlorine. 457. Chlorine as a disinfectant. 458. HYDROGEN PEROXIDE H_2O_2 bleaches and disinfects. 459. Open doors and windows, where possible, the best disinfectants. 460. MacDougall's disinfectant. Contains three agents: SULPHUROUS ACID H_2SO_3, in combination as a sulphite; CARBOLIC ACID OF PHENIC ACID HC_6H_5O in combination, and CALCIUM HYDRATE CaH_2O_2 as well as in combination. Dr. R. A. Smith's disinfectant known as MacDougall's.

432. SOAP, one of the commonest and, at the same time, most important articles of domestic use, is an extremely interesting substance in its chemical relations. A short account, therefore, of its com-

position, and the process of its manufacture, will not be out of place in a work professing to treat of applied Chemistry.

If cleanliness be, as the proverb has it, next to Godliness, the manufacture of Soap must needs be an important one. So much at any rate does the comfort of a community depend upon this useful article that the consumption of Soap has been considered by Liebig to afford a good test of the comparative civilization of nations. Our country, happily, will bear this test of its social condition, as Soap is not only manufactured in England in enormous quantities, but is also largely exported to foreign parts. In consequence of its universal use, it has long been regarded as a convenient source of national revenue, and from the days of Queen Anne to the year 1853, it was heavily taxed. This taxation naturally operated as a great check upon its use; but now that a sense of its importance for the health and comfort of the people has altered the views of our legislators, and effected the removal of the tax, its manufacture has greatly increased.

433. The earliest mention of Soap is made by Pliny, who declares it to be the invention of the Gauls, though he states his own preference for the German, over the Gallic Soap. The Latin word *Sapo*, from which we get our word Soap, is said by Beckmann to be derived from the German; the corresponding word, *Sepe*, being still in use in the Low German dialect. Both hard and soft Soaps were in use among the Germans; and Pliny describes not only the mode of manufacture from tallow and ashes, but also the use of Soap as a medicine, as well as a detergent.

434. In very remote times, clothes were cleansed simply by rubbing, or by stamping upon them in water, without the addition of any extraneous substance. In the Odyssey, Nansicaa and her attendants are represented as washing their clothes by treading upon them in pits which they had filled with water. The same practice still prevails in India.

435. At a later period, mention is made of the use of a lye of ashes in washing garments, and in cleansing oil and wine-jars; and there can be little doubt that a solution of wood-ashes or of impure Potassium carbonate, must have been employed for these purposes.

436. The cleansing property of Soap, or of Wood-ashes, depends

upon the power they possess, of rendering soluble in water the impurities adhering to the clothes or the person; and as these are not soluble in ordinary water, that agent cannot be effectively employed alone. To comprehend, however, the precise effects produced by Soap, it is necessary to be first acquainted with its nature and composition.

437. No substances are possessed of such powerful cleansing properties as POTASSIUM and SODIUM HYDRATES (pp. 168 and 173). Both of these are soluble in less than half their weight of water, and both have the power of dissolving the skin with very great rapidity. They possess further the remarkable property, of being able to enter readily into combination with every variety of oleaginous matter, to form compounds which are soluble in water. Every requisite to be desired in a cleansing agent, appears to be united in Potash and Soda: but their destructive action upon the skin would render their employment by themselves quite out of the question. The discovery, therefore, that they were capable of uniting with fatty matters, to form compounds possessing nearly all their cleansing powers, with little or none of their caustic properties, was a great step in advance towards their practical application as detergents.

438. For the understanding of the process of SAPONIFICATION, it is necessary to explain the nature and property of FATS somewhat more fully than has already been done under the head of the candle-manufacture (p. 123, par. 183).

Such animal and vegetable fats, as are fit for employment in soapmaking, are so similar in character that they may be considered together. All oleaginous bodies are divisible into two classes—FIXED and VOLATILE. Dropped upon paper or linen, both kinds produce a greasy stain, which, in the case of a Volatile Oil, disappears on the application of gentle heat, but is permanent in the case of a Fixed Oil. Volatile oils are alone odorous; and whenever a smell proceeds from a Fixed oil, it is always an indication of the presence in it of a Volatile oil.

439. The Fixed Oils and Fats all possess more or less affinity for OXYGEN; some to such an extent as to produce great heat, and, occasionally, to give rise to spontaneous combustion when mixed with inflammable substances, as cotton, jute, or flax. This property

of absorbing Oxygen, leads to a classification of the Fixed Oils into DRYING and NON-DRYING. Linseed, Rape, Poppy, and Walnut-oils become hard and resinous, in consequence of this absorption; hence their great use in paints. But Olive and Palm oils, and the animal fats become rancid on exposure to air, which renders them acid, but never dries them.

440. As a general principle, all Fats and Oils are INSOLUBLE IN WATER; how much, therefore, must their composition become changed, when they unite with Potash and Soda, to form Soaps readily soluble in water. Oils and Fats are slightly soluble in alcohol, but highly so in ether.

441. The consistency of these substances varies greatly, from that of Olive-oil to that of Mutton-suet; but from what has been previously stated, it will be readily seen, that this difference depends upon the proportions in which the fluid and solid Fatty Matters are mixed. A natural separation of these is often effected by cold, and even mechanically by pressure. Thus Olive oil, at a temperature slightly above the freezing point, separates into two distinct substances, liquid OLEINE $C_{57}H_{104}O_6$, and solid MARGARINE $C_{54}H_{104}O_6$. Again, if Suet be pressed between folds of blotting-paper, the latter will be found stained with Oleine, while the former will have become much harder, less fusible, and less tenacious; a mixture of STEARINE $C_{57}H_{110}O_6$ with MARGARINE.

442. Now, Margarine, Oleine, Stearine, and Palmitine (from palm-oil), consist of fatty acids in combination with GLYCERINE. Soaps are combinations of Potash and Soda with these Fatty Acids, which are known respectively as MARGARIC, OLEIC, STEARIC, and PALMITIC ACIDS.

GLYCERINE $C_3H_5H_3O_3$ is the *sweet principle of oils and fats*; a viscid, colourless liquid, soluble in water, it detracts from the hardness of Margarine, Stearine, and Palmitine, and is wanted neither for candles (p. 123, par. 183) nor soap-making. In Price's patent candles, the PALMITIC ACID $HC_{16}H_{31}O_2$ is separated from the Glycerine with which it is united as PALMITINE or PALMITATE of GLYCERINE $C_{51}H_{98}O_6$, by means of high-pressure steam. When a mixture of palmitine and water, is forced under strong pressure through tubes at a low red-heat, so as to raise the temperature to about

260° C., the PALMITINE is decomposed into GLYCERINE and PALMITIC ACID. The elements of three molecules of water enter into combination with those of the palmitine, and we obtain 3 molecules of Palmitic acid $3(HC_{16}H_{31}O_2)$ and 1 molecule of Glycerine $C_3H_5\mathit{H}_3O_3$. Thus: $C_{51}H_{98}O_6+3H_2O=3(HC_{16}H_{31}O_2)+C_3H_5\mathit{H}_3O_3$.

In common stearic acid candles, STEARIC ACID $HC_{18}H_{35}O_2$ is separated from the STEARINE $C_{57}H_{110}O_6$, by a similar process. Thus: $C_{57}H_{110}O_6+3H_2O=3(HC_{18}H_{35}O_2)+C_3H_5\mathit{H}_3O_3$. In the manufacture of soap, the same object is obtained by bringing these fatty matters together with Potash and Soda at high temperatures.

443. HARD SOAPS are made with Soda; SOFT, with Potash. Hard soaps may, however, be made with potash, if a hard fat is employed; but generally a fish or vegetable drying-oil is used with it.

444. Into a solution of caustic Soda, (p. 173, par. 310) called the SOAP-LYE, is introduced the fat, which may be Tallow, Lard, or bleached Palm-oil, and the mixture is then gently heated till it boils. After boiling two hours, the whole forms a viscid mass, capable of being drawn out into threads. To separate the Soap from the water in which it is dissolved, Salt or Sodium chloride is added, when the Soap, being insoluble in brine, rises to the surface. When cold, the Soap-paste is ladled into buckets, and spread on frames to dry. The mass of soap is then marked round with an iron rake, and divided by means of wire into slabs, which are again cut into bars, as seen exposed for sale.

445. This brief description of the manufacture of Curd-soap applies, in principle at least, to the making of all Soaps. Yellow Soap contains Resin and Palm oil, in addition to common Tallow; the Resin uniting with Soda to form a highly-cleansing compound. The only difference between Mottled and Curd-soaps is, that the more or less coloured impurities which, in the case of the latter, are allowed to subside, are, in that of the former, mixed up with the Soap, and transferred to the frames.

446. Soap made from Cocoa-nut oil has the property of retaining much more water than that made from tallow, a property more beneficial to the manufacturer than to the consumer. Many so-called improvements in the Soap-manufacture, and for which patents have been granted, have a similar recommendation, and are of benefit to the soap-makers only.

447. Marseilles-soap is made from Olive-oil and Soda: Naples-soap, from the pure Fatty acids of Mutton-suet, obtained in the same way as in the Candle-manufacture. The various toilet-soaps are far too numerous to admit of any lengthened description.

448. A good Soap should be a definite chemical compound—in fact a true Salt. The presence of any excess of water, or of any material foreign to its purpose as a cleansing agent, detracts from its value. Potash-soaps are, as a rule, more soluble than Soda-soaps. Both are said to be soluble in distilled water, but in reality this solution consists in a ready separation of the more soluble alkali, whose presence renders Soap of such value as a detergent, and so important an agent in the removal of fatty matters.

449. Used in a water containing Calcium sulphate, Soap is decomposed; the water becomes full of flakes, and the cleansing properties of the soap do not come into play, until the whole of the Calcium has been precipitated. The same result ensues when other earthy salts are present in the water, and hence the *waste* of Soap arising from the use of hard water, and the importance of soft water for domestic purposes.

450. The action is quite simple. Calcium sulphate is decomposed; Sodium sulphate is found in solution, while the Fatty acid in the Soap unites with the Calcium, to form an insoluble Calcium-soap. A solution of Soap is, in fact, one of the best tests of the hardness of waters; and such a solution in alcohol is most generally employed for that purpose by chemists. The impossibility of washing with common Soap in sea-water is sufficiently obvious, for not only does the Salt actually prevent solution, but the salts of Calcium and Magnesium so abundantly present, all aid in its decomposition.

451. It is also clear that when employed in cleansing our linen, &c., the Potash and Soda in the soap are expended in the formation with the grease contained in the linen, of a Soap soluble in the water.

452. The necessity of soap for cleansing the Skin depends on the same chemical fact. The effects of soap will be more apparent when the nature of the skin is described. If we examine the skin, we shall find its thickness vary in different parts of the body. The more we use our skin, the thicker it grows in the parts most exercised.

The skin is composed of three coats or layers, fitting close to each other, and covering every portion of the body. The innermost layer, which is the thickest, is called the DERMIS, CUTIS, or true Skin, and is the seat of touch and feeling. It consists of two Strata —a superficial or PAPILLARY STRATUM and a deeper or FIBROUS STRATUM, called the CORIUM. The outer surface of the Dermis is characterized by a number of small conical projections, called PAPILLÆ. It is these Papillæ which, by their erection, give rise to that peculiar rough appearance of the skin that is observed in winter, and which is known by the name of goose-skin. (According to Simon, warts are nothing more than these Papillæ elongated and covered with skin proper). Immediately upon the inner skin lies the Outer, or Scarf-skin, the EPIDERMIS, a peculiar horny membrane, thrown out upon the surface of the Cutis. With this, every-one is familiar. When in a state of health it is always undergoing renewal; the Dermis continually pushes up a number of little round cells, which become flattened and hard when in contact with the air, and lap one over another, something like the scales of a fish. These scales are not visible to the naked eye; but as there is a never-ceasing supply of the little cells from beneath, so there is a constant wear and rubbing-off of the outer surface. From head to foot this process is always going on; and a little attention to this fact will show the great benefit that must result from habitual cleanliness.

The Skin has a most important duty to perform, on which health is greatly dependent. This duty is the throwing off of the perspiration, and is accomplished by means of the little tubes or pores existing in every part of the skin. Commencing at the surface, the tubes pass inwards, twisted like a corkscrew, to the underside of the dermis, where they terminate in little balls or glands, in which the perspiratory matter is formed. The number of these pores is so great, that 3528 have been counted in a square inch, on the palm of the hand. Assuming each of these pores to be a quarter of an inch in length, and the surface of the body to contain 2500 square inches, the number of pores has been calculated at 7,000,000, and their length nearly 28 miles. This length of drainage, this breathing from the skin, is one of the means employed by nature for throwing off waste material which would be injurious to the body, and for preserving it at an equal temperature. The quantity of aqueous matter thrown off in the shape of insensible perspiration, amounts to thirty-three ounces in the twenty-four hours: a quantity so enormous, that, if checked by cold or impurity, must disturb the general health.

Besides the perspiratory tubes, the whole extent of the skin, except the palms of the hand and the soles of the feet, contains a second set, penetrating also through the inner skin, from whence they pour out oil to the surface, which, by this means, is kept smooth and pliant. Were it not for this oil, the outer skin would become harsh and dry, and peel off. The tubes through which it passes are most numerous on the face, head, and joints, as these parts are most exposed to external influences, or to friction while in exercise. This oil also assists in retaining the warmth of the body, and at the same time prevents injury from the occasional contact of noxious substances. (See also Erasmus Wilson on the Skin).

From these statements we learn, that the Skin, as the natural covering of the body, has the duty of throwing out two fluids, the one consisting of waste and noxious watery matter, the other an Oil, necessary to keep it in health. The simplest and best means of keeping the skin in health, must be by bathing and washing. But water by itself, be it ever so soft, is not always sufficient. The abundantly-secreted oil tends also to retain, with considerable tenacity, the particles of dust and dirt which come in contact with it, and which, therefore, cannot be effectually removed except by solution of the grease by means of SOAP.

453. FULLER'S EARTH, a grey clay, acts, partly by scouring, partly by absorbing the greasy dirt.

454. BLEACHING is an operation of such magnitude and importance, and is so intimately connected with cleansing, that it may, with propriety, be considered in connexion with soap.

The practice of bleaching by exposure to air and moisture, dates from a very remote period; but though this process in sunny lands is very efficient, the necessity of devoting large tracts of meadow land for this sole purpose, and the amount of labour and time expended upon it, render the operation at best a very expensive one.

455. The use of CHLORINE (p. 126, par 189) as a bleaching

agent, has effected such great changes in the process, and has conferred such vast benefits upon our manufacturers, as to render its discovery one of the greatest triumphs of modern chemistry.

CHLORINE is never found in a free state in nature, but requires to be separated by art from the SODIUM of SODIUM CHLORIDE (Common Salt) with which it is chemically combined (p. 127, par. 192).

The opposition made to the introduction of Chlorine as a bleaching agent may be readily imagined. Its disagreeable smell, and the injurious effects produced by it upon the respiratory organs, were far from being recommendations to its use. Owing to its unskilful application, workmen were killed by it, and the linens bleached were made rotten in the process. Even a dilute solution of Chlorine in water was found to be scarcely more manageable than the gas itself; while, when a stronger solution was used, the linens became yellow on exposure to air. And as water can only absorb twice its own volume of Chlorine, it is evident that for carrying on the process of bleaching on a large scale, the vats used must be of enormous size. At this time, when everything seemed to portend failure, the discovery was made that Chlorine would combine with Lime to form CHLORIDE OF LIME, and that this was one of the most manageable of compounds.

CHLORIDE OF LIME or CALCIUM OXY-CHLORIDE $CaOCl_2$ (p. 198, par 369), is generally manufactured, by allowing Chlorine and Lime to come together. The powdered Lime is placed in wooden trays arranged in layers one above the other, in chambers formed of flagstones or of lead, ând the Chlorine allowed to enter by openings from the top, whence it descends by its own weight to the bottom of the chamber.

As the only important part of Calcium oxy-chloride is soluble in water, a solution of it is employed in bleaching and in disinfecting, but the quantity of Chlorine evolved is insufficient without the aid of dilute Sulphuric acid.

456. The MODE OF BLEACHING BY CHLORINE is thus conducted. The fabrics to be bleached having been cleansed from all traces of greasy substances, are immersed in the bleaching-liquor. Having been allowed to soak for some hours, they are then transferred for a short time to the acid liquor, by which the Chlorine is set free in

contact with every part of them. When the Bleaching is completed (and to do this may require a repetition of the former process), the articles are well washed, and then boiled in dilute solution of Sodium carbonate, for the purpose of removing the colouring matter rendered soluble by the action of the Chlorine. Again they are passed through a solution of chloride of lime, and then through dilute acid, and finally they are washed in abundance of water, for the removal of every trace of free acid, which would, if allowed to remain, soon render the fabric rotten.

The chemistry of bleaching is not so clearly understood as could be desired. Chlorine discharges the colour from all vegetable dyes and, uniting with them, renders them easily soluble in water. The property of bleaching is intimately connected with that of oxydizing. When Chlorine gas is placed in contact with water in the sunshine, the gas disappears, Hydrochloric acid is found in the Water, and Oxygen is set free from the latter. Nor is this all; for the liberated oxygen possesses, in part, at least, the properties of ozone (p. 59, par. 63), with a power of bleaching and oxydizing not possessed by common Oxygen.

457. The important property of CHLORINE, as a DISINFECTANT, depends upon the power of decomposing every compound which elementary substances are capable of forming with HYDROGEN. Now, Hydrogen is a component of all the most noxious gases; thus, from sewers is generated SULPHURETTED HYDROGEN H_2S (p. 91, par. 109), a compound of Sulphur with Hydrogen; and, when Chlorine is brought into contact with it, the hydrogen is absorbed for the formation of hydrochloric acid, and the sulphur, a perfectly harmless substance of itself, is separated.

The best mode of employing Chlorine as a disinfectant is to soak a cloth in a weak solution of Chloride of lime (say one part in 10 of water), and to suspend it, after being wrung out, in the sick room. Except in cases of diseases of the chest, Chlorine may be always thus employed with safety; but as it acts injuriously upon steel and brass, it is desirable to remove articles made of those materials from any room where Chlorine is used.

458. HYDROGEN PEROXIDE, H_2O_2, (p. 62, par. 65,) excels even Chlorine as a bleaching agent. It would, doubtless, be more

frequently employed, if it were a more stable compound, and less expensive. It is so easily decomposed, that at any temperature above 20° C., Oxygen begins to escape. As an oxydizing agent, its powers are not excelled by any known substance. Indeed, it is more than probable, that Hydrogen peroxide is produced in the process of bleaching by Chlorine.

459. Important as disinfectants undoubtedly are, there is none equal to a thorough change of air. Open doors and open windows are most desirable, when the object is the removal of foul air. Yet there are cases in which a genuine disinfectant is invaluable.

460. As a preventive against putrefaction, no disinfectant can surpass that invented by DR. R. A. SMITH, F.R.S. and MR. A. MAC DOUGALL, of Manchester. It is a compound of CALCIUM SULPHITE with CALCIUM CARBOLATE, together with CALCIUM HYDRATE or LIME. The bleaching properties of SULPHUROUS ANHYDRIDE SO_2, have been already mentioned at p. 92, par. 111; its chief use in combination is, however, due to the fact, that it is in readiness to accept more Oxygen: so that the Oxygen of the air attacks it, instead of the putrescent body.

CARBOLIC OR PHENIC ACID HC_6H_5O possesses antiseptic properties similar to those of kreasote. A crystallized HYDRATE $2(HC_6H_5O)H_2O$, properly prepared, is much employed as a local application in surgical cases, and as an active agent against the spread of infection. Carbolic acid literally prevents the decomposition of many organic bodies, and its value has been fully proved by Mr. Crookes, F.R.S., who employed the impure acid, obtained from the distillation of oil of coal-tar, with the greatest success, in disinfecting cattle-lairs, &c., during the cattle-plague.

Nor is the LIME itself without value, as it unites with and renders harmless, Hydrogen Sulphide and Carbonic anhydride, both of which gases are extensively evolved in putrefactive processes.

CHAPTER XXI.

ON THE TRIAD ELEMENTS ALUMINUM, CHROMIUM, AND URANIUM.

461. XXVIII. **Aluminum** Al=27·5. Properties. Preparation. 462. ALUMINUM OXIDE OF ALUMINA Al_2O_3=103. ALUMINUM HYDRATE $Al_2H_6O_6$. The ruby and sapphire are alumina. Spinelle-ruby MgO,Al_2O_3. Gahnite ZnO,Al_2O_3. 463. ALUMINUM SULPHIDE Al_2S_3. 464. ALUMINUM CHLORIDE Al_2Cl_6. ALUMINUM SULPHATE $Al_23SO_4,18H_2O$. POTASSIUM-ALUM $KAl2SO_4,12H_2O$. Isomorphous with $NH_4Al2SO_4,12H_2O$; $KCr2SO_4,12H_2O$; $KFe2SO_4,12HO$, &c. [ALUMINUM PHOSPHATE $4Al_2O_3,3P_2O_5,18H_2O$. WAVELLITE $3(4Al_2O_3,3P_2O_5,18H_2O),Al_2F_6$. LAZULITE [$2(3$[CaMgFe]$O,P_2O_5),(4Al_2O_3,3P_2O_5),6H_2O$]. CLAY $Al_2O_3,2SiO_2,2H_2O$. [CYANITE or DISTHENE Al_2O_3,SiO_2. ANALCIME ($Na_2O,Al_2O_3,4SiO_2,2H_2O$. STILBITE ($CaO,Al_2O_3,6SiO_2,6H_2O$). PREHNITE ($2CaO,Al_2O_3,3SiO_2,H_2O$). FELSPAR, ORTHOCLASE or ADULARIA $K_2O,Al_2O_3,6SiO_2$. ALBITE $Na_2O,Al_2O_3,6SiO_2$. PETALITE $L_2O,Al_2O_3,6SiO_2$. LABRADORITE $CaO,Al_2O_3,6SiO_2$. GRANITE is a mixture of Felspar, Quartz and Mica. SYENITE contains Hornblende, instead of Mica. Gneiss is a granite containing mica in layers. GARNETS [$3(CaMgFeMn)O,(AlFe)_2O_3 3SiO_2$]. UNI-AXAL MICA [$4(MgKFe)O,(AlFe)_2O_3,4SiO_2$.] BI-AXAL MICA [$(KFe)O,3(AlFe)_2O_3,6SiO_2$.] CHLORITE [$4(MgFe)O,(AlFe)_2O_3 2SiO_2,3H_2O$]. Slates. AUGITE or PYROXENE [$(CaMgFeMn)O,SiO_2$]. HORNBLENDE or AMPHIBOLE [$5(CaMgFeMn)O,6SiO_2,x(KCa)F_2$]. Asbestos, a fibrous variety of Amianthus. Ultramarine ($2Al_2O_3 3SiO_2,Al_2O_3,4SiO_2,Na_2S_2O_3,3Na_2S$.)]. 465. Tests for Aluminum. 466. XXIX. **Chromium** Cr=52·5. Triad element, of sp. gr. 6·81. *Prep.* $Cr_2Cl_6+3K_2=6KCl+2Cr$. 467. CHROMOUS OXIDE CrO. CHROMIC OXIDE Cr_2O_3. CHROMIUM SULPHATE Cr_23SO_4. CHROME-ALUM $KCr2SO_4,12H_2O$. CHROMIUM CHLORIDE Cr_2Cl_6. CHLORO-CHROMIC ACID CrO_2Cl_2. CHROMIUM HEXA-FLUORIDE CrF_6. CHROMIUM NITRIDE Cr_3N_4. CHROMOUS CHROMIC OXIDE CrO,Cr_2O_3. CHROME-IRON-ORE FeO,Cr_2O_3. CHROMIC ANHYDRIDE CrO_3. CHROMIC ACID H_2CrO_4. POTASSIUM CHROMATE K_2CrO_4. POTASSIUM DI-CHROMATE $K_2Cr_2O_7$. [PERCHROMIC ACID $HCrO_4$. 468. CHROMIC SULPHIDE Cr_2S_3. 469. TESTS FOR CHROMIUM. [470. XXX.—**Uranium** U=120. A metal of sp. gr. 18·4. *Prep.* from URANOUS

CHLORIDE UCl_2, by Potassium. Pitchblende is chiefly DI-URANOUS URANIC OXIDE $2UO,U_2O_3$. URANITE $[Ca2(U_2O_2)2PO_4,8H_2O]$. CHALCOLITE $[Cu\,2(U_2O_2)2PO_4,8H_2O$. 471. URANOUS OXIDE UO. URANIC OXIDE U_2O_3. URANIC NITRATE $(U_2O_2)2NO_3,6H_2O$. DI-URANOUS URANIC OXIDE $2UO,U_2O_3$, used in painting china black. URANOUS URANIC OXIDE UO,U_2O_3, is green. URANOUS CHLORIDE UCl_2. URANIC CHLORIDE U_2Cl_3. URANIC OXY-CHLORIDE UOCl. 472. TESTS FOR URANIUM.]

XXVIII. Aluminum Al = 27·5.

461. ALUMINUM Al=27·5. A triad metal, never found native. Aluminum was discovered by Wöhler, and derives its name from alum. It is white, lustrous and malleable, very sonorous and a good conductor of electricity. Sp. gr. from 2·5 to 2·67. It melts at a temperature of about 450° C. Nitric acid has but little action upon it, but Hydrochloric acid dissolves it with extreme facility: $Al_2+6HCl=3H_2+Al_2Cl_6$. Sulphuric acid also dissolves it. The chief uses of Aluminum are: for purposes of ornament, the manufacture of small weights, and the formation of Aluminum-bronze, an alloy of a golden colour. *Prep.* of Aluminum. 1. From the mineral Bauxite $3Na_2O,Al_2O_3$, an ALUMINUM HYDRATE $Al_2H_6O_6$, is prepared. This is made up into balls with charcoal and sodium chloride, and dried; and these balls are heated to redness in earthen retorts, while a stream of dry Chlorine is passed through them. SODIUM-ALUMINIC CHLORIDE $NaCl,AlCl_3$ sublimes. Ten parts of this salt are mixed with 5 parts of the mineral Cryolite $3NaF,AlF_3$ and projected into a highly-heated reverberatory furnace, together with 2 parts of Sodium. ALUMINUM collects at the bottom, while Sodium chloride volatilizes. 2. From SODIUM ALUMINIC CHLORIDE $NaCl,AlCl_3$, either by electrolysis or by heating with Sodium; nothing but Sodium chloride and Aluminum result. Thus: $2(NaCl,AlCl_3)+3Na_2=8NaCl+Al_2$. 3. From Cryolite $3NaF,AlF_3$ by Sodium: $2(3NaF,AlF_3)+3Na_2=12NaF+Al_2$.

462. Aluminum has but one oxide; ALUMINUM OXIDE $Al_2O_3=103$. It is found nearly pure as "corundum" in 6-sided prisms of sp. gr. 3·95: "emery" is impure corundum. The "ruby" and the "sapphire" also consist of aluminum oxide, tinged with colouring matters.

Pure aluminum oxide is a snow-white, porous mass, infusible except in the oxy-hydrogen flame, and perfectly insoluble in water. When heated to redness, it becomes insoluble in acids. As the basis of CLAY, its important character and wide distribution will be easily appreciated. It is isomorphous with Ferric oxide, and is generally more or less discoloured owing to the presence of the latter. "Spinelle ruby" is MAGNESIUM ALUMINATE MgO,Al_2O_3; "Gahnite" is a ZINC ALUMINATE ZnO,Al_2O_3. ALUMINUM HYDRATE $Al_2H_6O_6$, is a translucent gum-like mass, insoluble in water, but soluble in acids and also in Potassium hydrate. The mineral "Diaspore" is also a hydrate Al_2O_3,H_2O.

Aluminum oxide Al_2O_3, is best prepared by heating AMMONIUM ALUM $NH_4Al2SO_4,12H_2O$, to bright redness. Or, as HYDRATE $Al_2H_6O_6$, in the form of a bulky-white, gelatinous mass.

463. Aluminum unites with Sulphur to ALUMINUM SULPHIDE Al_2S_3. In fact, Aluminum may be burnt in the vapour of Sulphur, but Aluminum hydrate is precipitated from Aluminum salts, by Ammonium sulphide.

464. SALTS OF ALUMINUM. ALUMINUM CHLORIDE $Al_2Cl_6=268$. Except in solution, or in crystals, as $Al_2Cl_6,12H_2O$, Aluminum chloride can only be obtained, by heating a mixture of aluminum oxide and charcoal to bright redness in an atmosphere of Chlorine. $Al_2O_3+3C+3Cl_2=3CO+Al_2Cl_6$. A yellow, semi-translucent mass; very volatile. It may be employed for making Aluminum, by passing its vapour over melted Sodium. [Al_2Br_6. Al_2I_6. Al_2F_6. Sodium aluminic fluoride $3NaF,AlF_3$, is largely found in Greenland as Cryolite.] ALUMINUM SULPHATE $Al_23SO_4,18H_2O$. In thin scales, soluble in two parts of water. Largely prepared by heating roasted shale or clay, with sulphuric acid. It is sold as "concentrated alum." POTASSIUM-ALUM $KAl2SO_4,12H_2O$. It crystallizes in octohedra, and, when massive, often presents a most beautiful appearance, as represented in Fig. 91.

It is soluble in 18 parts of cold and in less than its own weight of hot water, and its solution has an acid reaction and a strong astringent taste. When heated to 100° C., this alum loses 5 molecules of water, and other 5 at 120° C. At 200° C., it loses all its water, and at a high temperature much of the acid. Alum is much used in dyeing. "Roman alum" crystallizes in cubes and is entirely free from iron. Potassium alum may be prepared from aluminum sulphate, by the addition of Potassium sulphate. Or, from alum-shale.

Fig. 91.

On careful roasting, the Ferrous di-sulphide or iron-pyrites FeS_2, which is contained in the Whitby Alum-shale becomes oxydized; FERROUS SULPHATE $FeSO_4$, is obtained as well as SULPHURIC ANHYDRIDE SO_3. This latter acts upon the alumina of the shale, and forms Aluminum sulphate Al_23SO_4. On proper concentration, and addition of POTASSIUM CHLORIDE KCl, the necessary POTASSIUM SULPHATE is obtained for the production of POTASSIUM ALUM: $2KCl + FeSO_4 = FeCl_2 + K_2SO_4$. Then $K_2SO_4 + Al_23SO_4 = 2(KAl2SO_4)$. AMMONIUM-ALUM $NH_4Al2SO_4,12H_2O$, is isomorphous. So are also: SODIUM-ALUM $NaAl2SO_4,12H_2O$; IRON-ALUM $KFe2SO_4,12H_2O$; CHROME-ALUM $KCr2SO_4,12H_2O$ and MANGANESE-ALUM $KMn2SO_4,12H_2O$. [ALUMINUM PHOSPHATE $4Al_2O_3,3P_2O_5,18H_2O$, is a white, insoluble, gelatinous, precipitate, soluble in acids and in Potassium hydrate. *Wavellite* $3(4Al_2O_3,3P_2O_5,18H_2O),Al_2F_6$, is a mineral, crystallizing in radiating tufts. *Lazulite* $[2(3[CaMgFe]O,P_2O_5),(4Al_2O_3,3P_2O_5),6H_2O]$. ALUMINUM SILICATES. CLAY $Al_2O_3,2SiO_2,2H_2O$, is formed from felspar and other silicates. It is very absorbent of moisture, white when pure, and most retentive of alkalies. Marl is a species of clay, containing more or less Calcium carbonate and sulphate. Fuller's earth is a porous clay. *Cyanite* or *disthene* $Al_2O_3SiO_2$. [*Zeolites* are hydrated double silicates; they dissolve in acids, leaving ORTHO-SILICIC ACID H_4SiO_4 and they boil up when heated on charcoal before the blowpipe-flame, whence their name, from ζέω, *zeo*, I boil. *Analcime* ($Na_2O,Al_2O_3,4SiO_2,2H_2O$) crystallizes in cubes. *Stilbite* ($CaO,Al_2O_3,6SiO_2 6H_2O$) crystallizes in radiated needles. *Prehnite* ($2CaO,Al_2O_3,3SiO_2,H_2O$), in 6-sided prisms]. *Felspars*. POTASSIUM FELSPAR $K_2O,Al_2O_3,6SiO_2$, in oblique rhombic prisms, and massive, is the *orthoclase* or *adularia* of mineralogists. It is used as a glaze in porcelain. SODIUM FELSPAR $Na_2O,Al_2O_3,6SiO_2$, is *Albite*. LITHIUM FELSPAR $L_2O,Al_2O_3,6SiO_2$ is *Petalite*. CALCIUM FELSPAR $CaO,Al_2O_3,6SiO_2$, is known as *Labradorite*. All of these minerals yield by disintegration *kaolin* or china-clay. *Porphyry* is a felspar, with crystals of felspar interspersed. *Granite* is a mixture of felspar, with quartz and mica. [*Syenite* is a granite, which contains Hornblende instead of mica. In *Gneiss*, we have a granite, with the mica almost in layers. *Garnets*, which crystallize in rhombic dodecahedra, are double silicates, in which the Dyad metal Calcium is often replaced by the Dyads Mg, Fe and Mn, and the metal Aluminum in part by Ferrum. [$3(CaMgFeMn)O,(AlFe)_2O_3,3SiO_2$]. *Idocrase* is a similar double silicate, crystallizing in square prisms. *Uniaxal mica* is chiefly a double silicate of Mg and Al. [$4(MgKFe)O,(AlFe)_2O_3,4SiO_2$]. In *bi-axal mica* [$(KFe)O,3(AlFe)_2O_3,6SiO_2$], Potassium silicate predominates. *Chlorite* [$4(MgFe)O,(AlFe)_2O_3,2SiO_2,3H_2O$], is green; sometimes massive as *Chlorite slate*. *Roofing slate* is mainly an aluminum silicate. *Augite* or *Pyroxene* [$(CaMgFeMn)O,SiO_2$] is a crystalline mineral often found in basalt and lava. *Hornblende* or *Amphibole* [$5(CaMgFeMn)O,6SiO_2.x(KCa)F_2$]. *Asbestos* is a fibrous variety of Amphibole. *Ultramarine* is mainly: ($2Al_2O_3,3SiO_2,Al_2O_3 4SiO_2,Na_2S_2O_3,3Na_2S$)].

465. TESTS FOR ALUMINUM. The salts are colourless, and possess an acid reaction upon blue litmus paper. When heated before the blowpipe-flame with Cobalt nitrate, a blue mass is formed. Sodium carbonate precipitates mainly Aluminum hydrate, with evolution of CO_2. Ammonia precipitates Aluminum hydrate insoluble in excess.

Potassium hydrate precipitates also $Al_2H_6O_6$, easily soluble in excess. In an *acid* solution, Hydrogen sulphide gives no precipitate; but AMMONIUM HYDROGEN SULPHIDE occasions a white gelatinous precipitate of $Al_2H_6O_6$, soluble in Potassium hydrate. In this manner, Aluminum is separated quantitatively from the metals of the alkalies and also from Barium, Strontium, Calcium and Magnesium. The separation of Alumina from Phosphoric acid is best accomplished by dissolving the precipitate in nitric acid and adding Bismuthic nitrate; phosphoric acid is precipitated as BISMUTHIC PHOSPHATE $BiPO_4$. Any excess of Bismuth may be removed from the acid solution by a stream of Hydrogen sulphide, which, as already stated, does not precipitate Aluminum.

XXIX. Chromium Cr = 52·5.

466. 29.—CHROMIUM Cr=52·5. Is a triad metal in its chief combination, but never found native; it was discovered by Vauquelin, in 1797. Of pure Chromium, little is known, except in the form of a dark-grey powder of sp. gr. 6·81. It dissolves readily in acids, and when strongly heated, it burns brilliantly. It is most intractable in fire

Chromium may be prepared from the CHLORIDE Cr_2Cl_6, by treatment with Potassium or Sodium.

467. Chromium unites with Oxygen in four proportions. *a.* CHROMOUS OXIDE CrO=68·5. Is only known as HYDRATE CrH_2O_2. It forms a brown powder which rapidly absorbs Oxygen, and becomes CrO,Cr_2O_3,H_2O. *b.* CHROMIC OXIDE Cr_2O_3=153. A bright green powder, much used as a pigment in enamelling. It is readily prepared by heating the bluish-green HYDRATE $Cr_2H_6O_6,4H_2O$ to low redness; or, by heating AMMONIUM CHROMATE $(NH_4)_2CrO_4$, or MERCUROUS CHROMATE Hg_2CrO_4. There are two sets of chromic salts, green and violet-coloured; the latter are alone crystallizable. Chromic hydrate is, however, the only hydrate, and is very soluble in potassium hydrate with bright green colour, and with difficulty soluble in Ammonia with rose colour. Chromic oxide is the colouring matter of the emerald, greenstone, &c. CHROME-ALUM $KCr2SO_4$,

$12H_2O$, crystallizes in violet-coloured octohedra, isomorphous with potassium-alum. CHROMIUM SULPHATE $Cr_2 3SO_4$. CHROMIUM CHLORIDE Cr_2Cl_6, in pale violet-coloured scales, insoluble in water, but soluble on boiling. It is obtained by passing dry Chlorine over a red-hot mixture of Chromic oxide and charcoal. [CHLORO-CHROMIC ACID CrO_2Cl_2, is a dense red liquid, fuming on exposure. It is chromic anhydride, in which a molecule of Chlorine displaces an atom of oxygen, and it is immediately decomposed by water into Hydrochloric and Chromic acid. The formation of Chloro-chromic acid is explained by the following symbols: $K_2CrO_4 + 2NaCl + 2H_2SO_4 = K_2SO_4 + Na_2SO_4 + 2H_2O + CrO_2Cl_2$. CHROMIUM HEXAFLUORIDE CrF_6, is a deep-red vapour, which condenses to a blood-red fluid at low temperatures. $PbCrO_4 + 3CaF_2 + 4H_2SO_4 = PbSO_4 + 3CaSO_4 + 4H_2O + CrF_6$. By water it is decomposed into Hydrofluoric acid and Chromic anhydride. CHROMIUM NITRIDE Cr_3N_4, is a chocolate-brown powder. It is obtained by heating Chromic chloride in dry Ammonia.] CHROMOUS CHROMIC OXIDE CrO,Cr_2O_3,H_2O. CHROME-IRON-ORE FeO,Cr_2O_3, is the principal ore of chrome, isomorphous with FERROUS FERRIC OXIDE FeO,Fe_2O_3. It crystallizes in octohedra and contains 48·27 per cent. of Chromium. In order to obtain the Chromium, in the form of a Chromate, the roasted ore is fused with Potassium carbonate in a reverberatory furnace; soluble Potassium chromate is withdrawn by water, and nitric acid added. Thus we obtain: POTASSIUM DI-CHROMATE $K_2Cr_2O_7$. It crystallizes in 4-sided tables, of brilliant red-colour, soluble in 10 parts of water. POTASSIUM CHROMATE K_2CrO_4, in transparent, yellow prisms, soluble in 2 parts of water. *c.* CHROMIC ANHYDRIDE $CrO_3 = 100·5$. Crimson needles, most deliquescent. By water, into: CHROMIC ACID H_2CrO_4, dibasic, and forming salts which are isomorphous with those of Sulphuric acid. Chromic anhydride is most easily prepared, by mixing 5 volumes of sulphuric acid with 4 volumes of a saturated solution of Potassium di-chromate; in a solution of Sulphuric acid of sp. gr. 1·55, Chromic anhydride is nearly insoluble. BARIUM CHROMATE $BaCrO_4$, is yellow, and soluble in Nitric acid. [*d.* PERCHROMIC ACID $HCrO_4$, is purple. It is obtained from Chromic acid by the action of Hydrogen peroxide.]

[468. CHROMIC SULPHIDE Cr_2S_3, in black scales resembling graphite. By passing the vapour of Carbonic sulphide over Chromic oxide heated to redness.]

469. TESTS FOR CHROMIC SALTS. They are green or violet-coloured, and their solutions have an acid reaction. Hydrogen sulphide gives no precipitate, but Ammonium hydrogen sulphide precipitates a dirty green hydrate, soluble in Potassium hydrate with bright-green colour. Potassium hydrate precipitates a green hydrate soluble in excess with emerald-green colour. If any compound of Chromium be fused with Potassium nitrate and carbonate, a yellow soluble potassium chromate is obtained. CHROMIC ACID TESTS. The solutions are yellow or red. Hydrogen sulphide reduces acid solutions to the condition of salts of Chromic oxide. Thus: $K_2Cr_2O_7 + 8HCl + 3H_2S = 2KCl + 7H_2O + 3S + Cr_2Cl_6$. Alcohol brings about a similar reduction. LEAD CHROMATE $PbCrO_4$, is bright-yellow, and soluble in Potassium hydrate. In the reducing flame of the blow-pipe, Chromic acid colours the borax-bead green.

Chromium is weighed as CHROMIC OXIDE, which contains 68·63 per cent. of the metal.

[XXX. Uranium U = 120.

470. 30.—URANIUM U = 120. Is a metal, which is never found native. It was discovered by Klaproth in *Pitchblende*, which is chiefly a DI-URANOUS-URANIC OXIDE $2UO,U_2O_3$. Uranium is a steel-grey metal, of sp. gr. 18·4. It somewhat resembles Ferrum and Manganesium. Acids dissolve it.

Uranium is prepared from URANOUS CHLORIDE UCl_2 by heating it with Potassium.

Uranite [$Ca2(U_2O_2)2PO_4,8H_2O$] and *Chalcolite* [$Cu2(U_2O_2)2PO_4,8H_2O$] are the other two ores of Uranium.

471. Uranium has two principal oxides. URANOUS OXIDE UO, forms green salts, which rapidly absorb Oxygen. It may be prepared by strongly heating Uranic oxalate in absence of air. Its HYDRATE UH_2O_2, is reddish-brown. URANIC OXIDE U_2O_3, is a brick-red powder, capable of acting the part of an acid as well as of a base. Its salts are yellow. URANIC NITRATE $(U_2O_2)2NO_3,6H_2O$, in long,

striated prisms, of bright-yellow colour. The compounds of U_2O_3 with BaO, SrO and CaO communicate a bright-yellow colour to glass. DI-URANOUS URANIC OXIDE $2UO,U_2O_3$, is jet-black, and much used in painting china. It is made by heating Uranium nitrate to redness. URANOUS URANIC OXIDE UO,U_2O_3, is the green oxide. It is prepared from the di-uranous uranic oxide, by heating in air or oxygen. By an intense heat, it is re-converted into the black oxide. URANOUS CHLORIDE UCl_2, is green, volatile, and deliquescent. URANIC CHLORIDE U_2Cl_3, crystallizes in dark-brown needles. URANIUM OXY-CHLORIDE UOCl, is deliquescent.

472. TESTS FOR URANIUM. URANOUS SALTS have a green colour; they absorb oxygen readily, and become yellow. Hydrogen sulphide produces no change in their solutions, but Ammonium hydrogen sulphide precipitates black uranous sulphide. The alkaline hydrates throw down a dark-brown hydrate, which absorbs Oxygen and becomes yellow. URANIC SALTS. They are yellow. Hydrogen sulphide produces no reaction, but Ammonium hydrogen sulphide gives a yellowish-brown precipitate. Ammonia occasions a yellow precipitate, of ammonium uranate. Alkaline carbonates, although they precipitate, yet in excess they re-dissolve the yellow granular precipitate.]

CHAPTER XXII.

ON THE MANUFACTURE OF POTTERY.

473. THERE are few branches of human industry of greater antiquity than the manufacture of Pottery. The plastic qualities of CLAY and its power of hardening under exposure to heat, must have suggested at a very early period in the history of man its application to the making of utensils for the manifold requirements of daily life, while scarcely any art has made greater advances from its rude commencement, and probably none has been more indebted for its progress to the aid of science.

474. The BRICKS of which our houses are built, the SLATES and tiles with which they are roofed, the CHINA and EARTHENWARE which we use, are all of them but varieties of CLAYS, so abundantly distributed over the earth.

475. What then is clay, so extensively used for the purposes of civilized life, and what more particularly is the clay used for making pottery, china, and earthenware?

Pure clay (p. 244, par. 464) may be described as a compound of Silicate of Alumina, which has been already mentioned as a constituent of certain kinds of glass, with Water. It is purely white, soft, unctuous to the touch, and emits an earthy smell when breathed upon. It is perfectly insoluble in water, and forms with it a tenacious paste. When slowly dried, it shrinks considerably, and soon shows its unfitness, of itself, to form good utensils. It does not fuse, even under exposure to the greatest heat of an air-furnace, but shrinks and splits into hard pieces. When burnt, it is still perfectly white and adheres tenaciously to the tongue. It is greatly absorbent of water, acting like capillary tubes, and allowing the water to flow through,

476. The only clay suitable for the manufacture of porcelain, is that called CHINA-CLAY or KAOLIN, so named from two Chinese words signifying Porcelain-clay. But even this is not able to perform all the service required of it, without the assistance of some substance to obviate the two defects just mentioned as common to all clays; viz., its porosity, and its troublesome property of shrinking as it dries. These faults are entirely remedied by admixture of the clay with SILICA, the same substance which was found to be so essential in the manufacture of glass.

477. Clay, calcined flints, and some cement capable of partial fusion, are all the materials required for the various kinds of porcelain.

After the FLINTS have been calcined, they are transferred to the grinding vats, in which, by mechanical force, they are crushed to powder. In these vats the flints are ground in water, with which they form a thick, creamy fluid, which is then drawn off and conveyed to the washing chamber. There the coarser particles are separated from the finer, by their more rapid subsidence in the water, and are returned to the grinding vats. And this process of grinding and washing is continued, till the flint is reduced to the state of the finest possible powder.

The KAOLIN having also been stirred with water to a similar cream-like consistency, the two fluids are thoroughly mixed in the blending reservoirs. From these, the mixture is passed through fine sieves containing 300 threads to the square inch, into other receivers, and finally into a receiver of open kilns where the excess of water is evaporated by heat. When tolerably firm, the mixture is cut into blocks and reduced by pressure and by the rotating knives attached to the centre shaft of the "mill," into the perfectly-plastic state in which it is required for use by the potter.

478. FELSPAR, (p. 245, par. 464) if not already present in the kaolin in sufficient quantities, is added to the flint previous to its mixture.

479. The paste thus carefully prepared is now fit for making porcelain; but it has been found by experience that it improves greatly by being kept for some years in a damp place. When so kept, it undergoes the process of rotting, and gives out the unsavoury odour of Sulphuretted hydrogen. The evolution of this gas arises from the presence of small quantities of Sulphates contained in the clay, acted upon by the organic (chiefly vegetable) impurities contracted during the manipulation of the clay and flint. This improvement therefore may be ascribed to the separation of these impurities.

480. Before the paste is worked up, it is again mixed by hand, and squeezed into balls, which are violently thrown on to the table at which the potter is at work.

The process of throwing and turning on the POTTER'S WHEEL, must be witnessed to be understood. In executing his work, the Potter is guided not only by patterns, but also by various measures or moulds, for shaping the article in hand. If the work be on a large scale, the clay is added in successive batches, as required for working up.

481. In MOULDING, the ceramic paste (*κέραμος*, *keramos*, potter's clay, from *κέω*, *keo*, I burn, and *ἔρα*, *era*, earth) is applied to the mould, the form of which it is intended to assume. The mould is always made of a porous material, capable of absorbing the moisture from the clay, generally of plaster of Paris. When the pattern is very complex, the mould is made in several pieces. For round objects it is made in two parts, fitting each other exactly; the article being moulded in separate halves, which are united while the paste is yet moist. When the paste has been sufficiently dried by absorption of its moisture by the mould, the latter is removed by separating its two parts, leaving the work complete.

482. The moulded articles have next to be thoroughly DRIED. This is done by BAKING them in the so-called BISCUIT OVENS; and

porcelain, which has passed through this process, is called BISCUIT-WARE.

483. The very porous ware requires now the protection of the GLAZE: this is formed of some vitreous substance, analogous to that used in the making of the porcelain-paste itself. To be effectively applied, it must not be too fusible, and yet must melt below the temperature which would soften the porcelain, and destroy its form.

The GLAZING is generally accomplished by immersion. The glaze, formed of finely-powdered felspar and quartz, held in suspension by vinegar and water, is applied by very rapid dipping and removal of the biscuit ware. A skilled workman is said to be able to dip seven hundred dozen plates in a day.

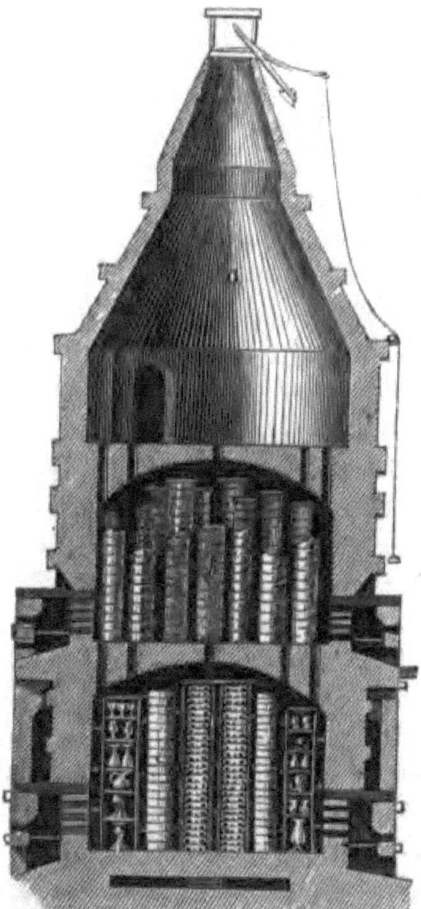

Fig. 92.

484. The PORCELAIN-KILNS are constructed of two or three stories. Fig. 92 represents a vertical section of a kiln of the usual construction, showing, also in section, several of the so-called SEGGARS and SUPPORTS. As the current of heated air carries with it quantities of fine ash, which would adhere to the glaze, the porcelain cannot be exposed to its direct action, nor can the articles be allowed to touch each other, or they would adhere by the fusing of the glaze. Each piece is, therefore, placed in a vessel, called a Seggar, *c*, formed of refractory fire-clay and furnished with supports of the same material. From the explanations previously given of the general character of clays, it

must be evident, not only that the articles should be carefully and completely dried before their introduction into the kiln, but also that the kiln itself should be heated gradually. It is also desirable that the kiln should be as nearly as possible filled, and yet so as to allow the fire to circulate freely between each pile of seggars.

The details of the manufacture of porcelain might be treated at much greater length, did the limits of this work permit. Enough, however, has been said, to convey a fair idea of its principle.

485. The materials used for colouring porcelain, are much the same as for Glass, (p. 190, par. 353). Most of them are Silicates, which after fusion are reduced to an impalpable powder, and ground into paint with oil of lavender. They are applied with a brush before the glazing, and the porcelain is heated sufficiently to fuse them. Black is produced by DI-URANOUS URANIC OXIDE (p. 249, par. 471;) brown, by ferric and manganese oxides. The pink colour used on china is prepared by heating a mixture of one part of potassium di-chromate (p. 247, par. 467), 10 parts of chalk and 30 parts of stannic oxide. CHROMIC OXIDE gives the green, and COBALT OXIDE the blue colours most in vogue. Common GILDING is effected, by mixing pulverulent gold, a little borax, and Bismuth oxide, with essence of lavender, or with some other volatile oil, the mixture being applied with a brush. After burning, the gold appears dead, and is polished, first with an agate, and afterwards with a bloodstone-burnisher.

486. STONE-WARE is a coarser description of porcelain, containing both Ferric oxide and Lime, to the presence of which it owes both its colour and fusibility. This ware is seldom GLAZED by dipping into a fusible glass, but by an ingenious substitute. When the Pottery has attained a high temperature, a handful of dry SALT, (Sodium chloride) is thrown into the furnace, when it is volatilized and decomposed, by the joint action of the Silica of the ware, and the vapour of Water. Hydrochloric acid and Soda are produced; the former passing off in vapour, and the latter being deposited as a glaze of Silicate of Soda upon the surface of the ware.

487. The Pottery commonly known as STAFFORDSHIRE WARE, is made from a description of white clay of less value than kaolin. It is always mixed with ground flints, which are prepared by the process

already described. This ware is first burnt at a high temperature, and then, after immersion in a highly-fusible Glaze, is exposed to a second fire at a less intense heat. The glaze itself consists of a mixture of Litharge (Lead oxide), Clay, and ground Flints, made up into a thin paste.

The COLOUR applied to this ware is generally blue, on account of the cheapness and easy application of oxide of cobalt. The patterns are printed off upon paper, which is applied while wet to the ware before it is glazed. When the ink is dry, the paper is washed off, and the glazing completed as has been described.

488. Such common pottery as Flower-pots, and the like, is manufactured from impure clays which contain Ferric oxide and Sand. The pieces are formed on the Potter's wheel, dried by exposure to air, and burnt without glazing.

BRICKS are made from similar clays, the clay being prepared in various ways according to its quality, or to the kind of brick desired. When bricks are well burnt, they are often superficially glazed by the fusion of certain Silicates, especially of silicate of Lime, contained in them.

489. Of late years, a description of Pottery, very strong but coarse, and generally highly glazed, has been made, as pipes for drainage of towns, and for other purposes chiefly of a sanitary nature. Its complete imperviousness to moisture, its great strength, and unlimited durability, render it an admirable material for house-drains and small sewers; and its manufacture is carried on at Lambeth, and elsewhere, on an enormous scale.

CHAPTER XXIII.

ON THE METALS MANGANESIUM, COBALT AND NICKEL, WITH NOTES ON GLUCINUM, YTTRIUM, ERBIUM, CERIUM, LANTHANUM AND DIDYMIUM.

490. XXXI.—**Manganesium** Mn=55. Metal of sp. gr. 8·013. 491. *a.* MANGANOUS OXIDE MnO; HYDRATE MnH_2O_2. *b.* MANGANIC OXIDE Mn_2O_3. MANGANESE ALUM $KMn2SO_4,12H_2O$. *c.* MANGANOUS-MANGANIC OXIDE MnO,Mn_2O_3. *d.* MANGANESE PEROXIDE MnO_2. A source of O_2. $3MnO_2 = Mn_3O_4 + O_2$. A source of Cl_2. $MnO_2 + 4HCl = MnCl_2 + 2H_2O + Cl_2$. Also: $MnO_2 + 2NaCl + 2H_2SO_4 = Na_2SO_4 + MnSO_4 + 2H_2O + Cl_2$. *e.* MANGANIC ACID H_2MnO_4. POTASSIUM MANGANATE K_2MnO_4. Mineral Chameleon. *f.* PERMANGANIC ACID $HMnO_4$. POTASSIUM PERMANGANATE $KMnO_4$. 492. Manganous salts. MANGANOUS CHLORIDE $MnCl_2,4H_2O$. MANGANOUS SULPHATE $MnSO_4,5H_2O$ and $7H_2O$. MANGANOUS CARBONATE $MnCO_3$. When precipitated, $2MnCO_3,H_2O$. 493. MANGANOUS SULPHIDE MnS, native as Manganese Blende. 494. Tests for manganese. 495. XXXII.—**Nickel** Ni = 59. Sp. gr. 8·82. Chief ores: *Kupfernickel* NiAs; *Arsenical nickel* $NiAs_2$ and *Nickel glance* NiSAs. Preparation of Nickel from Speiss. 496. *a.* NICKEL OXIDE NiO. Its HYDRATE NiH_2O_2. *b.* NICKEL PEROXIDE. Ni_2O_3. 497. Nickel salts. NICKEL CHLORIDE $NiCl_2,9H_2O$. NICKEL SULPHATE $NiSO_4,7H_2O$. NICKEL CARBONATE $NiCO_3$; BASIC CARBONATE $2NiH_2O_2,2NiCO_3,5H_2O$. NICKEL CYANIDE $NiCy_2$. POTASSIUM-NICKEL CYANIDE $2KCy,NiCy_2$. Alkaline solution decomposed by mercuric nitrate. 498. NICKEL SULPHIDE NiS, native as *Millerite*. NICKEL SUB-SULPHIDE Ni_2S and NICKEL DI-SULPHIDE NiS_2. 499. Alloys of nickel. *German Silver* $Cu_5Zn_3Ni_2$. *Tutenag*. 500. Tests for nickel. 501. XXXII.—**Cobalt** Co = 59. Sp. gr. 8·95. Chief ores: COBALT DI-ARSENIDE $CoAs_2$ or *tin-white cobalt* and COBALT-GLANCE CoSAs. Derivation of Cobalt from Kobold. 502. Cobalt and Nickel occur together. Separation. 503. *a.* COBALT OXIDE CoO; HYDRATE CoH_2O_2. Smalt is a blue glass. Preparation. Separation from Speiss. *b.* COBALT PEROXIDE Co_2O_3. HYDRATE $Co_2O_3,3H_2O$. 504. Cobalt salts. COBALT CHLORIDE $CoCl_2,6H_2O$. COBALT SUL-

PHATE $CoSO_4,7H_2O$. COBALT NITRATE $Co2NO_3,6H_2O$. COBALT ORTHO-PHOSPHATE Co_32PO_4. COBALT ARSENATE $Co_32AsO_4,8H_2O$ is Cobalt-bloom. COBALT CARBONATE $3CoCO_3,2H_2O$. 505. Behaviour of ammonia towards cobalt salts. 506. Tin-white cobalt is a DI-ARSENIDE $CoAs_2$; COBALT-GLANCE CoSAs, in dodecahedra, octohedra and cubes. 507. Tests for cobalt. 508. XXXIV.—**Glucinum** G = 9·5. Sp. gr. 2·1. 509. GLUCINUM OXIDE GO. Its HYDRATE GH_2O_2. GLUCINUM CHLORIDE $GCl_2,4H_2O$. GLUCINUM SULPHATE $GSO_4,4H_2O$. Chryso-beryl or GLUCINUM ALUMINATE GO,Al_2O_3. *Phenakite* is GLUCINUM SILICATE $2GO,SiO_2$. GLUCINUM and ALUMINUM SILICATE $3GO,Al_2O_3,6SiO_2$. 510. Tests for Glucinum. 511. XXXV.—**Yttrium** Y = 61·7. 512. YTTRIUM OXIDE YO = 77·7. 513. XXXVI.—**Erbium** E = 112·6. ERBIUM OXIDE EO = 128·6. 514. XXXVII.—**Cerium** Ce = 92. CEROUS OXIDE CeO and CERIC OXIDE Ce_2O_3. POTASSIUM CERIC SULPHATE K_2Ce2SO_4. 515. XXXVIII.—**Lanthanum** La = 92. LANTHANUM OXIDE LaO = 108. 516. XXXIX.—**Didymium** Di = 96. DIDYMIUM OXIDE DiO = 112. Salts are pink.

XXXI. Manganesium Mn = 55.

490. 31.—MANGANESIUM Mn=55. This metal is never found native, although its chief compound MANGANESE PEROXIDE MnO_2, is an abundant mineral. Gahn, in 1774, discovered the character of it. The metal is grey-white, lustrous and hard, but very brittle. Sp. gr. 8·013. It is feebly magnetic. It oxydizes somewhat readily, and even slowly decomposes water. Manganesium may be prepared by strongly igniting Manganous carbonate, mixed with charcoal, in a wind-furnace: it requires a high temperature for its fusion. In its chief combination, Manganesium is a dyad.

491. Manganesium unites with Oxygen in several proportions. *a.* MANGANOUS OXIDE MnO=71, is a powerful base, isomorphous with ZINC OXIDE ZnO, and MAGNESIUM OXIDE MgO. It may be obtained in the form of a green powder, by heating Manganous carbonate $MnCO_3$ in a stream of Hydrogen, so as to prevent all access of air. MANGANOUS HYDRATE MnH_2O_2, is white, but readily browns on exposure, owing to its conversion into a superior oxide. *b.* MANGANIC OXIDE Mn_2O_3, is found native as *Braunite*, crystallized in square-based octohedra. Its hydrate Mn_2O_3,H_2O, is also met with as *Manganite*. It has very feeble basic properties. A MANGANESE-ALUM $KMn2SO_4,12H_2O$ is well-known, proving the isomorphism of

Manganic oxide Mn_2O_3, and Alumina Al_2O_3. *c.* If manganese peroxide be strongly heated, we obtain: MANGANOUS MANGANIC OXIDE MnO, Mn_2O_3. It is not basic, and is also formed by strongly igniting manganous or manganic oxides. It is found native as *Hausmannite* crystallized in black, four-sided pyramids. *d.* MANGANESE PEROXIDE $MnO_2=87$. Is *Pyrolusite*, and found in steel-grey, rhombic prisms. It also occurs as *Psilomelane*, a stalactitic variety. It is black, lustrous, and a good conductor of electricity. In chemical processes, it is of great value. In making OXYGEN: $3MnO_2=Mn_3O_4+O_2$. In making CHLORINE: $4HCl+MnO_2=MnCl_2+2H_2O+Cl_2$. Also: $2NaCl+MnO_2+2H_2SO_4=Na_2SO_4+MnSO_4+2H_2O+Cl_2$. It has no basic properties; heated with sulphuric acid we obtain Manganous sulphate $MnSO_4$, and Oxygen: $2MnO_2+2H_2SO_4=2MnSO_4+2H_2O+O_2$. A black hydrate $4MnO_2,H_2O$ is also known. *e.* MANGANIC ACID H_2MnO_4 is very unstable. Its salts are green. It is isomorphous with sulphuric acid. POTASSIUM MANGANATE K_2MnO_4, is green. It is formed when Manganese peroxide and Potassium hydrate are fused together. SODIUM MANGANATE Na_2MnO_4, constitutes *Condy's* most admirable disinfecting fluid. It is one of the best oxydizing agents, because so very unstable. On addition of an acid, we obtain a mixture of a permanganate, with a manganous salt: $5K_2MnO_4+4H_2SO_4=3K_2SO_4+4H_2O+MnSO_4+4KMnO_4$. The change of colour from a manganate to a permanganate, gave the name of Mineral Chameleon. *f.* PERMANGANIC ACID $HMnO_4$. Is very unstable, as its solution decomposes at 40° C. Its salts are amethystine. POTASSIUM PERMANGANATE $KMnO_4$, crystallizes in dark-red needles, soluble in 16 parts of water. It is also a splendid oxydizing and disinfecting agent.

492. SALTS OF MANGANOUS OXIDE. MANGANOUS CHLORIDE $MnCl_2,4H_2O$. Tabular crystals of bright pink colour. MANGANOUS SULPHATE $MnSO_4,5H_2O$, in pale rose-red crystals; also with $7H_2O$, isomorphous with $FeSO_4,7H_2O$. MANGANOUS CARBONATE $MnCO_3$, native as *Manganese spar*, isomorphous with *Calcspar* $CaCO_3$. Precipitated CARBONATE is $2MnCO_3,H_2O$.

493. MANGANESIUM and SULPHUR. MANGANOUS SULPHIDE MnS, is found native as *Manganese blende*. It is steel-grey. The HYDRATE MnS,H_2O, is flesh-coloured, and is precipitated by Ammonium

hydro-sulphide from neutral solutions of manganous salts. It readily oxydizes and becomes brown.

494. TESTS FOR MANGANESE. All compounds, when heated with Sodium carbonate in the blowpipe-flame, furnish a bluish-green mass. Heated with Borax or Microcosmic salt, the resulting "bead" becomes violet in the oxydizing flame. MANGANOUS salts are not precipitated in acid solutions by Hydrogen sulphide; but, on addition of Ammonium hydrogen sulphide, MANGANOUS SULPHIDE MnS,H_2O of a flesh colour is precipitated. This sulphide browns by exposure to air, owing to oxydation. Potassium hydrate precipitates white Manganous hydrate, which browns also on exposure. Ammonia, in presence of ammoniacal salts, gives no precipitate in neutral solutions, but the liquid becomes brown from deposit of Manganic hydrate. When no ammoniacal salt is present, ammonia gives a white precipitate, soluble in a large excess. Manganese is estimated quantitatively as MANGANOUS-MANGANIC OXIDE MnO,Mn_2O_3, and is precipitated by Sodium carbonate as Manganous carbonate $MnCO_3$, which, by ignition, becomes Mn_3O_4.

XXXII. Nickel Ni = 59.

495. 32.—NICKEL Ni=59. This metal was discovered by Cronstedt, in 1751. It is found in meteoric stones. The chief ores are, however, *Kupfernickel* NiAs; *Arsenical nickel* $NiAs_2$ and *Nickel-glance* NiSAs.

Nickel is almost silver-white, hard, tenacious and ductile. Sp. gr. 8·82. It is more magnetic than Manganesium, and loses its magnetism temporarily when heated to 330° C. It melts at a somewhat higher temperature than Iron. Nitric acid is its best solvent, but it also dissolves slowly in dilute Sulphuric and Hydrochloric acids.

Nickel may be obtained by heating Nickel oxide in a stream of Hydrogen, or from Nickel oxalate at a white heat.

On a large scale, Nickel is obtained from Speiss (par. 512), a mixture chiefly of arsenides and sulphides of Nickel, Cobalt, Iron and Copper. It is first fused with Calcium fluoride and carbonate, and the mass subsequently roasted to drive off the Arsenicum as Arsenious anhydride. It is then dissolved in Hydrochloric acid, the Ferrous changed into a Ferric salt by means of Calcium oxy-chloride, and then the whole of the Iron, as well as any Arsenic acid, carefully precipitated by Calcium hydrate as Ferric hydrate and Ferric arsenate. After filtration, the Copper is precipitated by a stream of Hydrosulphuric acid, as Cupric sulphide; the Nickel and Cobalt remain in solution. The liquid is filtered, boiled, neutralized with Calcium hydrate, and the Cobalt precipitated by means of further Calcium oxy-chloride, as Cobalt peroxide Co_2O_3. The Nickel is finally thrown down by means of Calcium hydrate, in the form of NICKEL HYDRATE NiH_2O_2, and reduced.

496. Nickel forms TWO OXIDES, only one of which is basic. NICKEL OXIDE $NiO=75$. Is olive-green, and obtained by heating the nitrate strongly in absence of air. The HYDRATE NiH_2O_2 is light, unripe apple-green, soluble in ammonia with blue colour. It gives insoluble compounds with potash and soda. It forms salts. NICKEL PEROXIDE Ni_2O_3, is black. It unites, as HYDRATE, with three molecules of water. It is not basic, and it behaves towards Hydrochloric acid like Manganese peroxide. It is formed when Nickel oxide is suspended in a strong solution of Potassium hydrate, through which a stream of Chlorine is passed: $2NiH_2O_2+2KHO+Cl_2=2KCl+Ni_2O_3,3H_2O$.

497. SALTS OF NICKEL. They are green when they contain water, otherwise they are yellow. NICKEL CHLORIDE $NiCl_2,9H_2O$ in green crystals. NICKEL SULPHATE $NiSO_4,7H_2O$, in green rhombic prisms, soluble in 3 parts of cold water. NICKEL CARBONATE $NiCO_3$, forms a green powder. Precipitated in the ordinary manner from a Nickel salt by Sodium carbonate, a BASIC NICKEL CARBONATE $2NiH_2O_2, 2NiCO_3,5H_2O$, is obtained. NICKEL CYANIDE $NiCy_2$, is soluble in Potassium cyanide as POTASSIUM NICKEL-CYANIDE $2KCy,NiCy_2$. If an alkaline solution of this salt is boiled, and mercuric nitrate added, Hydrated nickel oxide is precipitated, and mercuric cyanide obtained in solution: $2KCy,NiCy_2+H_2O+HgO=2KCy,HgCy_2+NiH_2O_2$

498. NICKEL and SULPHUR. NICKEL SULPHIDE NiS, occurs in grey, capillary crystals, as *Millerite*. The hydrate NiS,H_2O is black, and precipitated from Nickel salts by Ammonium hydrogen sulphide. A SUB-SULPHIDE Ni_2S and a DI-SULPHIDE NiS_2, are also known.

499. ALLOYS OF NICKEL. *German Silver* $Cu_5Zn_3Ni_2$ is an alloy of Copper, Zinc and Nickel. *Tutenag* is a white alloy common among the Chinese.

500. TESTS FOR NICKEL. Heated with borax in the oxydizing flame, nickel compounds give a reddish-yellow bead. The salts are green. Hydrogen sulphide gives no reaction in acid solutions. Ammonium hydrogen sulphide, a black precipitate of hydrated sulphide. Alkaline carbonates produce an unripe apple-green precipitate, and so also does Potassium hydrate. Ammonia, however, redissolves the precipitate, forming a bright blue solution. Nickel is always estimated quantitatively as Nickel oxide NiO, 100 parts of which contain 78·67 per cent of the metal.

XXXIII. Cobalt Co = 59.

501. 33.—COBALT Co=59. This metal was discovered by Brandt in 1733. It occurs in meteoric stones, but its chief ores are: COBALT DI-ARSENIDE $CoAs_2$ or *tin-white cobalt*, and COBALT-GLANCE CoSAs. The ores are found principally in the primitive rocks, associated with Copper, Manganese, Iron and Nickel.

The name was given it by the German miners. In the Middle ages, Sprites or Kobolds were supposed to haunt the mines, and to occasion endless annoyance by substituting inferior for good ores. When the ores, supposed to be those of Silver, turned out to be dross, the fault was laid at the door of the Sprites, and the name Kobold (Cobalt) was given in derision.

Cobalt is a metal of reddish colour, hard, ductile, and, with as great difficulty, fusible as iron. It is magnetic. Sp. gr. 8·95. It has no application in the arts, except in an oxydized form. Cobalt may be obtained from its oxide by intensely heating it in a stream of Hydrogen; or from Cobalt oxalate at a white heat.

502. Cobalt and Nickel constantly occur together. On admixture of the two nitrates with Hydrocyanic acid in excess, and subsequent addition of Potassium hydrate, the liquor absorbs Oxygen by degrees and becomes pale-yellow. The Cobalt is now found as POTASSIUM

COBALTICYANIDE K_3CoCy_6, but the Nickel as POTASSIUM NICKEL CYANIDE $2KCy,NiCy_2$.

The separation of Nickel depends upon the fact, that Mercuric oxide decomposes solutions of Potassium Nickel cyanide, precipitating Nickel hydrate; but it does *not* decompose a solution of POTASSIUM COBALTICYANIDE K_3CoCy_6.

503. COBALT and OXYGEN. Cobalt forms TWO OXIDES, only one of which is basic. *a.* COBALT OXIDE $CoO=75$. It is greenish-grey, but becomes black when heated to dull redness in air, owing to the formation of COBALTOUS-COBALTIC OXIDE Co_3O_4; when more strongly heated, it loses Oxygen, becomes brown, and represents Cobalt oxide again. COBALT HYDRATE CoH_2O_2, is rose-coloured, and readily soluble in ammonium hydro-carbonate and in ammonia, especially in presence of ammoniacal salts. "Zaffre" is an impure oxide, much used in commerce for the production of a blue colour in glass and pottery. "Smalt" is a blue glass. In the preparation of Smalt, the ore of Cobalt is carefully roasted, so as only to oxydize the Cobalt and to leave the Nickel, Copper and Iron as sulphides and Arsenides. Then the roasted ore is mixed with its own weight of Potassium carbonate, and twice its weight of calcined flints, and the mixture carefully melted in glass pots. Whilst the Nickel, Copper and Iron arsenides and sulphides collect as SPEISS at the bottom, a true glass, coloured by Cobalt silicate, is formed and skimmed off. *b.* COBALT PEROXIDE Co_2O_3, is black, and readily obtained by careful heating of the HYDRATE $Co_2O_3,3H_2O$. This hydrate may be obtained in the same manner as Nickel peroxide, and also by simply passing Chlorine through Cobaltous hydrate. $3CoH_2O_2+Cl_2=CoCl_2+Co_2O_3,3H_2O$. If Cobalt peroxide be strongly heated, we obtain cobaltous-cobaltic oxide CoO,C_2O_3.

504. SALTS OF COBALT OXIDE. COBALT CHLORIDE $CoCl_2,6H_2O$ is a beautiful salt, crystallizing in octohedra of a ruby-red colour. A weak solution leaves so slight a pink stain upon paper, that it may pass unnoticed. But, when heated, writing comes out of a deep-blue colour. Hence it is used as *sympathetic ink.* COBALT SULPHATE $CoSO_4,7H_2O$, is isomorphous with Magnesium, Zinc and Cobalt sulphates. COBALT NITRATE $Co2NO_3,6H_2O$, is a red, deliquescent salt. COBALT ORTHO-PHOSPHATE Co_32PO_4. COBALT ARSENATE Co_3

$2AsO_4,8H_2O$ is a beautiful mineral of a red colour, known as *Cobalt-bloom*. COBALT CARBONATE $3CoCO_3,2H_2O$ may be obtained from any of the basic carbonates, which are generally precipitated from soluble salts, by digesting them in solutions of NH_4HCO_3.

505. The behaviour of ammonia towards cobalt salts is remarkable. It first precipitates a basic salt which is blue, then converts it into the red hydrate, and finally dissolves the latter, forming a brown solution. By the absorption of Oxygen, the brown liquid becomes red, and we can obtain by proper treatment different classes of salts.

506. The two arsenides of Cobalt have been mentioned. *Tin-white cobalt* or COBALT DI-ARSENIDE $CoAs_2$, loses Arsenicum when heated in absence of air, and is changed into ARSENIDE CoAs. COBALT-GLANCE CoSAs, crystallizes in dodecahedra, octohedra and cubes.

507. TESTS FOR COBALT. Anhydrous salts of cobalt are purple; crystallized and dissolved are red. They communicate a blue colour to the borax-bead in the oxydizing flame. Sulphuretted hydrogen does not precipitate Cobalt sulphide in presence of free mineral acids, but Ammonium hydrogen sulphide occasions a black precipitate of Cobalt sulphide. Potassium hydrate precipitates a blue basic salt, which becomes rose-coloured on the addition of an excess. Ammonia re-dissolves the blue precipitate first formed, and gives a brown solution which, by absorption of Oxygen, becomes red. Sodium carbonate precipitates a basic salt of a rose colour. Potassium ferro-cyanide gives a dirty-green, and Potassium ferri-cyanide a reddish-brown precipitate. Cobalt is best estimated as COBALTOUS COBALTIC OXIDE Co_3O_4, which contains 73·44 per cent. of the metal.

XXXIV. Glucinum G = 9·5.

508. 34.—GLUCINUM G=9·5. Is the Beryllium of German Chemistry. The name is derived from γλυκὺς, *glukus*, sweet, from the character of its salts. The metal is white, lustrous and malleable, of sp. gr. 2·1. It does not decompose water. It dissolves readily in Hydrochloric and Sulphuric acids, but scarcely in Nitric acid.

Glucinum is prepared from GLUCINUM CHLORIDE GCl_2 by heating with Sodium. $GCl_2 + Na_2 = 2NaCl + G$.

509. Glucinum has ONE oxide, GLUCINUM OXIDE $GO = 25{\cdot}5$. It is white, and its HYDRATE GH_2O_2, is bulky, and very readily soluble in alkaline hydrates and carbonates. The precious *beryl* or *emerald* contains 13·6 per cent. of the oxide: it is a SILICATE OF GLUCINA and ALUMINA $3GO,Al_2O_3,6SiO_2$. Phenakite is GLUCINUM SILICATE $2GO,SiO_2$. GLUCINUM CHLORIDE $GCl_2,4H_2O$, in crystals. The anhydrous salt is volatile. GLUCINUM SULPHATE $GSO_4,4H_2O$, crystallizes in octohedra. The *Chryso-beryl* is GLUCINUM ALUMINATE GO,Al_2O_3.

510. TESTS FOR GLUCINUM. The salts are colourless, sweet and slightly astringent. They form no alum, and the carbonate is soluble in alkaline carbonates.

[XXXV. Yttrium Y = 61·7.

511. 35.—YTTRIUM $Y = 61{\cdot}7$. It is very like Glucinum, but it burns readily and magnificently in Oxygen.

512. Yttrium has but ONE oxide. YTTRIUM OXIDE $YO = 77{\cdot}7$, is a white powder of sp. gr. 4·842, soluble in the alkaline carbonates and not in the alkalies. It is made from *Gadolinite*, a silicate of yttrium, glucinum, cerium and iron, found at Ytterby, in Sweden.]

[XXXVI. Erbium E = 112·6.

513. 36.—ERBIUM $E = 112{\cdot}6$. Is but little known, but is found in association with Yttrium. It has but ONE oxide. ERBIUM OXIDE $EO = 128{\cdot}6$, is pale rose-red, and its salts are of the same colour. When the oxide is heated in a Bunsen flame, it glows with a green light and affords a continuous spectrum crossed by brilliant streaks, when the light is seen through a spectroscope.]

[XXXVII. Cerium Ce = 92.

514. 37.—CERIUM $Ce = 92$. Is contained in the mineral *Cerite*, basic Cerium silicate. It has TWO oxides. CEROUS OXIDE CeO,

and CERIC OXIDE Ce_2O_3. POTASSIUM CERIC SULPHATE K_2Ce2SO_4, is nearly insoluble in Potassium sulphate. CEROUS OXALATE CeC_2O_4, $3H_2O$, is used in medicine.]

[XXXVIII. Lanthanum La = 92.

515. 38.—LANTHANUM La=92. Was discovered by Mosander, in 1841, and so-called from the Greek, λανθάνω, *lanthano*, I lie hid, because hidden in Cerite. LANTHANUM OXIDE LaO=108, is buff-coloured; its salts are colourless and astringent.]

[XXXIX. Didymium Di = 96.

516. 39.—DIDYMIUM Di=96. Derives its name from δίδυμος, *didymos*, twin, because associated with Lanthanum in Cerite. DIDYMIUM OXIDE DiO=112, is dark-brown. Its hydrate is insoluble in alkalies. With Potassium sulphate, like Cerium and Lanthanum, Didymium sulphate forms a double salt. Its salts are pink.]

CHAPTER XXIV.

THE CHEMISTRY OF THE METAL IRON.

517. XL.—FERRUM Fe=56. A dyad in Ferrous, a triad in Ferric salts. 518. The most widely diffused metal. 519. Known from earliest times. 520. Native in Meteorites. 521. Properties of Iron. Sp. gr. 7·844. Melts at 1530° C. Welding. Passive iron. 522. Ores of iron. MAGNETIC OXIDE FeO,Fe_2O_3. SPECULAR IRON-ORE Fe_2O_3. HAEMATITE Fe_2O_3. BROWN HAEMATITE $2Fe_2O_3,3H_2O$. SPATHIC IRON-ORE $FeCO_3$. CLAY-IRON STONE, principally Ferrous carbonate. 523. Manufacture of Pig-iron in the Blast-furnace. 524. Cast iron; grey, white and mottled. 525. Wrought iron. 526. Bessemer's steel. 527. Steel, a Carbide of iron. Cementation. 528. Iron the most magnetic of metals. 529. The magnetic needle. Galvanometer. 530. FERROUS OXIDE FeO. FERRIC OXIDE Fe_2O_3. FERROUS-FERRIC OXIDE FeO,Fe_2O_3. FERRIC ACID H_2FeO_4. 531. FERROUS SULPHIDE FeS. FERROUS DI-SULPHIDE FeS_2, is Iron-pyrites. MISPICKEL FeSAs. 532. FERROUS SALTS. FERROUS CHLORIDE $FeCl_2,4H_2O$. FERROUS IODIDE $FeI_2,4H_2O$. FERROUS CARBONATE $FeCO_3$. FERROUS SULPHATE $FeSO_4,7H_2O$. FERROUS NITRATE $Fe2NO_3,6H_2O$. [FERROUS HYDROGEN ORTHO-PHOSPHATE $FeHPO_4$. FERROUS ORTHO-SILICATE Fe_2SiO_4. POTASSIUM FERRO-CYANIDE $K_4FeCy_6,3H_2O$. Prussian blue $Fe_4Fe_3Cy_{18},18H_2O$.] FERRIC SALTS. FERRIC CHLORIDE Fe_2Cl_6. FERRIC SULPHATE $Fe_23SO_4,9H_2O$. FERRIC NITRATE $Fe3NO_4$. [FERRIC PHOSPHATE $FePO_4,2H_2O$. FERRIC CYANIDE Fe_2Cy_6. POTASSIUM FERRI-CYANIDE K_3FeCy_6. TURNBULL'S BLUE $Fe_3Fe_2Cy_{12}$. NITRO-PRUSSIC ACID $H_4Fe_2Cy_{10}N_2O_2$.] [533. IRON NITRIDES. Fe_2N. Fe_3N. Fe_5N_2.] 534. 4,970,206 tons of Iron made in Great Britain in 1868. 535. Tests for iron.

XL. Ferrum Fe = 56.

517. 40.—FERRUM Fe=56. The metal iron is somewhat anomalous in its combinations, being distinctly a DYAD, in its relations to Hydrogen, and in its compound Ferrous oxide; but a TRIAD, in its other basic oxide, Ferric oxide.

518. Of all metals, Iron is by far the most important. It possesses, indeed, neither the beauty of colour, nor the fineness of quality which distinguish Gold and Silver, but it surpasses all metals in its wonderful tenacity. As oxide, Iron is universally diffused throughout the earth, and there is no soil which does not contain more or less of it. In minuter quantities, it is equally universal in the vegetable and animal kingdoms, and forms a necessary constituent of every article of food capable of conversion into flesh.

519. Iron has been known and employed by man from the earliest ages of his existence; knives and weapons of war being frequently mentioned in the books of Moses, and artificers in Iron being alluded to as early in the history of man as the fourth Chapter of Genesis. The name and symbol of Mars ♂, the God of War, was given by the Alchemists to this metal.

520. Native Iron is found in small quantities in the Ural mountains, and at Canaan, in Connecticut, U.S. The large meteoric masses, known as METEORITES, consist of Iron, containing besides, Nickel, Cobalt, Manganese, Copper, Tin, Magnesium, Carbon, Sulphur and Phosphorus. A block computed to weigh 14 tons, has been found in S. America, and very good specimens are to be met with in the Mineral Department of the British Museum.

521. Pure Iron is a white, lustrous metal, very malleable and the most ductile of all metals. An iron wire of 0·787 of a line in thickness, supporting a weight of no less than 549·25 lbs. The machine used for wire-drawing is represented in Fig. 69, p. 160. A B is an iron plate, pierced with holes of various decreasing diameters, and supported by the frame-work D. The bar of metal coiled round the cylinder C, is forced by steam-power through the largest hole, and then wound round the reel F. When the whole of the wire has been drawn through the largest hole, it is then passed through that next in size, and the operation repeated until the wire is reduced to the desired diameter. During the process, it requires frequent annealing.

The sp. gr. of Iron is 7·844. It fuses at about 1530° C. In damp air it oxydizes readily, and becomes coated with *rust*, but it does not oxydize when covered with an alkaline liquid, or a paste of

Calcium hydrate and water. The surface of iron-plate is frequently "galvanized," or coated with zinc in order to prevent this corrosion; but, if imperfectly done, the coating hastens the destructive influence of atmospheric Oxygen, inasmuch as Iron is more electro-positive than Zinc. The presence of Solder similarly hastens oxydation, as may be seen daily in London, at the places where old iron-railings are found embedded in stone-sockets filled with Solder.

Iron is capable of being *welded*. It softens before it fuses, and so, when a bundle of Scrap-iron is strongly heated to redness, it is possible, by a few blows with a Nasmyth's hammer, to form it into a homogeneous mass.

The acids dissolve Iron with readiness. But strong Nitric acid renders Iron PASSIVE, by superficially oxydizing it, and thus preventing further chemical action.

On a small scale, Pure Iron may be obtained in the condition of fine powder, by heating FERRIC OXIDE Fe_2O_3, to redness in a stream of Hydrogen.

522. The most important ORES of Iron are OXIDES. The most valuable ore is *magnetic iron-ore* or loadstone, a FERROUS-FERRIC OXIDE FeO,Fe_2O_3. In Norway and Sweden, whole mountains of it are found, and it forms one of the most important of Indian ores. It exists only in primary rocks, sometimes crystallized in octohedra. Charcoal is employed in its reduction, hence is the resulting metal purer than that derived from any other source.

Specular iron-ore or *Fer oligiste* is another ore of iron found in primary rocks. It occurs in the Islands of Elba and Formosa, often beautifully crystallized in steel-grey rhomboids. It is a pure FERRIC OXIDE Fe_2O_3, and yields, when reduced with Charcoal a very pure Iron. *Haematite*, which so abounds at Ulverstone, in Lancashire, consists also of Ferric oxide. The reduction to the metallic state is effected with coal. From the metal obtained from this ore, some of the finest qualities of Iron-plate and Wire are manufactured.

Brown haematite is a HYDRATE of this Ferric oxide, represented by the symbol $2Fe_2O_3,3H_2O$. When roasted, it loses water and becomes very porous; consequently, it is easily reduced by Coal at a red-heat.

Spathic iron is a FERROUS CARBONATE $FeCO_3$. It is a most abundant ore, and is sometimes found in a state of purity, crystallized in rhomboids isomorphous with Calc-spar. When roasted, it loses Carbonic anhydride and becomes oxydized into the higher oxide, Ferric oxide. Thus: $4FeCO_3 + O_2 = 4CO_2 + 2Fe_2O_3$. *Clay-iron stone*, is the ore from which the greater portion of British Iron is manufactured. This consists of an impure Ferrous carbonate, mixed with Clay, Calcium carbonate, Magnesium carbonate, and small quantities of Calcium sulphate, &c.

523. The process of the manufacture of PIG-IRON from Clay-iron-

Fig. 93.

stone is simple enough in its broad features. The clay-iron-stone is first roasted by means of small coal in great heaps about 10 feet high, 15 feet wide, and many hundred yards in length. By this operation, Water and Carbonic anhydride are got rid of, the Ferrous becomes Ferric oxide, and the actual percentage of Iron is consequently increased. The ore is now ready for smelting. This is done in the BLAST-FURNACE, Fig. 93, which generally consists of a truncated, pyramidal mass of brickwork, about 15 metres or 50 feet high, and from 4·25 to 5 metres (14—17 feet) in diameter, in the widest part of the cavity, and with a double-coned hollow centre. The lowest portion, or neck of the funnel, is called the crucible, and is made of the most refractory stone. On the sides are the openings for the Tuyères, (or blast-pipes), *t*, through which hot air under pressure is introduced, these being the only openings for the supply of air. Into this Furnace a mixture of equal weights of roasted Ore and Coal, with one-fifth of Limestone (Calcium carbonate) are thrown from above. The Ore is reduced through the agency of the Carbon of the Coal, and the Silica, contained in the Clay-iron-stone, unites with the Lime and Alumina to form fusible Slag or Glass, which, being lighter than the metal, swims upon its surface. In quantity, this Slag is five times that of the iron, and is constantly run off from an opening left for the purpose. In the course of a day and night, the Iron is reduced to the metallic state and is drawn off into channels of sand. In this state it is known as "Pig-iron."

The quantity of air driven hourly into the furnace amounts to 6000 kilos, or nearly six tons, so that the weight of the air alone is greater than that of the Ore, Limestone and Coal put together. The temperature of the hot blast is about 325° C., or even 700° C., when Mr. Siemens' *Regenerator* is employed. The gases which escape at the top of the furnace consist principally of Nitrogen (55 per cent.), Carbonic oxide (26 per cent.), Hydrogen (7 per cent.), Light carbonetted hydrogen (4 per cent.), and Carbonic anhydride (8 per cent.) The slags are principally silicates of Lime and Alumina, with Ferrous silicate. The roasted ore yields about 35 per cent. of Iron, and each furnace from 8000 to 10,000 kilos in 24 hours.

524. CAST-IRON is of three kinds: GREY, WHITE and MOTTLED. None are pure metal, but contain in addition, Carbon, Sulphur, Phosphorus and Silicon. *White* cast iron contains 5 per cent. of

Carbon, and may be represented by the formula Fe_4C: it is the *Spiegel-eisen* of the Germans, and crystallizes in lustrous tables. *Grey* cast-iron consists principally of Fe_8C; it requires a much higher temperature for its fusion than the white variety. Grey and mottled cast iron always contain uncombined Carbon diffused through them. It is the grey cast-iron which is generally obtained from the furnace, and, although somewhat less fusible than the white, is more employed for making castings: it is also employed in making chilled shot, by sudden cooling. In white cast-iron, the whole of the Carbon is in a combined state; it is more fusible than the other sorts, contains less Silicon, and more Sulphur and Phosphorus.

525. WROUGHT IRON is made from cast-iron by the process of *de-carbonizing*. This is accomplished by two processes. 1. *Refining*. Pig-iron, in quantities of about 25 cwt. at a time, is strongly heated upon the hearth of a sort of forge, called the refinery. This operation lasts about two hours. The metal loses more than 10 per cent. in weight, for the Silicon oxydizes as well as a small portion of the Iron and Carbon, and a yet smaller quantity of Sulphur; a slag of FERROUS ORTHO-SILICATE Fe_2SiO_4, together with some Ferrous phosphate is formed; Sulphurous anhydride and Carbonic oxide escape. The molten mass is run off into flat cakes, and quickly cooled by water as soon as it begins to set. 2. *Puddling*. It is then introduced into the *puddling* furnace, in quantities of about 5 cwt. at a time, in order to free the Iron yet more from Carbon and Silicon. As the Carbon diminishes, the metal becomes less fusible; the temperature is then increased, while the air is, as far as possible, excluded from the furnace. The puddler, collecting the iron in balls (*blooms*) on the end of a rod, removes it from the furnace and subjects it, while intensely hot, to tremendous pressure. The melted slag, consisting of FERROUS ORTHO-SILICATE Fe_2SiO_4, is squeezed out, and the metallic particles brought closer together; after which it is re-heated and passed under grooved rollers, by which it is shaped into bars. Thus is the iron freed from the greater part of its Carbon and Silicon.

Wrought iron surpasses all metals in tenacity, hence its great value in the construction of roofs, bridges, chain-cables, girders, or other works required to resist great strains, or supporting heavy weights. In the form of wire-rope, the greatest possible strength is obtained from a given weight of metal.

526. Mr. BESSEMER burns off the greater portion of the Carbon by forcing cold air at a pressure of about 15lbs. per square inch through melted cast-iron. Although the process has not been as complete and as satisfactory as was first expected in the reduction of cost, its application in the manufacture of STEEL has proved of the utmost value. Bessemer steel is now made by adding a given quantity of white cast-iron to molten wrought-iron just before pouring out.

527. The carbon contained in STEEL, the most important alloy of iron, varies between 0·7 and 1·5 per cent. The formation of steel may be said to be intermediate between cast and wrought iron. The addition of manganese improves the quality of the steel. Spiegeleisen contains as much as 5 per cent. of manganesium.

The best kind of Steel is made by a process known as *Cementation*. If the very best steel is required, bars of Swedish iron, made from magnetic iron ore, are embedded in charcoal powder in a suitable furnace, and exposed for several days to a well-regulated heat. When the process is completed, the furnace is allowed slowly to cool down. The Steel has then the appearance of being covered with blisters, and is hence called "Blistered Steel." This description of Steel when melted is termed "Cast Steel." The texture of blistered Steel is much improved and rendered more uniform by welding together a number of bars, and forging the mass under a tilt-hammer into smaller ones. The most characteristic property of Steel is that of becoming very hard, and, within certain limits, perfectly elastic when suddenly cooled from a high temperature. If a piece of Steel heated to redness is plunged into cold water, it becomes sufficiently hard to cut glass, and it can no longer be acted upon by a file; again heated, and suffered to cool slowly, Steel becomes once more as soft as common iron. The process of "tempering" or "annealing" consists in a proper application of these principles. The greater the heat employed, the softer the Steel. The degree of heat attained, and the resulting quality of the steel, are judged of by the experienced workman from the COLOUR of the Steel, brought about by superficial oxydation. A faint straw-colour, obtained at a temperature of about 225° C., is indicative of the hardest tempered Steel, suitable for razors, lancets, &c.; a full yellow, (245° C.) or brown tint, (255° C.) of that fit for penknives or scissors. A tinge of purple shows itself at 265° C.; but the Steel for swords and watch-springs, which must be soft and elastic, has to be heated till its surface becomes deep blue, which happens between 280° and 300° C.

528. The phenomenon of MAGNETISM is best studied in the metal Iron, or rather in Steel, as Iron is the most magnetic of all metals. When a magnetized needle is suspended at its centre, so as to be free to move in a horizontal plane, it is always found to assume a particular direction with regard to the earth, one end pointing nearly North, the other nearly South. The pole which points to the Astronomical North is called the North pole of the magnet, and that which points Southward, the South pole. Poles of the same name repel each other, those of an opposite kind attract. The mariner's compass consists of a suspended needle attached to a circular card marked with the points.

529. If a wire conveying an electrical current be brought near a magnetic needle, the latter will instantly assume a position as nearly perpendicular to the wire, as the mode of suspension and the earth's magnetism will permit. When the wire is placed over the needle and parallel to its length while the current itself travels from North to South, the needle is deflected, and the North pole is moved to the Eastward; when the current is reversed, the same pole moves towards the West. Placing the wire below the needle produces the same effect as reversing the current.

When the magnetic needle is subjected to the action of two currents in opposite directions, the one above and the other below, they will obviously concur in their effects. The same thing happens when the wire carrying the current is bent upon itself, and the needle placed between the two portions; and since every time the bending is repeated, a fresh portion of the current is made to act in the same manner upon the needle, it is easy to understand how a current, too feeble to produce any effect when a straight wire is employed, may be made by this contrivance to exhibit a powerful action upon the needle. It is on this principle that GALVANO-

METERS are constructed. They serve not only to shew the existence of an electric current (p. 3, par. 7), but also the direction in which it is moving. The delicacy of the instrument can be immensely increased by the use of a very long coil of insulated wire, and two needles of equal strength with their poles opposed. The advantage gained is twofold: the needles are almost unaffected (*astatic*) by the earth's magnetism, and, being both acted upon in the same manner by the electric current, are urged with much greater force than one alone would be, all the actions of every part of the coil being strictly concurrent. A divided circle is placed below the upper needle, by which the angular motion can be measured; and the whole is enclosed in glass to protect the needles from the action of the air.

530. The OXIDES of iron are most important. They are four in number. 1. FERROUS OXIDE $FeO=72$. Is with difficulty obtained and impossible to preserve, except in vessels hermetically sealed. The HYDRATE FeH_2O_2, is white, insoluble in water, and oxydizes with rapidity, passing through shades of green, blue-green and ochre. It is soluble in ammoniacal salts and somewhat freely in ammonia. It may be obtained from any Ferrous salt by means of Potassium hydrate: $FeSO_4+2KHO=K_2SO_4+FeH_2O_2$. 2. FERRIC OXIDE $Fe_2O_3=160$. Occurs native as Haematite, Blood-stone, Fer oligiste, &c. Artificial Ferric oxide is known as Colcothar, Rouge, &c. It is of a deep red-brown colour, insoluble in water and in alkalies. There are several hydrates, besides the brown haematite already mentioned. The formula of one HYDRATE is $2Fe_2O_3, 3H_2O$; it is precipitated from any Ferric salt by Potassium hydrate: $2Fe_2Cl_6+12KHO=12KCl+3H_2O+2Fe_2O_3,3H_2O$. Ferric hydrate is now employed in the purification of Gas from Hydrogen sulphide. $2Fe_2O_3,3H_2O+6H_2S=9H_2O+2Fe_2S_3$. By exposure to atmospheric air, Sulphur is set free, and Ferric oxide re-produced. 3. FERROUS-FERRIC OXIDE $FeO,Fe_2O_3=232$, constitutes "magnetic iron-ore" or "load-stone." It is formed by burning Iron in Oxygen, and by passing steam over red-hot iron-filings. $3Fe+4H_2O=4H_2+Fe_3O_4$. As a black hydrate, it may be precipitated from a mixture of a Ferrous and Ferric salt by means of Ammonia. 4. FERRIC ACID $H_2FeO_4=122$. Is only known in combination. POTASSIUM FERRATE K_2FeO_4, is of a splendid amethystine colour, when in

solution. As a black powder, insoluble in a strong solution of Potassium hydrate, it may be obtained by passing Chlorine through Ferric hydrate suspended in strong Potassium hydrate.

531. Iron also unites with SULPHUR. 1. FERROUS SULPHIDE $FeS=88$, is prepared by heating Iron and Sulphur together, or by heating Iron-filings with Ferrous di-sulphide. It is precipitated as a black hydrate, when Ammonium Hydrogen sulphide is added to a Ferrous salt. Ferrous sulphide is much employed in making Hydrogen sulphide. $FeS+2HCl=FeCl_2+H_2S$. Or: $FeS+H_2SO_4,H_2O=FeSO_4+H_2O+H_2S$. 2. FERROUS DI-SULPHIDE $FeS_2=120$. Crystallizes in cubes and in dodecahedra, besides being found massive as *Iron pyrites*. It is much used in the manufacture of Sulphuric acid (p. 94, par. 113). $4FeS_2+11O_2=2Fe_2O_3+8SO_2$. When carefully roasted, it is a source of Ferrous sulphate, and also of Sulphuric acid in the manufacture of Alum (p. 243, par. 464). It is insoluble in Hydrochloric acid. [3. MAGNETIC PYRITES Fe_7S_8, is lustrous and brass-like in colour: it is soluble in HCl.] FERROUS ARSENICUM SULPHIDE FeSAs, is found native in Cornwall, and other places, as "Mispickel." It is steel-grey, lustrous and crystalline.

532. The SALTS of iron are most numerous. 1. The FERROUS SALTS. FERROUS CHLORIDE $FeCl_2$, is white. In the form of green crystals, $FeCl_2,4H_2O$, it may be obtained by dissolving Iron in Hydric chloride. FERROUS IODIDE $FeI_2,4H_2O$, in green crystals, is prepared by bringing Iodine and Iron-filings together with water. It is employed in making Potassium iodide. $FeI_2+K_2CO_3=FeCO_3+2KI$. FERROUS CARBONATE $FeCO_3=116$, is found native as Spathic iron-ore. When precipitated, it falls as a hydrate of greenish hue. It is contained in Chalybeate waters, which deposit red Ferric hydrate on exposure to the air. FERROUS SULPHATE $FeSO_4,7H_2O$, in sea-green rhomboids, soluble in 2 parts of cold and one-third of hot water. It is known as Copperas and Green Vitriol. When strongly heated, it gives off Sulphuric acid containing Sulphuric anhydride (Nordhausen sulphuric acid H_2SO_4,SO_3), and Sulphurous anhydride: Ferric oxide or Colcothar remains. Its solution absorbs Oxygen: BASIC FERRIC SULPHATE $2Fe_2O_3,SO_33H_2O$, is precipitated, while Fe_23SO_4 remains in solution. FERROUS NITRATE $Fe2NO_3,6H_2O$, crystallizes in pale-green rhomboids. FERROUS

HYDROGEN ORTHO-PHOSPHATE $FeHPO_4$ is white, and insoluble in water. It absorbs Oxygen and becomes blue. FERROUS ORTHO-SILICATE Fe_2SiO_4, constitutes the slag already mentioned in the making of wrought iron. [FERROUS CYANIDE $FeCy_2$, as a red-brown hydrate, is probably precipitated from a Ferrous salt by Potassium cyanide. POTASSIUM FERRO-CYANIDE $K_4FeCy_6.3H_2O$, in lemon-yellow tables, constitutes yellow Prussiate of Potash. It is an important metallic test. When added to a solution of a Ferric salt, *Prussian-blue* is precipitated. $3K_4FeCy_6+2Fe_2Cl_6=12KCl+Fe_4Fe_3Cy_{18}$. Prussian-blue contains 18 molecules of water. Hydrochloric acid precipitates HYDROFERROCYANIC ACID H_4FeCy_6. $K_4FeCy_6+4HCl=4KCl+H_4FeCy_6$. When mixed with Sulphuric acid and distilled, Potassium ferro-cyanide yields Prussic acid. $K_4FeCy_6+3H_2SO_4=3KHSO_4+KFeCy_3+3HCy$. A solution of Potassium ferro-cyanide when added to a Ferrous salt gives a greenish-white precipitate of so-called "Everitt's salt," $KFeCy_3$. $FeSO_4+K_4FeCy_6=K_2SO_4+2KFeCy_3$. Lastly, it may be mentioned that, when Potassium ferro-cyanide is fused with Potassium carbonate and Charcoal, we obtain POTASSIUM CYANIDE KCy. $K_4FeCy_6+K_2CO_3+2C=Fe+3CO+6KCy$.]

FERRIC SALTS. FERRIC CHLORIDE $Fe_2Cl_6=325$. Brown scales, very deliquescent. It crystallizes with $6H_2O$, soluble in water, alcohol and ether. A powerful tonic. FERRIC SULPHATE $Fe_23SO_4,9H_2O$, is found native in Chili and in India as *Coquimbite* in white, silken masses. It is made from Ferrous sulphate and Sulphuric acid by oxydation with Nitric acid. $6FeSO_4+3H_2SO_4+2HNO_3=4H_2O+2NO+3Fe_23SO_4$. FERRIC NITRATE $Fe3NO_4$, in prisms with water of crystallization. FERRIC ACETATE $Fe3C_2H_3O_2$, is of a deep-red colour, very soluble in water. [FERRIC PHOSPHATE $FePO_4,2H_2O$, is white and insoluble in water. FERRIC CYANIDE Fe_2Cy_6, is not known. POTASSIUM FERRI-CYANIDE $K_3FeCy_6=329$, crystallizes in ruby-red, right rhombic prisms, soluble in 2½ parts of cold water. It is prepared by passing Chlorine through Potassium ferro-cyanide: $2K_4FeCy_6+Cl_2=2KCl+2K_3FeCy_6$. It gives no precipitate with Ferric salts, but a bright blue precipitate of TURNBULL'S BLUE Fe_5Cy_{12} or $Fe_3Fe_2Cy_{12}$, with a Ferrous salt. $2K_3FeCy_6+3FeSO_4=3K_2SO_4+Fe_3Fe_2Cy_{12}$. HYDROFERRICYANIC ACID H_3FeCy_6, is tri-basic; it affords NITRO-PRUSSIC ACID $H_4Fe_2Cy_{10}N_2O_3$, and PRUSSIC ACID when treated with Nitrous anhydride.]

[533. Iron unites with NITROGEN in several proportions. Fe_3N. Fe_2N and Fe_5N_2. It also unites with HYDROGEN.]

534. The production of Iron in Great Britain alone, amounted in 1868 to 4,970,206 tons.

535. TESTS for iron. 1. FERRIC TESTS. Solutions are yellow or reddish-brown, and strongly acid to test paper. Hydrogen sulphide reduces them, in presence of free acid, to Ferrous salts, with deposit of Sulphur. $2Fe_2Cl_6+2H_2S+2HCl=6HCl+S_2+4FeCl_2$. The alkalies give a bulky, red-brown precipitate of Ferric hydrate. Sodium carbonate, a pale red-brown precipitate of Ferrous oxide with Ferric carbonate. Potassium ferrocyanide, Prussian blue. Tincture of galls, a bluish-black precipitate : *ink*. Potassium sulpho-cyanide, a deep port-wine-red colour. Iron is always weighed as Ferric oxide Fe_2O_3, 100 parts of which contain 70 parts of the metal. 2. FERROUS SALTS. The solutions are pale green, feebly acid. In acid solutions, Hydrogen sulphide occasions no reaction ; but in neutral solutions Ammonium hydrogen sulphide produces a black precipitate of hydrated Ferrous sulphide. The alkalies precipitate Ferrous hydrate, of whitish or greenish colour, passing into Ferric hydrate : the precipitate is soluble to some extent in Ammonia. Sodium carbonate gives a greenish-white precipitate of Ferrous carbonate. Potassium ferricyanide, a blue precipitate of Turnbull's blue. A ferrous salt is changed by Nitric acid into a Ferric salt, with temporary production of an olive-brown compound of Nitric oxide and Ferrous sulphate $2FeSO_4,NO$. Whether present in the form of Ferrous or Ferric, Iron must ever be weighed as FERRIC OXIDE.

CHAPTER XXV.

THE CHEMISTRY OF LEAD, THALLIUM, INDIUM AND SILVER.

536. XLI. **Plumbum** or **Lead** Pb=207. Sp. gr. 11·4. Melts at 325° C. Effects of air and Carbonic anhydride. Formation of DI-PLUMBIC HYDRATED CARBONATE $PbH_2O_2,PbCO_3$. 537. Chief ore, Galena or LEAD SULPHIDE PbS. 538. Alloys. 539. Action of acids. 540. LEAD OXIDES. SUB-OXIDE Pb_2O. LEAD OXIDE PbO, the only basic oxide. HYDRATE PbO,PbH_2O_2. LEAD PEROXIDE PbO_2, the Heavy Lead-ore of Mineralogists. RED LEAD $2PbO,PbO_2$. 541. LEAD SULPHIDE PbS. LEAD CHLORO-SULPHIDE $3PbS,2PbCl_2$. 542. LEAD SALTS. LEAD CHLORIDE $PbCl_2$. LEAD OXY-CHLORIDE $PbO,PbCl_2$. LEAD IODIDE PbI_2. LEAD NITRATE $Pb2NO_3$. LEAD SULPHATE $PbSO_4$, native as Lead Vitriol. LEAD HYDRO-CARBONATE $PbH_2O_2,2PbCO_3$ is "White Lead." LEAD ACETATE $Pb2C_2H_3O_2$. Goulard's Extract PbH_2O_2, $Pb_22C_2H_3O_2$. LEAD CHROMATE $PbCrO_4$. DI-CHROMATE $PbO,PbCrO_4$. LEAD PHOSPHATES. $Pb2PO_3$. $Pb_2P_2O_7$. Pb_32PO_4. Pyro-morphite $PbCl_2$, $3(Pb_32PO_4)$. 543. Tests for lead. Lead tree. [XLII. **Thallium** Tl=204. 544. Properties of the metal. Sp. gr. 11·9. Melts at 294° C. 545. One basic oxide. THALLIOUS OXIDE Tl_2O. 546. THALLIUM SULPHIDE Tl_2S. 547. Salts. THALLIUM CHLORIDE TlCl. THALLIUM SULPHATE Tl_2SO_4. THALLIUM CARBONATE Tl_2CO_3. THALLIUM NITRATE $TlNO_3$. 548. Tests.] XLIII. INDIUM In=72. 549. Properties. Sp. gr. 7·36. Very fusible. 550. INDIUM OXIDE. SULPHIDE. CHLORIDE.] XLIV. **Argentum** Ag=108. 551. Properties of Silver. Sp. gr. 10·53. Melts at 1023° C. Spitting. 552. "Silver-glance" or SILVER SULPHIDE Ag_2S, the chief ore. Metallurgy of Silver. 553. Silver, from Silver-lead by Pattinson's process. 554. Alloys of Silver. 555. Sheffield-plating. 556. Electro-plating. Use of Potassium Silver cyanide AgCy,KCy. 557. Three oxides. SUBOXIDE Ag_4O. SILVER OXIDE Ag_2O, the only basic oxide. SILVER PEROXIDE Ag_2O_2. 558. SILVER SULPHIDE Ag_2S. Red antimony-ore $3Ag_2S$, Sb_2S_3. 559. SALTS OF SILVER. SILVER CHLORIDE AgCl, native as "Horn-silver." SILVER BROMIDE AgBr. IODIDE AgI. SILVER CYANIDE AgCy. SILVER NITRATE $AgNO_3$. "Lunar caustic." Marking-ink. SILVER CARBONATE Ag_2CO_3. SILVER SULPHATE Ag_2SO_4. SILVER PHOSPHATES. $AgPO_3$. $Ag_4P_2O_7$. Ag_3PO_4. 560. TESTS for SILVER. Cupellation.

XLI. Plumbum Pb = 207.

536. 41.—Plumbum Pb=207. A dyad metal, never native. It is bluish-white in colour, soft and lustrous; so soft, indeed, that it may be cut with the nail, and it leaves a mark upon paper. Malleable and ductile, it may be beaten into leaves or drawn into wire, but its tenacity is very slight. Lead wire is so flexible, as to be used for gardening purposes instead of string. Sp. gr. 11·4. When heated to 325° C. it melts, but as it contracts on cooling, it is ill adapted for making castings. It may be obtained crystallized in cubes and in octohedra. From its softness and durability, it is much employed for roofing over flat surfaces, for gutters and rain-water-pipes, and for the lining of wooden cisterns.

On exposure to air, Lead speedily tarnishes, owing to the formation of a minute quantity of Lead-oxide, which protects the surface of the metal from further action. Dry air, and perfectly chemically pure water, free from salts as well as from Oxygen and Carbonic anhydride, exert no action upon Lead, but it is powerfully affected by the united action of air and water. The Lead becomes slowly oxydized, and, almost as fast as formed, precipitated by Carbonic anhydride as di-plumbic hydrated carbonate $PbH_2O_2,PbCO_3$ which, most fortunately, is one of the least soluble of salts. Sodium chloride is also a great foe to lead, and so are also nitrates and nitrites in certain spring-waters. The presènce of Calcium sulphate in a water diminishes the probability of the metal being attacked, on account of the insolubility of Lead sulphate.

As lead salts, if at all soluble, are very injurious to health, it may be confidently affirmed that no soft water can be safely stored in a leaden cistern.

537. The chief ore of Lead is the Sulphide PbS, technically known as *Galena.* It resembles metallic Lead, but is easily distinguished by its brittleness. The process of obtaining the metal is sufficiently simple. The Galena, having been broken up and washed, is mixed with $\frac{1}{10}$ its weight of Lime and roasted in a reverberatory furnace with open doors. The free admission of air pro-

duces oxydation of a considerable portion of the Lead Sulphide, which becomes changed partly into LEAD OXIDE PbO, with loss of SULPHUROUS ANHYDRIDE SO_2, and partly into LEAD SULPHATE $PbSO_4$. Thus: $2PbS+3O_2=2PbO+2SO_2$. And: $PbS+2O_2=PbSO_4$. After the lapse of some hours, the furnace-doors are closed and the unaltered Lead sulphide re-acts on the oxydized portion, allowing the sulphur to volatilize as Sulphurous anhydride, and leaving metallic Lead at the bottom of the furnace. $PbS+2PbO=SO_2+3\mathbf{Pb}$. And: $PbS+PbSO_4=2SO_2+2\mathbf{Pb}$.

The metal is drawn off into cast-iron basins, whence it is poured into oblong moulds of Iron. When removed from the Moulds, it constitutes the Pig-lead of commerce.

538. The ALLOYS of lead are very important. *Solder* is of three kinds; fine solder consists of 2 parts of Tin and 1 part of Lead, coarse solder of 2 parts of Lead with 1 part of Tin, and common solder of equal parts of Lead and Tin. *Pewter* consists of 4 parts of Tin to 1 part of Lead. An alloy of Lead and Antimony expands at the moment of solidification, hence its use in making *type-metal.* Gun-shot contains a small portion of Lead arsenide, in order to harden it.

539. Sulphuric acid and Hydrochloric acid have no action upon Lead at ordinary temperatures. When boiled with Lead, Sulphuric acid forms Lead sulphate, with evolution of Sulphurous anhydride: $Pb+2H_2SO_4=PbSO_4+2H_2O+SO_2$. Nitric acid is the chief solvent, yielding Lead nitrate, Water and Nitric oxide: $3Pb+8HNO_3=3(Pb2NO_3)+4H_2O+2NO$.

540. Lead has several OXIDES, but only ONE BASIC OXIDE. 1. LEAD SUBOXIDE Pb_2O, is black. 2. LEAD OXIDE $PbO=223$. Is yellow in colour, and the only basic oxide. It is known in the fused state as "Litharge"; otherwise as "Massicot," and it is readily formed when Lead is melted and heated in air. LEAD HYDRATE PbO,PbH_2O_2, is white, and sufficiently soluble in water to turn red litmus paper blue. In solutions of Potassium and Sodium hydrate, it is much more soluble, and the liquid is an admirable test for organic sulphur, as all such compounds blacken when boiled with Potassium plumbate. 3. LEAD PEROXIDE PbO_2, is of a puce-colour. It constitutes the "Heavy Lead-ore" of Mineralogists. In its

relations towards acids, it behaves like Manganese peroxide. Sulphurous anhydride is immediately absorbed and oxydized by it: $PbO_2+SO_2=PbSO_4$. It is prepared from Red Lead and Nitric acid. 4. RED LEAD $2PbO,PbO_2$, or "Minium" is of a brilliant red colour, much used as a paint known as "priming," from its being the *first* coat of paint applied to the surface of new wood or plaster. It is also employed in making Glass, owing to its great oxydizing powers. Red lead is made by carefully heating finely levigated Litharge at a temperature of 320° C., in a reverberatory furnace. Nitric acid dissolves out Lead oxide, with which it forms Lead nitrate, and leaves Lead peroxide. $2PbO,PbO_2+4HNO_3=2(Pb2NO_3)+PbO_2+2H_2O$.

541. LEAD SULPHIDE $PbS=239$. Has been mentioned as the ore of Lead. "Galena" is one of the most valuable of the many mineral products of Great Britain: there are mines in Derbyshire known to have been worked by the Romans. In Cornwall, Galena is met with in veins which traverse the primitive rocks, mixed with Quartz, Fluorspar, Heavy-spar, and other minerals. In Derbyshire and in Cumberland it occurs in the Mountain-Limestone. It is found sometimes in the form of lustrous, lead-like cubes, of singular boldness and beauty. Sulphuretted hydrogen precipitates Lead salts in the form of black Lead sulphide; hence are Lead-salts tests for the presence of H_2S (p. 91, par. 109). When an acid solution of Lead Chloride is precipitated by Hydro-sulphuric acid, a CHLORO-SULPHIDE $3PbS,2PbCl_2$, of red colour is first formed, and subsequently entirely changed into black Lead sulphide.

542. LEAD SALTS. LEAD CHLORIDE $PbCl_2=278$. Crystallizes in long, white needles, soluble in 33 parts of boiling water. It is much more soluble in strong Hydrochloric acid. LEAD OXYCHLORIDE $PbO,PbCl_2$, is known in Commerce as "Pattinson's White." Turner's yellow is $7PbO,PbCl_2$. LEAD IODIDE PbI_2. in golden scales or as a yellow powder, sparingly soluble. LEAD NITRATE $Pb2NO_3$, forms milk-white octohedra, soluble in 8 parts of cold water. When heated to redness, it evolves Nitric peroxide NO_2 (p. 75, par. 85). $2(Pb2NO_3)=2PbO+O_2+4NO_2$. LEAD SULPHATE $PbSO_4$, is found native as "Lead vitriol," as well in prisms, as in octohedra. It forms a white powder, insoluble in water, but to a

certain extent soluble in Sulphuric and other acids, in alkalies, in ammonium acetate and tartrate. LEAD SULPHITE $PbSO_3$, is also a white powder, insoluble in water. LEAD CARBONATE $PbCO_3=267$, is found native as "Cerussite" or "Lead spar."

The most important carbonate, is LEAD HYDRO-CARBONATE PbH_2O_2, $2PbCO_3$, familiarly known as *White Lead.* It is commonly prepared in the following manner: The Lead employed should be of the purest kind. If it contains iron, the resulting White lead has a tawny hue; and if a trace of silver, it becomes dingy on exposure to light. Derbyshire lead contains traces of cobalt, and white lead prepared from it always has a pinky hue. The metal is first cast into the form of flat, circular stars or gratings, so as to present as much surface as possible. Several of these gratings are placed in small glazed earthenware vessels, containing a small quantity of vinegar, or ACETIC ACID $HC_2H_3O_2$. These pots are each covered with a sheet of Lead, and, embedded in a mixture of new and spent tan, are arranged side by side in a large chamber. Tier upon tier of these vessels are similarly placed, and separated from each other only by loose boards, to a height of 20 to 30 feet. The tan heats, and gives out large quantities of Carbonic Anhydride; the Acetic acid rises in vapour, and, together with the Oxygen of the air, effects a solution of the Lead. The LEAD ACETATE $Pb2C_2H_3O_2$, first formed, unites with Lead oxide to form a BASIC LEAD ACETATE $2PbO,Pb2C_2H_3O_2$. This latter salt absorbs Carbonic anhydride, and is converted into Lead carbonate and Lead acetate. $2PbO,Pb2C_2H_3O_2+2CO_2=2PbCO_3+Pb2C_2H_3O_2$. The Lead carbonate becomes Lead hydro-carbonate, and the Lead acetate again unites with Lead oxide, to be once again precipitated. On unpacking the stacks, the Lead is found thickly covered with, or entirely converted into, White lead. This is removed from any remaining metal, crushed between heavy rollers, ground to fine powder, and by repeated washings with water, cleansed from all admixture of metallic lead. The white lead, at last obtained in the form of a white paste, is dried and packed into barrels.

LEAD ACETATE $Pb2C_2H_3O_2,3H_2O$ constitutes "Sugar of lead," on account of its resemblance to Loaf-sugar. It is soluble in two parts of cold water. *Goulard's Extract* is a basic lead acetate PbH_2O_2,

$Pb_2 2C_2H_3O_2$. LEAD CHROMATE $PbCrO_4$, or "Chrome Yellow," is much used in Calico-printing, and in painting coach-panels. A BASIC CHROMATE $PbO,PbCrO_4$, is a splendid scarlet pigment. LEAD META-PHOSPHATE $Pb2PO_3$; LEAD PYRO-PHOSPHATE $Pb_2P_2O_7$ and LEAD ORTHO-PHOSPHATE $Pb_3 2PO_4$, are white salts, insoluble in water, and interesting as supplying the respective Phosphoric acids by treatment with Sulphuretted hydrogen. The mineral "Pyromorphite" is a CHLOR-ORTHO-PHOSPHATE $PbCl_2,3(Pb_3 2PO_4)$. It crystallizes in yellow, six-sided prisms.

543. TESTS FOR LEAD The soluble salts are for the most part colourless, and redden litmus. They are sweet to the taste and poisonous. Hydrochloric acid gives a white precipitate of Lead chloride, soluble in excess, as well as in boiling water. Even in acid solutions, Hydrosulphuric acid throws down black Lead sulphide; in solutions of the Chloride, the precipitate is red. Sodium carbonate, a white precipitate of basic carbonate with effervescence. Potassium hydrate, a white precipitate of Lead hydrate, soluble in large excess. Sulphuric acid, a white precipitate of Lead sulphate. Potassium chromate, a yellow precipitate of Lead chromate; and Potassium iodide, of yellow Lead iodide.

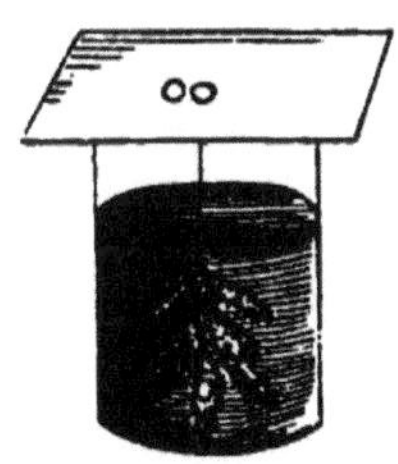

Fig. 94.

Metallic lead is easily precipitated from solutions of its salts by means of Iron or Zinc. The lead crystals may be deposited in an arborescent form, as in the so-called Lead-tree, Fig. 94.

Lead is best estimated as Lead sulphate, inasmuch as it is the only metal precipitable by Hydrosulphuric acid or by Ammonium Hydrogen sulphide which is thrown down by Sulphuric acid. LEAD SULPHATE $PbSO_4$, contains 68·31 per cent. of Lead.

[XLII. Thallium Tl = 204.

544. 42.—THALLIUM Tl=204. A monad metal, never found native, was discovered by Mr. Crookes, F.R.S., in 1861. The name was given to it because of the occurrence of a green line in its Spectrum, from the Greek θαλλὸς, *thallos*, a budding twig. It is

found in Pyrites obtained from Spain and Belgium; in various mineral waters, and also in certain kinds of Lepidolite.

Thallium somewhat resembles Lead, but oxydizes far more readily. It melts at 294° C., and has a sp. gr. of 11·9. When strongly heated in air, it burns with a beautiful green light. Both Nitric and Sulphuric acids dissolve it; Hydrochloric acid, however, possesses but little action, on account of the sparingly soluble Chloride. The chief source of the metal is from the thalliferous dust deposited in the flues of the Oil of vitriol works. This is boiled with water, filtered, and the Thallium precipitated as Thallium chloride by means of Hydrochloric acid; the latter is converted into Sulphate by means of Sulphuric acid, and from the Sulphate the metal is obtained by precipitation with Zinc.

545. Thallium forms ONE basic oxide. THALLIUM OXIDE Tl_2O= 424, is readily soluble in water, has an alkaline reaction, and precipitates Magnesium-hydrate, and others, from their respective solutions. There is also a SUBOXIDE and a PEROXIDE Tl_2O_3.

546. THALLIUM SULPHIDE Tl_2S, is brown-black.

547. SALTS OF THALLIUM. THALLIUM CHLORIDE $TlCl$, is yellowish-white and with difficulty soluble. THALLIUM SULPHATE Tl_2SO_4, crystallizes in six-sided prisms. It forms an ALUM $TlAl2SO_4,12H_2O$. THALLIUM CARBONATE Tl_2CO_3, in needles, requiring 25 parts of water for solution. THALLIUM NITRATE $TlNO_3$, in needles.

548. TESTS. The solutions are colourless, and precipitated by metallic zinc. In solutions of the sulphate, nitrate and chloride, Hydro-sulphuric acid produces no reaction, but Ammonium hydrogen sulphide gives a brown precipitate of Thallium sulphide. The alkalies give no precipitate. Hydro-chloric acid, however, throws down the Chloride. Platinic chloride forms an insoluble Thallium Platinic chloride $2TlCl,PtCl_4$.]

[XLIII. Indium In = 72.

549. 43.—INDIUM In=72. A white, malleable metal, of sp. gr. 7·36. It is very fusible. In the spectroscope, Indium gives two bright lines in the blue and indigo, not shown by any other body.

It is readily soluble in Hydrochloric acid. Whether a dyad, as assumed, is at present undetermined.

550. INDIUM OXIDE is white. SULPHIDE yellow, and not precipitated by Hydrosulphuric acid. INDIUM CHLORIDE is deliquescent.]

XLIV. Argentum Ag = 108.

551. 44.—ARGENTUM or SILVER Ag=108. A monad metal, frequently found native; generally in irregular masses, but often crystallized in cubes and octohedra. In 1478, a rich vein of Silver was discovered at Schneeberg, in Saxony, from which a mass of metal was cut out, sufficiently large to serve as a dining table to Duke Albert. When smelted, it yielded 44,000 lbs. of metal.

Silver is the most lustrous of all the metals. By the Alchemists it was accepted as the emblem of the pale "silvery" moon, under the name of Luna or Diana, and some of the salts of Silver are called "lunar" to this day. It melts at 1023° C., and has a sp. gr. of 10·53. It is the best conductor of heat and electricity. As it does not oxydize in dry or moist air, and as its oxide parts with its Oxygen below a red-heat, it is called a Noble Metal. When melted, Silver absorbs Oxygen, which it loses on solidifying, causing often a serious loss by what is technically called *spitting*. It tarnishes rapidly in Sulphuretted hydrogen, owing to its great affinity for Sulphur; it is from the presence of Sulphur in eggs and in mustard, that it is necessary to gild the bowls of silver spoons that are used. Nitric acid is the best solvent for silver; Argentum nitrate $AgNO_3$, is formed, together with Water and Nitrous anhydride. $4Ag+6HNO_3=4AgNO_3+3H_2O+N_2O_3$ (p. 74, par. 82). Sulphuric acid has no action upon it until boiled, then a Sulphate is formed: $Ag_2+2H_2SO_4=Ag_2SO_4+2H_2O+SO_2$. Hydrochloric acid scarcely injures it. Finally, the alkalies and their nitrates may be fused in vessels of Silver, without injury to the latter.

552. The chief ORE of Silver is "Silver-glance" or "Argentite," a SULPHIDE Ag_2S. The metal is generally obtained by the process of amalgamation with Quicksilver. At Freiberg, the ores contain about 0·24 per cent. of Argentum sulphide, the chief constituent as regards quantity being Iron-pyrites or FERROUS DI-SULPHIDE FeS_2,

with a little COPPER-PYRITES $FeCuS_2$. The mass is roasted carefully with one-tenth of its weight of Sodium chloride, the chlorine of which unites with the Silver, Copper, and Iron, to Chlorides of the respective metals; some of the Sulphur oxydizes to the fullest, and remains in combination with Sodium and Copper as Sulphates, and the rest of that element volatilizes as Sulphurous anhydride. $Ag_2S + 2NaCl + 2O_2 = 2AgCl + Na_2SO_4$. $2FeS_2 + 6NaCl + 7O_2 = Fe_2Cl_6 + 3Na_2SO_4 + SO_2$. $2CuS + 4NaCl + 4O_2 = Cu_2Cl_2 + Cl_2 + 2Na_2SO_4$. $CuS + 2O_2 = CuSO_4$. The roasted mass is pounded, mixed with water, and placed in casks which are made to revolve upon horizontal axes, and scrap-iron is then introduced, in order to convert the Silver and Copper salts into metallic Silver and Copper, and the Ferric into Ferrous chloride. Mercury is now added, which forms an Amalgam with the Silver and Copper, and the mercury is separated by heat, taking care to collect it as it volatilizes. The residual mass consists of about 70 per cent. of Silver and 30 per cent. of Copper. The Silver may now be separated by dissolving the metals in Nitric acid, precipitating by Hydrochloric as Silver chloride, and fusion with Chalk and Charcoal. $2AgCl + 2CaCO_3 + C = CaO,CaCl_2 + CO + 2CO_2 + Ag_2$.

553. Most ores of Lead contain also Silver, and nearly all samples of Lead, until purified, contain Silver. The separation of Silver from its combination with Lead, is one of the genuine triumphs of science. The late Mr. Pattinson, of Newcastle, discovered that an Alloy of Silver and Lead melted at a lower temperature than Lead alone, a very curious fact, when it is remembered that the melting point of Silver is three times higher than that of Lead. By a very simple process, therefore, when Lead containing Silver is melted, and set by to cool, the pure Lead is removed as fast as it solidifies; while the Silver-lead remaining, is separated from the Lead by a process of cupellation. The further separation depends on the fact, that Lead, being a base metal, oxydizes readily; while Silver, as a noble metal, does not. When, therefore, the Silver-lead is exposed to the red-heat of a reverberatory furnace, the Lead oxydizes and the light Oxide flows off the molten metal, exposing a fresh surface to the action of the air. When the chief part of the Lead is thus removed, the residue, rich in Silver, is placed in shallow oval basins, 'cupels,' made of a mixture of bone and wood-ashes, which,

when heated in a reverberatory furnace, absorb the Lead-oxide as fast as it is formed, leaving behind the Silver perfectly pure and brilliant.

554. The ALLOYS of Silver are important. Our English Silver coin is alloyed with 7·5 per cent. of Copper and has a sp. gr. of 10·3. When Silver so alloyed is heated to redness, it blackens from the formation on its surface of Cupric oxide. If this oxide be removed by immersion in hot dilute Sulphuric acid, a film of pure Silver, known as blanched or dead Silver, remains. The blanks for coins are thus treated before striking, and hence the whiteness of new silver money.

555. Silver is often used for plating and silvering the base metals. The better descriptions of SHEFFIELD-PLATE are prepared by placing a plate of Silver upon a somewhat larger plate of Copper, and heating them to just below the melting-point of the Silver. By hammering at this temperature, the two metals become inseparably united and may be rolled out to almost any extent of tenuity. Steel can be plated only by first lining its surface, and attaching the sheet-silver by soldering.

The application of a thin plating of Silver to kitchen utensils, now far from uncommon, is a very happy idea, improving both their appearance and utility; it is, however, necessary, that such articles should be kept scrupulously clean.

556. Sheffield plating has been almost superseded by ELECTRO-PLATING. A solution of POTASSIUM SILVER CYANIDE AgCy,KCy, containing a large excess of Potassium cyanide, is employed, as it is readily decomposed by electrolysis, Silver being deposited at the Zincode, in connection with which, the articles to be plated are placed.

557. Silver has THREE oxides, but only one basic oxide. SILVER SUB-OXIDE Ag_4O, is brown. It is quite unimportant. SILVER OXIDE $Ag_2O=232$. Is brown. Its hydrate is also brown, insoluble in Potassium hydrate and soluble in Ammonia. The ammoniacal solution, on exposure, deposits a black *fulminating silver*, most dangerous to handle. Silver oxide is the basis of Silver salts. It is precipitated from Silver salts on addition of Potassium hydrate. SILVER PEROXIDE Ag_2O_2, in grey needles.

558. SILVER SULPHIDE $Ag_2S=248$, is black and readily preci-

pitated from its salts by means of Sulphuretted hydrogen. It is found native as *Silver-glance*, sometimes massive, occasionally in cubes and octohedra. The mines of Schemnitz in Hungary, of Kongsberg in Norway, of Freyberg in Saxony, and of Kolyvan in the Altai mountains, are the most important in Europe and Asia. But the greatest quantities are derived from South America, chiefly from the celebrated mines of Potosi in Peru. The mountain was opened in 1545, and is now traversed in every direction in search of the ore. *Red antimony ore* is a Sulphide of Antimony and Silver $3Ag_2S, Sb_2S_3$.

559. SALTS OF SILVER. SILVER CHLORIDE $AgCl=143{\cdot}5$. Is known to Mineralogists as *Horn-silver*. When precipitated by Hydrochloric acid, it forms a dense, white, curd-like mass, which darkens on exposure to light. It is insoluble in water and in Nitric acid, but soluble in Ammonia. At 260°C it melts into a horn-like mass. By means of Zinc or Iron, Silver chloride is reduced to the metallic state. Silver is also readily obtained from it, by fusion with Sodium carbonate. Silver chloride is dissolved by Sodium hyposulphite. SILVER BROMIDE $AgBr=188$. It is yellowish, insoluble in water and in Nitric acid, but soluble in a large quantity of Ammonia. SILVER IODIDE $AgI=235$. Is pale yellow, insoluble in water, in Nitric acid, and in Ammonia.

SILVER CYANIDE $AgCy=134$. Is precipitated as well by Hydrocyanic acid, as by Potassium cyanide from Argentum nitrate. It is dense, white, curd-like: insoluble in water and in dilute Nitric acid, but soluble in Ammonia. Hydrochloric acid decomposes it, liberating Hydrocyanic acid. It dissolves in Potassium cyanide, and forms the double salt $AgCy,KCy$.

SILVER NITRATE $AgNO_3=170$. In square, colourless tables soluble in its own weight of water. It fuses at 219°C., and, when cast into moulds, forms the *Lunar caustic* of Surgeons. When Argentum nitrate is made from Silver containing Copper, the Copper can be easily separated by keeping the mass fused for a short time, as with the first rise of temperature the Cupric salt is entirely decomposed (leaving insoluble Cupric oxide) before the Silver salt is acted upon.

It is much employed in making *Marking ink*, as Silver nitrate is readily decomposed by organic matter, giving a black deposit of metallic silver. SILVER CARBONATE Ag_2CO_3. White, and readily soluble in Ammonia. SILVER SULPHATE Ag_2SO_4. In small rhombic prisms, requiring 90 parts of water for solution. SILVER ORTHOPHOSPHATE Ag_3PO_4. Yellow, insoluble in water, but soluble in Ammonia and in Nitric acid. When precipitated from neutral silver nitrate, by alkaline Di-sodium Hydrogen phosphate, the solution is found with an acid reaction on account of the liberation of Nitric acid. $3AgNO_3 + Na_2HPO_4 = Ag_3PO_4 + 2NaNO_3 + HNO_3$. SILVER PYROPHOSPHATE $Ag_4P_2O_7$ is white. So also is SILVER METAPHOSPHATE $AgPO_3$, and equally soluble in Nitric acid, and in Ammonia. SILVER ACETATE $AgC_2H_3O_2$, in silky needles, sparingly soluble in water.

560. TESTS. The salts of Silver are colourless, except those with coloured acids. Hydrochloric acid throws down Silver quantitatively as Silver chloride, soluble in Ammonia, but insoluble in Nitric acid. Hydrosulphuric acid gives a black precipitate of Silver sulphide. The alkalies throw down the brown Oxide, which Ammonia redissolves; an acid solution of a Silver salt is therefore not precipitated by Ammonia. Sodium carbonate precipitates the white Silver carbonate. Potassium chromate, a crimson-red Silver chromate. Silver is estimated either as Silver chloride, 100 parts of which contain 75·27 of the metal, or in the metallic state by *cupellation*. The *cupel* consists of a porous crucible made of bone-earth. The principle consists in the fact, that the metal Lead readily oxydizes when heated, and imparts its Oxygen to such metals as yield oxides which are not reduced by heat alone. These oxides form with Lead-oxide a fusible slag, which is readily absorbed by the Cupel, whilst the Silver, as a noble metal, remains behind in a state admitting of being weighed. The samples of Silver to be assayed are carefully weighed and wrapped in pure sheet-lead, the quantity of which is increased in proportion to the quantity of Copper, or other base metal present. It is then placed in the cupel and fused in the assay-furnace. When the whole of the lead is oxydized, the Silver is seen to flash, and the operation, all but the removal from the cupel, cleansing and careful weighing, is complete. Care must be taken to prevent loss from spitting, which is accomplished by allowing the cooling to take place slowly.

CHAPTER XXVI.

THE CHEMISTRY OF THE DYAD METALS COPPER AND MERCURY.

XLV. **Copper** Cu=63·5. 561. Antiquity of Copper. 562. Properties of the metal. Sp. gr. 8·952. Melts at 1091° C. 563. Employed to cover the bottoms of ships. 564. Action of acids upon Copper. 565. Salts readily reduced by Iron, Copper and Zinc. Cementation. 566. Alloys of copper. BRASS Cu_2Zn. MUNTZ METAL Cu_3Zn_2. 567. Ores of Copper. Ruby-copper-ore is CUPREOUS OXIDE Cu_2O. Redruthite or Vitreous Copper is CUPREOUS SULPHIDE Cu_2S. Malachite $CuH_2O_2,CuCO_3$, and Chessylite CuH_2O_2, $2CuCO_3$, are basic Carbonates. Copper-pyrites $FeCuS_2$, the chief ore. 568. Metallurgy of Copper. 569. Two oxides; both basic. CUPREOUS OXIDE Cu_2O. CUPRIC OXIDE CuO. [570. CUPREOUS HYDRIDE CuH or Cu_2H_2.] 571. CUPREOUS SULPHIDE Cu_2S. CUPRIC SULPHIDE CuS. 572. SALTS OF COPPER. CUPREOUS CHLORIDE Cu_2Cl_2. CUPREOUS BROMIDE Cu_2Br_2. CUPREOUS IODIDE Cu_2I_2. CUPREOUS CYANIDE Cu_2Cy_2. CUPRIC CHLORIDE $CuCl_2,2H_2O$. CUPRIC OXY-CHLORIDE $3CuO,CuCl_2,4H_2O$. CUPRIC CYANIDE $CuCy_2$. CUPRIC FERRO-CYANIDE Cu_2FeCy_6. CUPRIC SULPHATE $CuSO_4$, $5H_2O$. Blue vitriol. CUPRIC NITRATE $Cu2NO_3,6H_2O$. Malachite and Chessylite. Mineral Green $CuH_2O_2,CuCO_3$. CUPRIC ORTHO-PHOSPHATE Cu_32PO_4. Scheele's Green is CUPRIC HYDROGEN ARSENITE $CuHAsO_3$. [CUPRIC ACETATE $Cu2C_2H_3O_2,H_2O$: Verditer. Verdigris $CuO,2(Cu2C_2H_3O_2)$, $6H_2O$. CUPRIC TRI-BASIC ACETATE $(4CuO,2[Cu2C_2H_3O_2],3H_2O)$. Schweinfurth green $3(Cu2AsO_2),Cu2C_2H_3O_2$.] 573. TESTS FOR COPPER. XLVI. **Hydrargyrum or Mercury** Hg=200. 574. Derivation of the name Mercury. Long known. Chief mines. 575. Properties of the metal. Sp. gr. 13·56. Boils at 350° C. Freezes at—38°·8 C. 576. Cinnabar, the chief ore. A sulphide. Metallurgy of Mercury. 577. Mercury used as Medicine. 578. Amalgams. Preparation of Mirrors. 579. Two oxides, both basic. MERCUROUS OXIDE Hg_2O. MERCURIC OXIDE HgO: used in making Oxygen. 580. MERCURAMINE $Hg_4N_2O,5H_2O$. MERCURAMINE SULPHATE Hg_4N_2O,H_2O,H_2SO_4. MERCURAMINE CHLORIDE Hg_4N_2O,H_2O, 2HCl. 581. MERCUROUS SULPHIDE Hg_2S. MERCURIC SULPHIDE HgS.

MERCURIC CHLORO-SULPHIDE $2HgS,HgCl_2$. 582. SALTS. 1. MERCUROUS CHLORIDE Hg_2Cl_2. Calomel. MERCUROUS BROMIDE Hg_2Br_2. MERCUROUS IODIDE Hg_2I_2. MERCUROUS SULPHATE Hg_2SO_4. MERCUROUS NITRATE $Hg_22NO_3,2H_2O$. By *water* into: Hg_2O,Hg_22NO_3,H_2O. MERCUROUS ACETATE $Hg_22C_2H_3O_2$. MERCUROUS BASIC CHROMATE $Hg_2O,3Hg_2CrO_4$. 2. MERCURIC SALTS. MERCURIC CHLORIDE $HgCl_2$. Corrosive Sublimate. [MERCURIC OXY-CHLORIDES $2HgO,HgCl_2$. $3HgO,HgCl_2$. $4HgO,HgCl_2$.] Sal Alembroth $6NH_4Cl,HgCl_2,H_2O$. White precipitate HgH_2NCl. MERCURIC BROMIDE $HgBr_2$. MERCURIC IODIDE HgI_2. Nessler's test. MERCURIC CYANIDE $HgCy_2$. Cyanogen and Paracyanogen. MERCURIC SULPHATE $HgSO_4$. MERCURIC DI-BASIC SULPHATE $2HgO,HgSO_4$. MERCURIC NITRATE $[2(Hg2NO_3),H_2O]$. MERCURIC BASIC NITRATE $HgO,Hg2NO_3,2H_2O$. 583. Tests for mercury. 1. Mercurous. 2. Mercuric.

XLV. Cuprum Cu = 63·5.

561. 45.—COPPER or CUPRUM Cu=63·5. This beautiful metal has been known and used from the earliest times. The Phœnicians could scarcely have worked the Tin-mines of Cornwall, without discovering Copper. It probably derived its name from the Isle of Cyprus, where, according to Pliny, it was first discovered. The Copper from this island was known in the Roman market as Cyprian Copper. The alchemists called copper Venus ♀, the protecting goddess of Cyprus.

Copper frequently occurs native, in quantities varying from an ounce to a ton. Immense masses of the metal, covered with metallic Silver, have been found at Lake Superior in N. America: and small quantities of Copper are met with in all the Cornish mines. It is sometimes found in cubes and octohedra, or in dendritic masses.

562. The metal is of a red-colour, very hard, lustrous, malleable, and ductile. It is so tenacious, that it is the most useful of all metals for the construction of wire. It melts at 1091° C., and has a sp. gr. of 8·952. In power of conductivity for heat and electricity, Copper ranks next to Silver. At a high temperature, it can be volatilized: it burns with a green flame.

563. Copper is largely employed for coating the bottoms of ships, as a protection to the wood from the ravages of the "seaworms."

Unfortunately, the metal is attacked and corroded by the sea-water itself. To remedy this evil, Sir H. Davy suggested the employment of strips of Iron, or Zinc, at intervals across the Copper. These, oxydizing much more readily than Copper, protect the more expensive metal. So far, then, as the action of sea-water is concerned, the cure was complete. But a further inconvenience, almost as serious, remains, that the copper forms a nidus for vegetable growths, and rapidly becomes so foul, as materially to impede the vessel in its passage through the water.

564. Copper is not readily attacked by moist air, but in process of time a deposit of Cupric basic carbonate (falsely called Verdigris) forms upon the surface. Nitric acid is its best solvent; CUPRIC NITRATE $Cu2NO_3$, is formed, and Nitric oxide is set free. $3Cu+8HNO_3=3Cu2NO_3+4H_2O+2NO$. Sulphuric acid has no action until heated with Copper; then a SULPHATE $CuSO_4$, is formed. $Cu+2H_2SO_4=CuSO_4+2H_2O+SO_2$. Hydrochloric acid does not dissolve Copper. Chlorine attacks the metal violently, insomuch that Copper-leaf takes fire in the gas..

565. Like most of the reducible metals, Copper is readily separated from solutions of its salts in the metallic form. A bar of Iron, Zinc, or Lead in such solution, soon becomes coated with a deposit of Copper. The process of *cementation* consists in precipitating Copper from a dilute solution of Cupric sulphate by means of metallic Iron. $CuSO_4+Fe=FeSO_4+Cu$. But, on the other hand, a bar of Copper precipitates Silver from its solution, so that Silver from the sea is deposited on ships' bottoms. It was indeed the examination of these deposits by Mr. FIELD, F.R S., which led to the discovery of the existence of Silver in the sea.

566. The ALLOYS of copper are most useful. Ordinary BRASS Cu_2Zn, is a compound of Zinc with Copper. MUNTZ METAL Cu_3Zn_2. is employed in the sheathing of Ships' bottoms. Hard solder consists of 2 parts of Brass with 1 part of Zinc. The Brass of the Ancients with which they made their implements, both of agriculture and of war, was mainly an alloy of Copper and Tin. It had been often questioned whether such an alloy as our Brass was known to them; but after the able disquisition on the point in Dr. Percy's work on Metallurgy, no doubt of the fact can remain. Other alloys

containing Tin, such as Bronze, Gun-metal, Bell-metal and Speculum metal, will be mentioned under the head of Alloys of Tin.

567. The ORES of Copper are numerous. *Ruby-Copper-ore* is CUPREOUS OXIDE Cu_2O. It is met with principally in Chili, both massive and beautifully crystallized in octohedra. By means of Charcoal it is readily reduced to the metallic state. *Redruthite* or *Vitreous Copper* is a CUPREOUS SULPHIDE Cu_2S, a mineral of dark-grey colour, often found crystallized in 6-sided prisms. It is the "fine metal" of the Smelter. *Malachite* $CuH_2O_2,CuCO_3$, is one of the most beautiful of minerals. Besides its value as an ore, it is much used by the Lapidary as an article of "vertu." For the latter purpose, the Malachite of the Ural Mountains is most available, that from the Burra Burra mine in Australia being more important as a source of Copper. When roasted, it loses Water and Carbonic anhydride, and when the remaining Cupric oxide is heated with Charcoal, Copper is left. $CuH_2O_2,CuCO_3$ heated, leaves $2CuO$. —$CuO + C = CO + Cu$. *Chessylite* $CuH_2O_2,2CuCO_3$ is a blue mineral, which crystallizes in oblique rhombic prisms, but is of little importance in a metallurgical point of view. *Copper-pyrites* or *Towanite* $CuFeS_2$, is the most important of all the ores. It is a compound of Iron, Copper, and Sulphur. There are mines of it in Anglesea and Cumberland, but the most extensive and the richest are those of Cornwall and Devonshire. The chief of these lie between the Land's End and Truro, the most important being in the neighbourhood of Redruth. This ore occurs in the primitive rocks, and chiefly in the Clay-slate, or Killas. It is of a yellow colour, with a bright brass-like lustre. It is sometimes crystallized in tetrahedra, but it is more generally massive and less hard than Iron-pyrites. *Peacock-copper ore* contains more Cupric sulphide, and owes its name to its varied and beautiful hues.

568. The METALLURGY of COPPER is somewhat complicated. The *copper-pyrites* is *first* carefully calcined, or roasted for some 12 hours. During this process the Iron becomes oxydized; a portion of the Sulphur volatilizes as SULPHUROUS ANHYDRIDE, and the Copper remains in the form of CUPREOUS SULPHIDE Cu_2S. Inasmuch as *Mispickel* $FeSAs$, and other compounds containing Arsenicum, are nearly always present in the ore, besides *Fluorspar* CaF_2, fumes of Arsenious anhydride, Sulphuric and Hydrofluoric acids

accompany the escaping Sulphurous anhydride and form the so-called "Copper-smoke" which hangs over the town of Swansea, where the chief smelting is carried on. The *second* part of the process consists in ridding the Cupreous sulphide of the Ferric oxide. This is accomplished in the "Ore furnace." Siliceous minerals are melted with the ore; a slag of Ferrous silicate is formed, which is raked off, whilst the "Coarse metal," consisting of Cupreous sulphide with some Ferrous sulphide, remains. The *third* process consists in a re-calcination as at the first, so as to oxydize nearly all the Ferrous sulphide. And now again the admixture of a siliceous mineral is resorted to, so as to remove the remaining Ferric oxide. In this *fourth* operation, a copper-ore rich in quartz and poor in iron-pyrites is preferred, so as to increase the quantity of Cupreous sulphide, or "Fine metal" as it is now called. The *fifth* process consists in oxydizing the Cupreous sulphide Cu_2S in part, and subsequently reducing it to the metallic state. $Cu_2S + 2O_2 = SO_2 + 2CuO$. Then: $2CuO + Cu_2S = SO_2 + 4Cu$. The metal is run off into moulds made of sand, and thus cast into ingots of "blistered copper." Lastly, the blistered copper is "refined." Such is a brief sketch of the operation. For further particulars, the reader is referred to larger Works on Chemistry. Swansea produces annually 20,000 tons of Refined Copper.

569. The OXIDES of copper are principally *two*. Both of them are basic oxides. 1. CUPREOUS OXIDE $Cu_2O = 143$. Is of a red colour, much valued because of the ruby tint it imparts to glass. Its HYDRATE $4Cu_2O,H_2O$ is deep-yellow. Cupreous oxide has already been mentioned as an Ore of Copper. Artificially, it is easily prepared, by reducing a solution of a cupric salt by means of Glucose. In presence of Glucose or Grape-sugar, Potassium hydrate re-dissolves the precipitate of Cupric hydrate, and on heating such alkaline solution, Cupreous hydrate first falls,—and then Cupreous oxide by the loss of water. It may be obtained in Octohedra, by boiling a solution of Cupric di-basic acetate with Sugar. "Dry" or brittle Copper contains Cupreous oxide. Cupreous hydrate forms a colourless solution with Ammonia, which is a most delicate test of the presence of Oxygen, as it immediately becomes blue by its absorption. 2. CUPRIC OXIDE $CuO = 79{\cdot}5$. Is black, insoluble in water and soluble in acids. On account of the facility with which it

imparts its Oxygen, Cupric oxide is much employed in Organic analysis for oxydizing Carbon. It is readily prepared by dissolving Copper in nitric acid and heating the Cupric nitrate to redness. When Copper is heated to redness, its surface becomes covered with Cupric oxide. And so, also, when a Cupric carbonate is strongly heated, nothing but Cupric oxide remains. CUPRIC HYDRATE CuH_2O_2, is light blue, insoluble in Potassium hydrate, and soluble in Ammonia with bright blue colour. When the hydrate is heated, it loses water and becomes Cupric oxide: $CuH_2O_2 = H_2O + CuO$. Cupric salts are blue or green. Cupric oxide communicates a green colour to Glass.

[570. Copper unites with HYDROGEN. CUPREOUS HYDRIDE Cu_2H_2. It is obtained by the spontaneous decomposition of Cupric hypophosphite (p. 141, par. 233). With Hydrochloric acid it immediately evolves Hydrogen, with formation of CUPREOUS CHLORIDE Cu_2Cl_2. Thus: $Cu_2H_2 + 2HCl = 2H_2 + Cu_2Cl_2$. In the same manner is O_2 evolved from Hydrogen peroxide, when mixed with Manganese peroxide, because the Oxygen in the one case, and the Hydrogen in the other, is contained in a different electrical condition.]

571. The chief compounds of Copper and SULPHUR have been already mentioned as *Ores* of Copper. CUPRIC SULPHIDE $CuS = 95{\cdot}5$, may be prepared by burning Copper in Sulphur vapour. It is found native as *Covelline*, in thin flexible leaves of indigo-blue colour. As a dark-brown hydrate, very readily oxydizable, it is obtained by precipitating a Cupric salt by Hydrosulphuric acid.

572. SALTS OF COPPER. [1. CUPREOUS SALTS. CUPREOUS CHLORIDE Cu_2Cl_2, is white, and insoluble in water. It can be made by dissolving Copper in a solution of Cupric chloride. $Cu + CuCl_2 = Cu_2Cl_2$. CUPREOUS BROMIDE Cu_2Br_2 and CUPREOUS IODIDE Cu_2I_2. are also insoluble. CUPREOUS CYANIDE Cu_2Cy_2 is a white powder.] 2. CUPRIC SALTS. CUPRIC CHLORIDE $CuCl_2,2H_2O$, in green needles, highly deliquescent. The anhydrous salt is red-brown, and obtained by burning Copper in dry Chlorine. "Brunswick Green" is CUPRIC OXY-CHLORIDE $3CuO,CuCl_2,4H_2O$. CUPRIC CYANIDE $CuCy_2$, as a brownish-yellow hydrate, is precipitated by Potassium cyanide. It is soluble in excess of Potassium cyanide as Cupreous cyanide, with

evolution of Cyanogen. CUPRIC-FERROCYANIDE Cu_2FeCy_6, is deep red-brown. It is formed on the addition of Potassium ferrocyanide to a Cupric salt. Thus: $2CuSO_4+K_4FeCy_6=2K_2SO_4+Cu_2FeCy_6$. CUPRIC SULPHATE $CuSO_4.5H_2O$. Is called "Blue vitriol." It crystallizes in doubly-oblique rhombs of a splendid blue colour, soluble in four parts of cold water. It is insoluble in alcohol. At 200° C. the salt loses all its water. Cupric sulphate is readily prepared by careful roasting of cupreous sulphide. When mixed with Ferrous sulphate, it crystallizes with 7 molecules of water, and forms a salt isomorphous with Ferrous sulphate. CUPRIC NITRATE $Cu2NO_3,6H_2O$, crystallizes in blue, deliquescent rhombs. By heat it is changed into Cupric oxide. CUPRIC CARBONATE $CuCO_3$, has not been prepared. Chessylite and Malachite have already been mentioned as Ores of Copper. "Mineral Green" has the same composition as Malachite $CuH_2O_2,CuCO_3$. The BASIC CARBONATE $CuH_2O_2,CuCO_3,H_2O$, precipitated from Cupric salts by means of a solution of Sodium carbonate, contains one molecule more water than Malachite. CUPRIC ORTHO-PHOSPHATE Cu_32PO_4 is blue, and insoluble in water. CUPRIC HYDROGEN ARSENITE $CuHAsO_3$, is "Scheele's Green." [CUPRIC ACETATE $Cu2C_2H_3O_2,H_2O$. In oblique rhombic prisms. Known as "Verditer," a beautiful pigment. "Verdigris" is CUPRIC BASIC ACETATE $CuO,2[Cu2C_2H_3O_2],6H_2O$. CUPRIC TRI-BASIC ACETATE $(4CuO,2[Cu2C_2H_3O_2],3H_2O)$ forms a green crystalline powder, insoluble in water. "Schweinfurth Green" is a compound of Cupric arsenite and acetate $3(Cu2AsO_2)$, $Cu2C_2H_3O_2$.]

573. TESTS FOR COPPER. Copper is always weighed as CUPRIC OXIDE CuO, 100 parts of which contain 79·85 per cent. of the metal. In cases of poisoning, the material must be heated to redness in a crucible, and the remaining mass carefully but vigorously rubbed in a mortar, and well washed. Unless the quantity be very small, red flattened spangles will be found in the mortar. The solutions of Cupric salts are either blue or green. Even in acid solutions Hydro-sulphuric acid precipitates brown-black Cupric sulphide. Potassium hydrate, a blue hydrate insoluble in excess; but, when sugar is present, the precipitate re-dissolves and is thrown down as Cupreous oxide of a red colour. Ammonia gives a similar preci-

pitate, but re-dissolves it with dark-blue colour as $CuO,2NH_3$. Sodium carbonate, a pale blue basic carbonate. Ammonium hydrocarbonate, re-dissolves the precipitate with blue colour. Potassium ferrocyanide a claret-coloured precipitate of Cupric ferrocyanide. Metallic iron precipitates metallic Copper, and especially in presence of free Hydrochloric acid.

XLVI. Hydrargyrum Hg = 200.

574. 46.—MERCURY OR HYDRARGYRUM Hg=200. Atomic and Molecular volume 2. Mercury affords the only example of a metal, fluid under the temperatures of most climates. It is of great importance both in medicine and the arts. It owes its name (from the Greek ὕδωρ, *hydor*, water, and ἄργυρον, *argyron*, silver) to its silver-white colour and fluidity, while the name of the heathen god Mercury was given to it, on account of its mobile, restless character. In the word mercurial, applied to an individual, the ancient meaning of the word is still retained. The metal has been known from very remote ages. Pliny describes the Greeks as importing Cinnabar, the chief ore of mercury, from Almaden in Spain, 700 years before Christ, and states, that from this source alone they obtained an annual supply of some 300 tons.

At the present day, considerable quantities of mercury are obtained from California, Peru, China and Japan. But the chief mines are those of Almaden in Spain, and Idria in Carniola. In the latter mine, the metallic vein occurs at a depth of 720 feet from the surface. The mineral products of this mine are so varied, that the modes of carrying on operations differ from any other. In some parts, pure mercury distils in globules from the rock. The ore is raised in square boxes, by means of a water-wheel. Besides metallic mercury, obtained from the ore by distillation, all the preparations of mercury, used in medicine and the arts, are either found naturally, or are manufactured at Idria. From the great heat of the mines, and the unwholesome vapour of the metal, the miners suffer severely. They soon lose their teeth from salivation, and few survive their fortieth year. Cattle cannot be reared, nor will fruit or grain ripen in the neighbourhood.

575. Mercury is a highly lustrous metal, solid at $-38°\cdot 8$ C. Its sp. gr. is 13·56. At 350° C., it boils. Owing to its low specific heat, and the low temperature at which it freezes, Mercury is admirably adapted to the making of philosophical instruments. It does not tarnish on exposure to the air, and its Oxide parts with its Oxygen at a low red-heat. Like Silver and Gold, it is accounted a noble metal. Nitric acid dissolves it rapidly, and we obtain Mercurous or Mercuric nitrate, dependent upon the circumstance of temperature, and the relative quantity of Mercury to the acid. $3Hg_2 + 8HNO_3 = 4H_2O + 2NO + 3(Hg_2 2NO_3)$. Or: $3Hg + 8HNO_3 = 4H_2O + 2NO + 3(Hg2NO_3)$. Sulphuric acid forms a sulphate, when heated with Mercury: $Hg + 2H_2SO_4 = 2H_2O + SO_2 + HgSO_4$. Hydrochloric acid has no action. Chlorine and Bromine unite at once with Mercury, and, indeed, so do many of the metals.

576. The only ORE of Mercury is *Cinnabar* or MERCURIC SULPHIDE HgS. It occurs both massive and in 6-sided prisms. One mode of obtaining Mercury from the sulphide is by heating the ore with Calcium oxide and collecting the Metal in receivers, Fig. 95. The action is represented by the following symbols: $4HgS + 4CaO = CaSO_4 + 3CaS + 4Hg$. Again, it may be obtained by heating the Sulphide with Iron-filings; Ferrous Sulphide FeS is formed and Mercury sublimes. $HgS + Fe = FeS + Hg$. But, at Almaden, the Mercury is obtained by simply burning off the sulphur and distilling the metal. When Mercuric sulphide is roasted, the Sulphur alone oxydizes. $HgS + O_2 = SO_2 + Hg$.

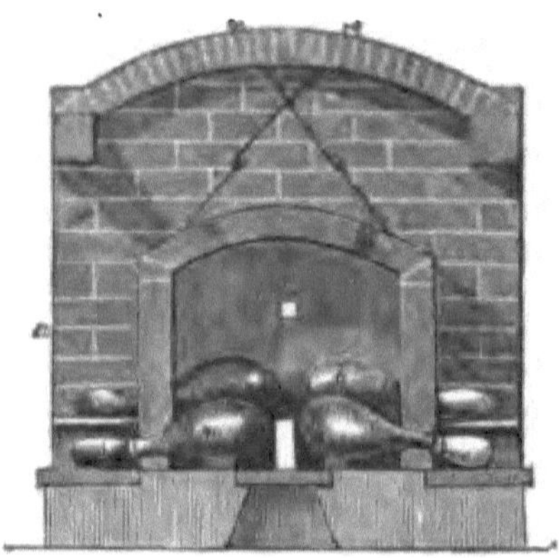

Fig. 95.

577. Mercury is much used in medicine. It is contained in the metallic form as well in Blue pill, as in Mercurial ointment and in Pulvis Hydrargyri cum cretâ.

578. The alloys of Mercury are called AMALGAMS. We have AgHg, CuHg, FeHg, Zn_2Hg, Pb_2Hg, $PtHg_2$, and others. The

amalgam with tin, Sn_7Hg is very important. The process of "silvering" mirrors is thus accomplished: A perfect sheet of tin-foil, somewhat larger than the plate-glass, is placed upon an even table of slate or stone: Mercury is then poured upon it, and rubbed uniformly upon its surface with a hare's foot, or a ball of flannel or cotton, until a clean and bright amalgam is formed. Upon this, Mercury in excess is poured, till the metal has a tendency to run off; the plate of glass, previously made quite clean, is then brought horizontally towards the table, and its edge so adjusted, as by gradually and steadily sliding it forward, to displace some of the excess of Mercury, and float the plate, as it were, over the amalgam, the dross upon its surface being pushed on by the edge of the glass, so that the Mercury appears beneath it with a perfectly uniform, clean, and brilliant reflecting surface. A number of square weights, of 10 or 12 lbs. each, are now placed side by side upon the surface of the plate, so as entirely to cover it, and press it down upon the amalgamated surface of the tin: in this way, the excess of Mercury is partly squeezed out, and the amalgam is made to adhere firmly to the glass. The Mercury, as it runs off, is received into a channel on the side of the table, (which is slightly inclined, to facilitate the draining), and in about forty-eight hours the weights are taken off, and the plate carefully lifted from the table and set nearly upright, by which the adhering Mercury gradually drains off, and the solid crystalline amalgam remains perfectly and uniformly adhering to the glass.

Amalgams of silver, gold and cadmium, are also employed for filling teeth.

579. There are TWO OXIDES of Mercury. 1. MERCUROUS OXIDE $Hg_2O=416$. Is black and most unstable. At a gentle heat, it changes into Mercury and Mercuric oxide. It is prepared by triturating Mercurous chloride with Potassium hydrate. $Hg_2Cl_2 + 2KHO = 2KCl + H_2O + Hg_2O$. 2. MERCURIC OXIDE $HgO=216$. Forms a yellow powder, when precipitated from Mercuric salts by Potassium hydrate, and red scales when obtained by heating Mercury at a temperature of 400° C. When heated, it blackens, but returns to its original condition on cooling. Mercuric oxide can also be

prepared by carefully heating Mercuric nitrate. It is employed for making Oxygen. $2HgO$ by heat $= 2Hg + O_2$.

580. When Ammonia is poured upon Mercuric oxide which has been precipitated from a mercuric salt, the yellow colour becomes paler by degrees, and at length a yellowish-white powder remains. This, when carefully dried over quick-lime, is found to be the Hydrate of the base MERCURAMINE $Hg_4N_2O,5H_2O$. It may be regarded as the TETRHYDRARGAMMONIUM HYDRATED OXIDE. It decomposes solutions of Ammonium salts, and combines with their acids. When dried over Sulphuric acid, it loses 2 atoms of Water, and yet another atom at 100° C. MERCURAMINE SULPHATE Hg_4N_2O,H_2O,H_2SO_4. MERCURAMINE CHLORIDE $Hg_4N_2O,H_2O,$ $2HCl$.

581. SULPHUR and MERCURY unite in *two* proportions. MERCUROUS SULPHIDE $Hg_2S=432$. A black powder, very unstable. It may be obtained by passing Hydrogen sulphide through Mercurous nitrate, or by trituration of the two elements. MERCURIC SULPHIDE $HgS=232$. Is also black, when obtained by precipitation. But when sublimed, it is known and valued as "Vermilion." When Hydrogen Sulphide is carefully added to a solution of Mercuric chloride, rendered acid by Hydrochloric acid, a white MERCURIC CHLORO-SULPHIDE $2HgS,HgCl_2$, is precipitated, which passes through shades of yellow and orange, to black sulphide.

582. SALTS OF MERCURY. 1. MERCUROUS SALTS. MERCUROUS CHLORIDE $Hg_2Cl_2=471$. A yellowish-white powder, perfectly insoluble in water. It is highly valued as a medicine and well-known as "Calomel." It may be prepared by heating MERCUROUS SULPHATE with SODIUM CHLORIDE. $Hg_2SO_4+2NaCl=Na_2SO_4+Hg_2Cl_2$. Or, by precipitating MERCUROUS NITRATE with Sodium chloride. $Hg_22NO_3+2NaCl=2NaNO_3+Hg_2Cl_2$. Or, by triturating Mercury with Mercuric chloride. $HgCl_2+Hg=Hg_2Cl_2$. Mercurous chloride is decomposed by Potassium hydrate, with separation of Mercurous oxide. $Hg_2Cl_2+2KHO=2KCl+H_2O+Hg_2O$. Lime-water produces a similar effect. The result is known as Black-wash. Ammonia decomposes Mercurous chloride, with formation of a black compound Hg_2H_2NCl. Thus: $Hg_2Cl_2+2NH_3=NH_4Cl+Hg_2H_2NCl$.

MERCUROUS BROMIDE Hg_2Br_2, is white and insoluble in water. MERCUROUS IODIDE Hg_2I_2, a green powder, precipitated by Potassium iodide from a mercurous salt. It is very unstable. MERCUROUS SULPHATE Hg_2SO_4. MERCUROUS NITRATE $Hg_22NO_3,2H_2O$; in prisms. It is decomposed by water into acid nitrate, and MERCUROUS BASIC NITRATE Hg_2O,Hg_22NO_3,H_2O. There are other Mercurous nitrates. MERCUROUS ACETATE $Hg_22C_2H_3O_2$, in silvery scales, on addition of Sodium acetate to mercurous nitrate. MERCUROUS BASIC CHROMATE $Hg_2O,3Hg_2CrO_4$. is orange-red, and insoluble in water. 2. MERCURIC SALTS. MERCURIC CHLORIDE $HgCl_2=271$. In crystalline needles soluble in 16 parts of cold, and 3 of boiling water. Its solution reddens litmus and possesses a burning taste. It is more generally known by its trivial name "Corrosive sublimate." The best antidote to poisoning with this substance is Albumen, as White of egg is coagulated and precipitated by it. Hence its use as an antiseptic. It is decomposed by Potassium hydrate, with precipitation of Mercuric oxide. $HgCl_2+2KHO=2KCl+H_2O+HgO$. Ammonia throws down *White precipitate* HgH_2NCl. $HgCl_2+2NH_3=NH_4Cl+HgH_2NCl$. Mercuric chloride is best prepared by subliming a mixture of Mercuric sulphate and Sodium chloride: $HgSO_4+2NaCl=Na_2SO_4+HgCl_2$. [There are several MERCURIC OXY-CHLORIDES. $2HgO,HgCl_2$. $3HgO,HgCl_2$. $4HgO,HgCl_2$.] Mercuric chloride forms double salts with Potassium chloride and with Ammonium chloride. *Sal Alembroth* $6NH_4Cl,HgCl_2,H_2O$ is an illustration of the latter fact. MERCURIC BROMIDE $HgBr_2$, soluble and crystallizable. MERCURIC IODIDE HgI_2, is scarlet. It fuses at 200° C. and sublimes in the form of yellow tables, which become scarlet when scratched. It is soluble in Potassium iodide, and forms a double salt $HgI_2,2HgCl_2$. Such a solution to which Potassium hydrate has been added, forms NESSLER's test for Ammonia (p. 179, par. 323); the latter throws down a brown precipitate of TETRARGAMMONIUM IODIDE Hg_2NI,H_2O. This *test* is prepared as follows:—Potassium iodide is added to a solution of Mercuric chloride until the scarlet precipitate of Mercuric iodide is nearly re-dissolved. Potassium hydrate is now added and when the liquid has become quite clear, it is decanted. On addition of a drop or two of the test to any water or liquid containing Ammonia, a brown colouration at least will take place, where none of the ordinary tests for Ammonia would give any indications. MERCURIC CYANIDE $HgCy_2=252$. In rectangular prisms, soluble in 8 parts of cold water. When heated, it yields CYANOGEN C_2N_2, MERCURY, and a residue of PARACYANOGEN. Mercuric cyanide is made either by saturating Hydrocyanic acid with Mercuric oxide, or by adding a solution of Mercuric sulphate to a solution of Potassium ferrocyanide. $3HgSO_4+K_4FeCy_6=2K_2SO_4+FeSO_4+3HgCy_2$. MERCURIC SULPHATE $HgSO_4$, is a white crystalline salt, much employed in making mercuric chloride. $Hg+2H_2SO_4=2H_2O+SO_2+HgSO_4$ It is decomposed by water, into Sulphuric acid which dissolves a portion of the salt, and into MERCURIC DI-BASIC SULPHATE $2HgO,HgSO_4$. Thus: $3HgSO_4+2H_2O=2H_2SO_4+2HgO,HgSO_4$. This basic salt is called *turpeth minerale*. MERCURIC NITRATE $[2(Hg2NO_3),H_2O]$ and MERCURIC BASIC NITRATE $(HgO,Hg2NO_3,2H_2O)$ are easily formed. Water decomposes them into a yellow basic salt $(2HgO,Hg2NO_3,H_2O)$.

583. TESTS FOR MERCURY. 1. MERCUROUS SALTS. When soluble, they possess an acid reaction. All are volatile, except those with fixed acids, with or without decomposition. Hydrochloric acid precipitates Mercurous chloride, insoluble in water and blackened by ammonia. Hydrogen sulphide even in acid solutions, gives a black precipitate of Mercurous sulphide. Potassium hydrate, a black precipitate of Mercurous oxide. Potassium chromate, an orange mercurous basic chromate: Potassium iodide, a green Mercurous iodide. Copper precipitates metallic mercury in silvery globules.

2. MERCURIC SALTS. When soluble, are colourless, and they are all volatile with or without decomposition, except in the case of fixed acids. Hydrochloric acid, no visible reaction. Hydrosulphuric acid, even in acid solutions, gives a black precipitate of Mercuric sulphide. With Sodium carbonate, a yellow precipitate of Mercuric oxide; but with Sodium Hydrogen carbonate, a red precipitate of Mercuric oxy-chloride. Potassium hydrate produces a yellow precipitate of Mercuric oxide. Ammonia, in solutions of Mercuric chloride, a white precipitate. Potassium iodide, a scarlet precipitate of Mercuric iodide, soluble in excess. Copper reduces mercuric salts to the metallic state. Mercury is ESTIMATED in the metallic form.

CHAPTER XXVII.

ON GOLD AND BISMUTH.

584. **Aurum or Gold** Au=196·6. In minute quantities very widely diffused. 585. Found in alluvial soils. 586. In the Ural Mountains and in the Brazils. 587. In California. 588. In Australia. 589. In Queensland, British Columbia, New Zealand. 590. Amalgamation. 591. Gold, a noble metal. Sp. gr. 19·34. Melts at 1102°·2 C. 592. Uses of Gold. 24-Carat Gold. 593. Making pure Gold. 594. Alloys and Amalgams. 595. Quartation process. 596. Two oxides. AUROUS OXIDE Au_2O. AURIC OXIDE Au_2O_3. 597. AUROUS-AURIC SULPHIDE Au_2S,Au_2S_3. 598. AUROUS CHLORIDE AuCl. AURIC CHLORIDE $AuCl_3$. POTASSIUM-AURIC CHLORIDE $2(KCl,AuCl_3),5H_2O$. [AURIC BROMIDE $AuBr_3$. AUROUS IODIDE AuI. AURIC IODIDE AuI_3.] AUROUS CYANIDE AuCy. POTASSIUM AUROUS CYANIDE KCy,AuCy. Electro-gilding. 599. Tests for Gold. PURPLE OF CASSIUS $SnAu_2Sn_2O_6,4H_2O$. **Bismuth** Bi=210. 600. Properties of Bismuth. Melts at 264° C. Sp. gr. 9·799. 601. Found native in Saxony and Transylvania. Metallurgy. 602. Used in medicine. Alloys. Fusible metal $PbBi_2Sn_2$. 603. Three oxides. BISMUTHOUS OXIDE BiO. BISMUTH OXIDE Bi_2O_3. BISMUTH OXY-HYDRATE $BiHO_2$. BISMUTHIC ANHYDRIDE Bi_2O_5. BISMUTHIC ACID $HBiO_3$. 604. BISMUTH SULPHIDE Bi_2S_3. 605. Bismuth salts. BISMUTH CHLORIDE $BiCl_3$. OXY-CHLORIDE BiOCl. Pearl-white $2BiOCl,H_2O$. BISMUTH NITRATE $Bi3NO_3,5H_2O$. Sub-nitrate $Bi_2O_3,2HNO_3$. 606. Tests.

XLVII. Aurum Au = 196·6.

584. 47.—AURUM AU=196·6 Although traces of this beautiful metal are to be found in every country in Europe, yet the labour of obtaining it is so great, as to render it the most expensive of metals. It always occurs native, and is usually associated with Quartz, so that the process of manufacture is generally a mechanical one.

585. In the time of Queen Elizabeth, Gold was found in the alluvial soil near the lead-hills of Scotland, and near the tin-works of Cornwall. It is now met with in Sutherlandshire, in the county Wicklow in Ireland, and in several parts of Wales. The richest mines in Europe are those of Hungary and Transylvania, which produce annually about 5000 ounces of gold.

The Gold-dust of commerce was formerly derived almost exclusively from Africa, where it was obtained by the natives from the washings of the sands of certain rivers.

586. Much Gold was, and is still, obtained from the Asiatic provinces of Russia.

In the Brazils, Gold was known to exist in 1543, but the first ore found in that country by a white man, was in 1693. This discovery led to the colonization of the Minas Geraes, and to all those evils resulting from the lust of Gold, with details of which the history of South America abounds. Dr. Walsh mentions that at a very early period, "two parties meeting on the banks of the river, where San José was afterwards built, instead of agreeing in their objects, and pursuing together their operations, set upon each other like famished tigers. A bloody encounter ensued, in which many were killed on both sides, and the river was from thenceforth called the Rio das Mortas, or the River of Death. "The vicinity of this river," proceeds our authority, "everywhere attests the extensive search for Gold formerly pursued here, as it was for a length of time considered one of the richest parts of Brazil, from the profusion of the precious metal found on its surface. All the banks of the stream are furrowed out in a most extraordinary manner, so as to be altogether unaccountable to one unacquainted with the cause. The whole of the vegetable mould was washed away, and nothing remained but a red earth, cut into square channels, like troughs, with a narrow ridge interposed between them. Above was conducted a head stream of water, let down through these troughs, which were all on an inclined plane. The lighter parts of the clay were washed away, and the Gold remained behind."

587. But all these sources of the precious metal sink into utter insignificance, when compared to the Gold-discoveries of the last ·hteen years. The discovery of Gold in California, attracted to

that place a vast population of adventurers from all parts of the globe, and great quantities have been derived from thence.

588. The knowledge of the geological feature of certain parts of Australia, had led Sir Roderick Murchison to suggest the probability of the existence of Gold in those regions. Between October 1851, and October, 1866, the known amount of Gold collected in Victoria was 36,514,361 ozs., of the value of £146,057,444.

The first discovery was made on the banks of the Summer-Hill creek and the Lewis Ponds river, small streams which run from the northern flank of the Conobalas down to the Macquarrie. The gold was found in the sand and gravel, accumulated, especially on the inside of the bends of the brook, and at the junction of the two watercourses, where the stream of each would be often checked by the other. It was coarse Gold, showing its parent site to be at no great distance. Shortly afterwards Gold was discovered in several other localities, especially on the banks of the Turon, some distance north-east of the Conobalas. At the head of the Turon river, among the dark glens and gullies in which it collects its head-waters, in the flanks of the Blue Mountains, the Gold got "coarser," occurring in large lumps or nuggets, but these being more sparingly scattered. The further discovery of gold ensued as a matter of course.

It is found usually in the drift-clay, sand, and gravel, or lying loose on the surface of the earth. The superficial drift in which the diggings have been carried on, varies in thickness from a few inches to 20 or 30 feet.

For the purpose of separation, a quantity of the auriferous sand or clay is taken and washed with water, when the gold, by its own weight, sinks to the bottom, and thus, by repeated cleansing, the Gold-dust of commerce is obtained.

As this appears a very simple process, the few words of advice given by Mr. Jukes to intending emigrants, are subjoined: "Gold digging is very hard work—just such work as you see navigators at in a railway cutting, or brickmakers in a brick-pit. You must work hard all day, lie hard all night, with but little shelter, often with scanty food, and with nothing of what you have probably been accustomed to consider necessary comfort."

A single nugget of Gold found in the Kingower diggings, Australia, weighed as much as 1743 ounces.

589. Still more recently, Gold has been found in British Columbia, the East Indies, Queensland, and in New Zealand, and always under similar conditions to those described as existing in Australia.

590. When Gold is found embedded in stone, it is stamped to powder and shaken, in a properly-constructed apparatus, with Mercury and water. An Amalgam is formed, from which the Mercury is removed by distillation.

591. Aurum or Gold is a noble metal, because it does not tarnish in air, and because of its small affinity for Oxygen. It is of a bright yellow colour, very lustrous, and the most malleable of all metals. It is soft and ductile, but not very tenacious. The sp. gr. of Gold is 19·34, and its melting-point 1102°·2 C. Gold is a triad in its chief combination, AURIC CHLORIDE $AuCl_3$.

The great solvent of Gold is Chlorine, with which it forms AURIC CHLORIDE $AuCl_3$. None of the ordinary mineral acids attack it, and the alkalies may be fused in it without injury. Its great solvent is *aqua regia*, which, indeed, is so-called because it attacks the King of the Metals; AURIC CHLORIDE $AuCl_3$ results.

592. Gold is much employed for various purposes. It is too soft to be used when unalloyed, hence in Gold coin it is with us in England united with 8·33 per cent. of Copper. Wrought Gold is said to be of so many "Carats"; pure Gold being 24 Carat. Gold-leaf is an illustration of the malleability of Gold, when it is considered that a single leaf is only $\frac{1}{250,000}$ of an inch in thickness. Gold-foil is much used in Dentistry, and the metal for communicating the ruby-red colour to Bohemian Glass.

593. To obtain pure Gold, several processes may be adopted. Advantage may be taken of the fact, that Auric chloride is readily reduced to the metallic state by means of a solution of FERROUS SULPHATE $FeSO_4$. $6FeSO_4 + 2AuCl_3 = 2Fe_2 3SO_4 + Fe_2Cl_6 + 2Au$. The metal is precipitated as a brown powder in a state of purity.

Or a solution of ANTIMONOUS CHLORIDE $SbCl_3$, may be employed: $3SbCl_3 + 2AuCl_3 = 3SbCl_5 + 2Au$. Or, lastly, OXALIC ACID $H_2C_2O_4$, $2H_2O$, may be used as the reducing agent: $3H_2C_2O_4 + 2AuCl_3 = 6HCl + 6CO_2 + 2Au$.

594. The ALLOYS of Gold, except for the purpose of hardening, are unimportant. An amalgam of Gold Au_8Hg, may be obtained in crystals.

595. The separation of Gold from Silver is often accomplished by the *Quartation-process*. It depends upon the fact that Nitric acid easily attacks an alloy containing *one-fourth* of Gold; it dissolves the Silver, and leaves the Gold as a brown powder, which cau be fused under Borax after proper washing with water. In an *assay* of Gold, the sample is weighed and wrapped up in a piece of paper with about three times the quantity of Silver as the Gold is supposed to contain. This is then treated exactly as in the assay of Silver (p. 288, par. 559). The Lead and Cupric oxides are absorbed by the Cupel, and a combination of Silver-Gold is obtained, which, after proper treatment, is submitted to the action of Nitric acid of sp. gr. 1·18. The Silver is dissolved as Argentum nitrate, and the Gold remains in the form of a brown mass. After careful washing, it is heated in a crucible, half-fused and weighed.

596. Gold has TWO oxides, the higher playing the part of the electro-negative element in salts, quite as much as the electro-positive. 1. AUROUS OXIDE $Au_2O=409{\cdot}2$. A dark-green powder, obtained by treating Aurous chloride with Potassium hydrate. 2. AURIC OXIDE Au_2O_3, a brown powder. It is best obtained by precipitating Auric chloride with Magnesia, washing and dissolving out the excess of Magnesia by means of Nitric acid, again washing and drying. At 245°, it is resolved into Oxygen and Gold. With the alkalies, auric hydrate forms salts. POTASSIUM AURATE $KAuO_2,3H_2O$ crystallizes in yellow needles. With Ammonia, it forms a fulminating powder, in the same manner as it does with Silver-oxide.

597. Gold sulphide appears to be an AUROUS-AURIC SULPHIDE Au_2S,Au_2S_3. It is precipitated from Auric chloride by Hydrogen sulphide. It is *soluble* in Ammonium sulphide.

598. AUROUS CHLORIDE $AuCl=232{\cdot}1$. A pale-yellow powder, very unstable. It is best obtained by heating Auric chloride to about 175° C. $AuCl_3=Cl_2+AuCl$. AURIC CHLORIDE $AuCl_3=303{\cdot}1$. In dark-yellow crystals, or as a red deliquescent mass. It is obtained by dissolving Gold in Aqua regia, and careful evaporation. It behaves like an acid. With Potassium chloride it forms crystals of

POTASSIUM AURIC CHLORIDE $2(KCl,AuCl_3),5H_2O$; also with Hydrogen chloride, and Sodium chloride. Auric chloride is readily reduced by many metals, organic matter, organic acids, &c., &c. Also by Phosphorus and hypo-phosphites. [AURIC BROMIDE $AuBr_3$. AUROUS IODIDE AuI. AURIC IODIDE AuI_3.] AUROUS CYANIDE AuCy=222·6. A lemon-yellow powder, soluble in Potassium cyanide, forming POTASSIUM AUROUS CYANIDE KCy,AuCy. It is much used in ELECTRO-GILDING. It may be obtained from Auric chloride by careful precipitation with Potassium cyanide, and dissolving the Aurous cyanide in excess of Potassium cyanide.

599. TESTS FOR GOLD. The salts of Gold are yellow or red. They stain the skin purple. All are readily reduced by heat. Even in an acid solution, Hydrogen sulphide gives a brown precipitate, *soluble* in Ammonium sulphide. Ferrous sulphate reduces them to the metallic state. On the addition of a mixture of Stannous-Stannic chloride *Purple of Cassius* $SnAu_2Sn_2O_6,4H_2O$ is precipitated. Gold is estimated in the form of metal.

XLVIII. Bismuth Bi = 210. Bi_4 = 840.

600. 48.—BISMUTH Bi=210. A metal of reddish hue, possessed of considerable lustre. It is very hard and brittle, and may be easily pounded in a mortar. Its sp. gr. is 9·799 and it fuses at 264° C.

Bismuth crystallizes in forms of extraordinary beauty, due not more to the regularity of the crystals, than to the variety of colour diffused over their surface. Large hollow cubes, coloured with yellow, blue and green, in many shades, gave rise to the name "Wiesmatte" or blooming meadow; whence the name of Bismuth.

Bismuth readily oxydizes at a red-heat to BISMUTH OXIDE Bi_2O_3. In a state of powder, it burns in Chlorine gas to BISMUTH CHLORIDE $BiCl_3$. It is a triad. Nitric acid dissolves it with vehemence as BISMUTH NITRATE $Bi3NO_3$. Hydrochloric acid has no action upon Bismuth. Sulphuric acid attacks it when boiled with it.

601. The metal is somewhat rare. It occurs almost always native, diffused through Quartz, and is met with chiefly in Transyl-

vania and Saxony. The broken ore is deposited in iron tubes, *A*, laid in a furnace on an incline, so as to allow the molten metal to flow into the iron basius *b*, placed to receive it (Fig. 96).

Fig. 96.

602. Bismuth is used in medicine, and somewhat in the arts. An alloy of $PbBi_2Sn_2$ is known as Fusible metal, which melts below 100° C. This alloy expands on cooling.

603. Bismuth has several oxides, but only one basic oxide at all well-known. 1. BISMUTHOUS OXIDE BiO. Is a velvet-black powder, which burns into Bismuth oxide when heated in air. 2. BISMUTH OXIDE $Bi_2O_3=468$. Is a yellow powder, insoluble in water and fusible at a red-heat. Its HYDRATE or OXY-HYDRATE $BiHO_2$ or Bi_2O_3,H_2O, is white, and precipitated readily from the nitrate by means of Ammonia. 3. BISMUTHIC ANHYDRIDE Bi_2O_5, is brown, and may be obtained by heating BISMUTHIC ACID $HBiO_3$ to 130° C. BISMUTHIC ACID $HBiO_3$, is a red powder. This is easily obtained, by passing Chlorine through a solution of Potassium hydrate containing Bismuth oxide in suspension. $Bi_2O_3+2Cl_2+4KHO=4KCl+H_2O+2HBiO_3$.

604. BISMUTH SULPHIDE $Bi_2S_3=516$. Is found native as "Bismuth Glance" in needles. As a brown powder when dried, it may be obtained by precipitating Bismuth nitrate with Hydrogen sulphide.

605. BISMUTH SALTS. BISMUTH CHLORIDE $BiCl_3=316{\cdot}5$. A white, deliquescent substance. Although soluble in Hydrochloric

acid, it is decomposed by water into a white OXY-CHLORIDE BiOCl. Thus; $BiCl_3+H_2O=2HCl+BiOCl$. "Pearl-white," so much used as a Cosmetic, is $2BiOCl,H_2O$: it is not soluble in Tartaric acid. BISMUTH IODIDE BiI_3, in brilliant scales. BISMUTH NITRATE $Bi3NO_3, 5H_2O=486$. In large transparent, colourless prisms, obtained by dissolving Bismuth in nitric acid. Water decomposes the solution, with separation of the so-called "Magistery of Bismuth"; it is BISMUTH BASIC NITRATE $Bi_2O_3,2HNO_3$.

606. TESTS The salts of Bismuth are colourless, except in presence of a coloured acid They have a strong acid reaction. Their solutions are decomposed by water, and the precipitate is greatly increased on addition of Tartaric acid. Even in acid solutions, Bismuth sulphide is precipitated with brown colour. The alkalies give a white precipitate, insoluble in excess. Sodium carbonate, a white precipitate. Potassium chromate, a bright-yellow precipitate of Bismuth chromate, insoluble in excess. Copper, Zinc and Iron precipitate the metal, and the salts are easily reduced in the blowpipe-flame. Bismuth oxide, with yellow colour, is deposited upon the charcoal around the bead of metal. Bismuth is always estimated as BISMUTH OXIDE Bi_2O_3, 100 parts of which contain 89·74 per cent. of the metal.

CHAPTER XXVIII.

ON PLATINUM AND THE ASSOCIATED METALS PALLADIUM, RHODIUM, OSMIUM, RUTHENIUM AND IRIDIUM.

XLIX.—**Platinum** Pt=197. 607. A tetrad, found native, as well as in certain alloys. 608. Fusion in Oxy-hydrogen. 609. Properties. Sp. gr. 21·5. 610. Action of aqua regia upon Platinum. 611. Spongy Platinum and Platinum black. 612. Use for making crucibles. 613. Alloys. 614. Two oxides. PLATINOUS OXIDE PtO. PLATINIC OXIDE PtO_2. 615. PLATINOUS SULPHIDE PtS. PLATINIC SULPHIDE PtS_2. 616. PLATINOUS CHLORIDE $PtCl_2$. PLATINIC CHLORIDE $PtCl_4$. POTASSIUM PLATINIC CHLORIDE $2KCl, PtCl_4$. AMMONIUM PLATINIC CHLORIDE $2NH_4Cl,PtCl_4$. [617. AMMONIA-BASES. Gros's salts. DI-PLATOSAMINE $PtN_4H_{10},2H_2O$, is Reiset's first base. PLATOSAMINE PtN_2H_4,H_2O. PLATINAMINE $PtN_2H_2,4H_2O$.] 618. Tests for Platinic salts. [L. **Palladium** Pd=106·5. 619. Metal of sp. gr. 11·4 to 11·8. Tincture of iodine leaves a blue stain, when heated, upon Platinum. 620. Alloys. 621. PALLADIOUS OXIDE Pd_2O. PALLADIC OXIDE PdO. PALLÁDIC PEROXIDE PdO_2. 622. PALLADIC SULPHIDE PdS. 623. PALLADIC CHLORIDE $PdCl_2$. PALLADIC PERCHLORIDE $PdCl_4$. PALLADIC IODIDE PdI_2. PALLADIC CYANIDE $PdCy_2$. 624. Preparation of Palladium.] [LI. **Rhodium** R=104·3. 625. Properties. Sp. gr. 12·1. 626. Two oxides. RHODIUM SUB-OXIDE RO. RHODIUM OXIDE R_2O_3. 627. RHODIUM SUB-SULPHIDE RS. RHODIUM SULPHIDE R_2S_3. 628. Salts. RHODIUM CHLORIDE RCl_3. POTASSIUM RHODIUM CHLORIDE $3KCl,RCl_3$. SODIUM RHODIUM CHLORIDE $3NaCl,RCl_3,9H_2O$. 629. Separation of Rhodium.] [LII. **Osmium** Os=199. 630. Found in Iridium-Osmium. 631. Properties. Sp. gr. 21·4. 632. Preparation of the metal. 633. Five oxides. OsO. Os_2O_3. OsO_2. OsO_3. OSMIUM PEROXIDE OsO_4. Volatile. Peculiar odour. 634. Sulphides. OsS. Os_2S_3. OsS_2. OsS_3. OsS_4. 635. OSMIUM CHLORIDE $OsCl_2$. OSMIUM PERCHLORIDE $OsCl_4$. POTASSIUM OSMIC CHLORIDE $2KCl, OsCl_4$. 636. All Osmic salts evolve OsO_4, when boiled with Nitric acid.] [LIII. **Ruthenium** Ru=104·2. 637. Properties of the metal. Sp. gr. 11 to 11·4. Oxydizes to RUTHENIC OXIDE Ru_2O_3. 638. Four oxides. RuO. Ru_2O_3. RuO_2. RuO_3. RUTHENIC OXIDE Ru_2O_3. Salts yellow. Its HYDRATE $Ru_2H_6O_6$. RUTHENIUM PEROXIDE RuO_2. 639. Ruthenium chlorides. $RuCl_2$. $RuCl_3$. $RuCl_4$. RUTHENIC CHLORIDE $RuCl_3$. RUTHENIUM PEROXIDE RuO_2. 640. Preparation of RUTHENIUM.] [LIV. **Iridium** Ir=197. 641. Metal of sp. gr. 21·15. Not soluble in mineral acids. 642. Three oxides. IrO. Ir_2O_3. IrO_2. Three sulphides. IrS. Ir_2S_3 and IrS_2. 643. Three chlorides. $IrCl_2$. $IrCl_3$. $IrCl_4$. POTASSIUM-IRIDIUM CHLORIDE 2KCl,$IrCl_4$.

XLIX. Platinum Pt = 197.

607. 49.—PLATINUM Pt=197. This tetrad metal always occurs native, and is met with usually in the form of flattened grains of iron-grey lustre. It is found more especially in the Ural Mountains, in Mexico and Brazil, and, more recently, in California.

Platinum was proved to be a distinct metal in 1741. It is always associated with Palladium, Ruthenium, Iridium, Rhodium and Osmium, for which reason alone they are treated of in the same Chapter.

608. Until quite recently, all Platinum articles were made by welding the metal, at a high temperature, in a wind-furnace. For, although it was well known that Platinum was fusible under the heat of the oxy-hydrogen blowpipe-flame, the knowledge was not available for the purposes of manufacture, until M. Deville invented his form of furnace, in which the most intractable metals can be worked in a liquid state. By a jet of mixed Coal-gas and Oxygen, in a furnace lined with Lime, as much as 11·595 kilos. or 25·4lbs. of Platinum may be melted in 42 minutes: it is then cast into an ingot in a mould made of graphite.

609. Platinum is a white, lustrous metal, very ductile and tenacious. It melts at a very high temperature, and resists even the heat of the forge. Its sp. gr. is 21·5. It is an inferior conductor of heat and electricity.

610. Platinum is not acted upon by the mineral acids. Its great solvent is Aqua regia, with the Chlorine evolved from which it forms PLATINIC CHLORIDE $PtCl_4$. With Ammonium chloride NH_4Cl, Platinic chloride forms an insoluble salt, AMMONIUM PLATINIC CHLORIDE $2NH_4Cl,PtCl_4$.

611. When Ammonium Platinic chloride is ignited, pure Platinum, in the form of SPONGY PLATINUM, is left behind. It was by hammering

Spongy Platinum, that Wollaston was able to work the metal into utensils fitted for use. Spongy Platinum has the power, owing to the amount of surface which it presents, of condensing Oxygen in its pores, to such an extent, as to become red-hot when brought into Hydrogen. In the form of PLATINUM BLACK, it possesses still greater oxydizing powers; and in this form Platinum may be obtained by boiling a solution of PLATINOUS CHLORIDE $PtCl_2$ in Potassium hydrate, with alcohol.

612. Platinum is much valued by Chemists for making crucibles and evaporating dishes. Care must be taken not to employ them for melting the alkalies in them, or heating in them the Oxides of Lead, or other easily reducible metals.

613. Platinum forms ALLOYS with many of the metals. With Phosphorus and Silicon it forms very brittle compounds. It also unites readily with Carbon and Boron.

614. Platinum has TWO oxides. PLATINOUS OXIDE $PtO = 213$. Can be obtained in the form of a black hydrate, by dissolving Platinous oxide in Potassium hydrate, and neutralizing the green solution by means of Sulphuric acid. Its salts are most unstable. PLATINIC OXIDE $PtO_2 = 229$. Is very dark-brown, almost as ready to play the part of an acid, in its HYDRATED state PtH_2O_2,H_2O, as that of a base. Owing to this property, it cannot well be obtained from Platinic chloride by means of Potassium hydrate, but by the addition of Sodium carbonate, in quantity far from sufficient to precipitate the whole. The hydrate is brown. At a comparatively low temperature, Platinic oxide loses its oxygen, as, indeed, do all the oxides of the noble metals.

615. There are TWO sulphides. PLATINOUS SULPHIDE PtS, is black. PLATINIC SULPHIDE PtS_2, is dark-brown, and somewhat soluble in Ammonium sulphide. It is obtained by precipitating a Platinic salt with Hydrogen sulphide.

616. Platinum has also TWO chlorides. PLATINOUS CHLORIDE $PtCl_2 = 268$. An olive-brown salt, insoluble in water, is obtained by carefully heating Platinic chloride $PtCl_4$ to 235° C. PLATINIC CHLORIDE $PtCl_4 = 339$. A salt of deep-red colour, highly deliquescent. If any Platinous chloride be present, the solution is brown; otherwise orange-red. By heat it is changed, first into Platinous chloride,

and then into Platinum and Chlorine. It forms double salts. POTASSIUM-PLATINIC CHLORIDE $2KCl,PtCl_4$, a yellow salt, very sparingly soluble in water, and still has in a mixture of ethylic alcohol and ether. It is used for estimating the quantity of Potassium in a compound. 100 parts contain 16·02 per cent. of Potassium. AMMONIUM-PLATINIC CHLORIDE $2NH_4Cl,PtCl_4$. Also a yellow salt, nearly insoluble. Similarly employed to estimate Ammonia quantitatively. 100 parts of the salt contain 7·62 per cent. of NH_3. PLATINIC IODIDE PtI_4. The solution of Platinic chloride becomes deep claret-red, on addition of Potassium iodide.

[617. AMMONIA-BASES. Platinous chloride forms bases with ammonia, in which a part of the Hydrogen is displaced by Ammonia. When Platinous chloride is dissolved in Hydrogen chloride and Ammonia is added, brilliant green needles are obtained of a compound $2PtCl_2,N_4H_{12}$. On addition of Nitric acid, a new salt, in flattened prisms, is obtained, $PtCl_2,N_4H_{10},2HNO_3$: in this salt, neither the Platinum nor the Chlorine can be detected by the ordinary tests. These salts, and the others of a similar kind, are known as *Gros' salts*. The base of these salts has not yet been isolated. DI-PLATOSAMINE $PtN_4H_{10},2H_2O$. An alkaline base, obtainable in needles. Absorbs Carbonic anhydride. Can be separated from the sulphate by means of Barium hydrate. DI-PLATOSAMINE HYDROCHLORATE $PtN_4H_{10},2HCl,H_2O$, is made by dissolving Platinous chloride in warm Ammonic hydrate and evaporation. PLATOSAMINE PtN_2H_4,H_2O, a grey compound, insoluble in water. It is made by heating Di-platosamine carefully at 110° C., as long as Water and Ammonia are evolved. Also called *Reiset's second base*. When Di-platosamine hydrochlorate is carefully heated, Water and Ammonia are liberated, and PLATOSAMINE DI-HYDROCHLORATE $PtN_2H_4,2HCl$ is obtained. PLATINAMINE $PtN_2H_2,4H_2O$, is *Gerhardt's* base. In brilliant rhombohedral prisms of yellow colour. Is obtained from PLATINAMINE NITRATE $PtN_2H_2,4HNO_3$, by Ammonia.]

618. TESTS for PLATINIC salts. The solutions are yellow or deep orange-red, with an acid reaction. They are decomposed by heat, and leave a residue of Platinum. Even in acid solutions, Hydrogen sulphide gives a brown-black precipitate of Platinic sulphide, soluble in a large quantity of Ammonium sulphide. Potassium hydrate, a

brown hydrate, soluble in excess. Or, if free Hydrochloric acid be present, a yellow precipitate of Potassium Platinic chloride. Ammonia, also. Potassium iodide, a deep-claret colour, with brown precipitate of Platinic iodide.

Platinum may be estimated either as PLATINUM, or as AMMONIUM PLATINIC CHLORIDE $2NH_4Cl,PtCl_4$, which contains 44·17 per cent. of Platinum.

[L. Palladium Pd = 106·5.

619. 50.—PALLADIUM Pd=106·5. Was discovered by Wollaston in 1803, in Platinum-ores. It is a white lustrous metal, very ductile and tenacious. It is found native as well in cubes as in 6-sided scales. It is very difficult of fusion, except in Oxy-hydrogen. Sp. gr. from 11·4 to 11·8. It undergoes no change on exposure to air. It is soluble in nitric acid. If Tincture of Iodine be evaporated upon Palladium, it leaves a blue stain, and may thus be distinguished from Platinum. In its chief combination, Palladium is a Dyad.

620. Palladium forms a white ALLOY with Gold. It has a great affinity for Carbon, and for Hydrogen. The late Prof. GRAHAM has made an alloy of Hydrogen with Palladium, containing many hundred times its volume of Hydrogen. That Hydrogen is therefore to be considered rather as a metal, than as a non-metallic element, has already been mentioned under the heading Hydrogen.

621. Palladium has three oxides. PALLADIOUS OXIDE Pd_2O, forms salts resembling those of Cupreous oxide. PALLADIC OXIDE PdO. A black powder, obtained by heating the nitrate to low redness; more strongly heated, it becomes Palladious oxide. PALLADIC PEROXIDE PdO_2. Is also black. It forms a double salt with Potassium chloride $2KCl,PdCl_4$, which, by treatment with Potassium hydrate, yields Palladium peroxide.

622. PALLADIC SULPHIDE PdS, is black, and precipitated from Palladic salts, even in acid solutions, by Hydrogen sulphide.

623. SALTS. PALLADIC CHLORIDE $PdCl_2$=177·5. Black when anhydrous, otherwise brown. It is obtained by evaporating PALLA-

DIC PERCHLORIDE $PdCl_4$, which is obtained by dissolving Palladium in Aqua regia, to dryness. PALLADIC IODIDE $PdI_2=360{\cdot}5$. Is black and insoluble. A solution of Palladic nitrate may be employed, to separate an iodide from a chloride and bromide. PALLADIC CYANIDE $PdCy_2$, is yellow. It is insoluble in water, and readily separated from neutral Palladic salts by Potassium cyanide.

624. Palladic cyanide is of the greatest importance. When the Platinic ores have been treated with Aqua regia, and the Platinum has been separated by means of Ammonium chloride, the Palladium is precipitated by the careful addition of Mercuric Cyanide. When the native alloy of Gold and Palladium is employed, it is first fused with Silver in sufficient quantity, and then treated with Nitric acid. Palladic and Silver nitrate dissolve, whilst the Gold remains. The Silver may be separated by means of Sodium chloride. If, as is usually the case, iron, copper, and lead are associated, Ammonia in excess is added, which leaves Ferric and Lead oxides undissolved, and dissolves Cupric and Palladic oxides. On the addition of Hydrochloric acid, a yellow salt of Palladium, PALLADAMINE HYDROCHLORATE $PdN_2H_4,2HCl$, is precipitated, which on heating leaves Palladium.]

[LI. Rhodium R = 104·3.

625. 51.—RHODIUM $R=104{\cdot}3$. A hard, white metal, of sp. gr. 12·1. It was discovered by Wollaston in 1803, constituting about one-half per cent. of Platinum ores. When pure, Rhodium is not soluble in the mineral acids, not even in Aqua regia: but in its alloys with Platinum and other metals, it is readily dissolved by aqua regia. It may be oxydized by fusion with Potassium nitrate and carbonate. The salts of Rhodium are rose-coloured, whence the name, from ῥόδον, *rhodon*, a rose.

626. Rhodium forms TWO oxides. RHODIUM SUBOXIDE RO. Is but little known. RHODIUM OXIDE $R_2O_3=256{\cdot}6$. Its hydrate is grey. The oxide forms with Potassium oxide an insoluble compound, which is decomposed by Hydrochloric acid.

627. RHODIUM SUB-SULPHIDE RS; bluish-grey. By heat into

metallic rhodium. RHODIUM SULPHIDE R_2S_3. Hydrate, brown. It is slowly formed, when Hydrogen sulphide is added to a solution of Rhodic chloride.

628. SALTS. RHODIUM CHLORIDE RCl_3. Is obtained from POTASSIUM RHODIUM CHLORIDE $3KCl,RCl_3$, by means of Hydrogen Silico-fluoride, and evaporation. It is rose-red and forms important double salts with the alkaline chlorides. SODIUM RHODIUM CHLORIDE $3NaCl,RCl_3,9H_2O$, in cubes, *insoluble in ethylic alcohol* of sp. gr. 0·837.

629. The separation of Rhodium depends upon the fact, that Sodium rhodium chloride is insoluble in ethylic alcohol. After the Palladium has been removed from the solution, as described in par. 630, Hydrogen chloride is added, together with Sodium chloride, and the mixture evaporated to dryness. Ethylic alcohol of sp. gr. 0·837 re-dissolves all the double chlorides, except that of Rhodium.]

[LII. Osmium Os = 199.

630. 52.—OSMIUM Os=199. Was discovered by Tennant in 1803 in Platinum ores. It is always associated with Iridium, Rhodium, and Ruthenium, and is found more especially in flattened, metallic scales, insoluble even in Aqua regia, better known as Osmium-Iridium.

631. Osmium is a bluish metal, of sp. gr. 21·4. It is the most infusible of all metals. When roasted in air, it oxydizes and volatilizes as OSMIUM PEROXIDE OsO_4. From the irritating vapour thus produced, OSMIUM derives its name (ὀσμή, *osme*, odour).

632. Osmium is best prepared from Osmium peroxide, by treatment with Mercury and Hydrogen chloride. $OsO_4+8Hg+8HCl=4H_2O+4Hg_2Cl_2+\mathbf{Os}$.

633. Osmium has FIVE oxides. OsO. Os_2O_3. OsO_2. OsO_3. OSMIUM PEROXIDE OsO_4=263. Colourless needles, very soluble in water and fusible. It stains the skin black. Its vapours are very poisonous and irritating. It does not unite with acids.

634. There are FIVE sulphides. OsS. Os_2S_3. OsS_2. OsS_3.

OsS_4. OSMIUM PERSULPHIDE OsS_4, is precipitated black by Hydrogen sulphide from a solution of the OsO_4, in water.

635. The best known CHLORIDES are: OSMIUM CHLORIDE $OsCl_2$. Is green, and obtained by heating Osmium in Chlorine gas. It sublimes in green needles. OSMIUM PERCHLORIDE $OsCl_4$. A red powder. It is formed by treating Osmium or Osmium chloride, with excess of Chlorine. POTASSIUM OSMIC CHLORIDE $2KCl,OsCl_4$, a red salt sparingly soluble in water.

636. All Osmium salts evolve vapours of Osmium peroxide when boiled with nitric acid.]

[LIII. Ruthenium Ru = 104·2.

637. 53.—RUTHENIUM Ru=104·2. Was discovered by Claus in 1845 in the so-called "Osmium-Iridium." It is a hard and brittle metal, of sp. gr. 11· to 11·4. It is almost as infusible as Osmium. At a red-heat it oxydizes and forms RUTHENIC OXIDE Ru_2O_3.

638. Ruthenium has FOUR oxides. RuO. Ru_2O_3. RuO_2. RuO_3. RUTHENIC OXIDE Ru_2O_3, forms yellow salts. It is the most stable. Its HYDRATE $Ru_2H_6O_6$, is blackish-brown. RUTHENIUM PEROXIDE RuO_2, is not volatile by itself.

639. RUTHENIUM CHLORIDES are THREE in number. $RuCl_2$. $RuCl_3$. $RuCl_4$. RUTHENIC CHLORIDE $RuCl_3$, is greenish-blue and deliquescent. Metallic zinc reduces $RuCl_3$ to $RuCl_2$, which is blue, and then to the metallic state.

640. Osmium and Ruthenium are separated from the ore containing them, by carefully roasting the so-called Iridium-Osmium in a stream of air. The Osmium is oxydized as OSMIUM PEROXIDE OsO_4, which volatilizes and carries with it mechanically RUTHENIUM PEROXIDE RuO_2, which is formed at the same time. Ruthenium peroxide is deposited in the form of square prisms.

Ruthenium peroxide is easily reduced by heat in a stream of Hydrogen.]

[LIV. Iridium Ir = 197.

641. 54.—IRIDIUM IR=197. Was discovered by Tennant. It is sometimes found native, but more usually as an alloy with Osmium. It is a hard, white, brittle metal, fusible in Oxy-hydrogen. Sp. gr. 21·15. In a finely-divided state, Iridium undergoes oxydation when heated.

Iridium is not soluble in the mineral acids, nor even in aqua regia, unless alloyed. From the facility with which the oxides pass into one another, giving rise to various rainbow-tints, (*Iris*), the name Iridium was given.

642. Iridium has THREE oxides. IrO. Ir_2O_3. IrO_2. And THREE sulphides IrS, Ir_2S_3, and IrS_2.

643. The CHLORIDES are also three in number. $IrCl_2$, $IrCl_3$ and $IrCl_4$. POTASSIUM-IRIDIUM CHLORIDE $2KCl,IrCl_4$ is reddish-black. The metal Iridium is best prepared by heating this salt in a stream of Hydrogen.]

CHAPTER XXIX.

THE CHEMISTRY OF THE TETRAD METAL TIN, AND THE TRIAD ANTIMONY.

LV.—**Stannum** or **Tin** Sn=118. 644. Properties of the metal. Sp. gr. 7·292. Melts at 228° C. 645. Tin-foil. Tin-plate. 646. Moirée métallique. 647. Alloys of tin. Solder. Pewter. Speculum-metal. Bronze. Gun-metal. Amalgams. 648. Gilt-paper. 649. Tin known from the earliest times. 650. Tinstone or Stannic oxide SnO_2, the ore of tin. 651. Mode of preparing the ore at Redruth and of smelting. 652. Three oxides. STANNOUS OXIDE SnO. HYDRATE $2SnO,H_2O$. STANNIC OXIDE SnO_2. Two acids. STANNIC ACID H_2SnO_3. METASTANNIC ACID $H_2Sn_5O_{11},4H_2O$. STANNOUS-STANNIC OXIDE SnO,SnO_2. Purple of Cassius $SnAu_2Sn_2O_9,4H_2O$. 653. Sulphides. STANNOUS SULPHIDE SnS. STANNIC SULPHIDE SnS_2, Mosaic Gold. STANNOUS STANNIC SULPHIDE SnS,SnS_2. 654. Salts of tin. STANNOUS CHLORIDE $SnCl_2,2H_2O$. OXY-CHLORIDE $SnO,SnCl_2,2H_2O$. STANNIC CHLORIDE $SnCl_4,5H_2O$. 655. Tests for stannous and stannic salts. LVI. **Stibium** or **Antimony** Sb=122. 656. Properties. Sp. gr. 6·71. Melts at 621° C. Action of acids. 657. Alloys of antimony. Type-metal. Britannia metal. Pewter. 658. Antimony-preparations used in medicine. TARTAR EMETIC $[2(KSbC_4H_4O_7)H_2O]$. 659. Grey-antimony-ore is an ANTIMONOUS SULPHIDE Sb_2S_3. The ore of antimony. Metallurgy. 660. ANTIMONIETTED HYDROGEN SbH_3. 661. Three oxides. ANTIMONOUS OXIDE Sb_2O_3, the only basic oxide. Native as White antimony-ore. ANTIMONIC ANHYDRIDE Sb_2O_5. ANTIMONOUS ANTIMONIC OXIDE Sb_2O_4. ORTH-ANTIMONIC ACID H_3SbO_4. MET-ANTIMONIC ACID $HSbO_3$. PYR-ANTIMONIC ACID $H_4Sb_2O_7$. POTASSIUM DI-HYDROGEN PYRANTIMONATE $K_2H_2Sb_2O_7,6H_2O$ precipitates SODIUM DI-HYDROGEN PYR-ANTIMONATE $Na_2H_2Sb_2O_7,6H_2O$. 662. ANTIMONOUS SULPHIDE Sb_2S_3. Native. Red antimony ore is an OXY-DI-SULPHIDE $SbOS_2$. ANTIMONIC SULPHIDE Sb_2S_5. 663. ANTIMONOUS CHLORIDE $SbCl_3$. OXY-CHLORIDE SbOCl. ANTIMONIC CHLORIDE $SbCl_5$. ANTIMONIC OXY-CHLORIDE $SbOCl_3$, by more water into: ORTH-ANTIMONIC ACID H_3SbO_4. 664. Tests. 665. Antimony weighed as Sb_2O_4. Separation of antimony by tin.

LV. Stannum or Tin Sn = 118.

644. 55.—STANNUM Sn=118. This metal is never found native. It is somewhat soft, malleable and lustrous, very white, with considerable ductility above 100° C, but little tenacity. Its sp. gr. is 7·292; it melts at 228° C. When tin is bent, it emits a cracking sound, and, like Copper, communicates a peculiar odour to the damp hand. On exposure to damp air, it tarnishes but slowly; but at a red-heat, it rapidly burns to STANNIC OXIDE SnO_2. In its chief combination, Tin is a Tetrad.

Hydrochloric acid dissolves the metal as STANNOUS CHLORIDE $SnCl_2$, with evolution of Hydrogen. Nitric acid converts tin into a white, insoluble HYDRATE of META-STANNIC ACID $H_2Sn_5O_{11},4H_2O$. When boiled with Sulphuric acid, STANNIC SULPHATE $Sn2SO_4$, is obtained. Aqua regia dissolves the metal as STANNIC CHLORIDE $SnCl_4$.

645. Tin-foil is laminated tin. Owing to its lustre, and the slight action on it both of water and air, Tin is largely employed as a coating to articles of Copper and Iron, and for culinary purposes. *Tin-plate* is thin sheet-iron coated with Tin. It is prepared by dipping the Iron, when its surfaces are perfectly clean, into a bath of molten Tin: when the plate is withdrawn, the excess of Tin is detached by a sharp blow.

646. The beautiful appearance of Tin-plate, known as *Moirée Métallique*, is produced by bringing out the crystalline character of the metal. Tin-plate is washed over with a mixture of equal parts of Nitric and Hydrochloric acids, diluted with their own bulk of water; as soon as the crystals appear, the plate is washed, dried, and varnished.

647. The ALLOYS of Tin are very important. *Solder*, an alloy of Tin and Lead, has been already mentioned. *Pewter* is an alloy of 4 parts of Tin with 1 part of Lead. *Speculum-metal* Cu_4Sn, is employed for the mirrors of telescopes. *Bell-metal* Cu_6Sn; *Bronze* contains less Tin, and about 4 per cent. of Lead. The Bronze in coin consists of 95 parts of Copper, 4 of Tin, and 1 part of Zinc. *Gun-metal* contains only half as much Tin as Bell-metal. An amal-

gam of Tin and Mercury is used for silvering Mirrors (p. 298, par. 578).

648. In the manufacture of so-called *Gilt paper*, no Gold is used. The paper is covered with Tin-foil, upon which a transparent lacquer, coloured yellow with turmeric, is afterwards spread.

649. Tin has been known and employed from very early times. Pliny relates that the Phœnicians were in the habit of trading both with Spain and Great Britain for Tin. In this country, it still occurs abundantly in Cornwall; and it is found in considerable quantities in Malacca, India and Mexico. The Alchemists worked incessantly upon Tin, in the hopes of being able to transmute it into Silver. They imagined it to be under the influence of Jupiter, and symbolized it accordingly by the name, and sign ♃ of that planet.

650. Its only important ore is TIN-STONE or STANNIC OXIDE SnO_2, which is found occasionally crystallized in square prisms, possessed of high lustre, and hard enough to cut glass. More generally it occurs in veins, diffused through primitive rocks of granite and porphyry. The principal Tin-mines of Cornwall are in the neighbourhood of St. Anstell and the Land's End. The Tin-ore obtained in alluvial soils, where it has been washed from its bed by running water, is called Stream-tin. It is purer, and more free from arsenic than when found in veins.

651. The largest Stream-tin works are at Carnon, near Redruth. The mode practised there of reducing the ore is briefly as follows. The ore is pounded, or stamped on the spot, in the stamping-mill, which consists of heavy beams of wood, shod with masses of iron, weighing 1½ cwt. each. These beams are raised by means of water-wheels, and fall by their weight on the ore, which is placed in troughs, through which a small stream of water runs, in order to wash away the ore as it is pounded fine. One end of the troughs is formed of iron plates pierced with holes, through which the ore is sifted as it were, and passes into other troughs or pits, in which it settles from the water in various degrees of fineness.

The ore thus pounded is washed in vats called Keeves, to free it from all soluble earthy matters, and from dirt, previous to being smelted. It is now ready for reduction. It is mixed with $\frac{1}{6}$ of its weight of charcoal, and with a small quantity of Lime, and heated carefully in a reverberatory furnace. The Carbon unites with the Oxygen of the Stannic oxide to Carbonic anhydride, and the metal Tin is separated. Then the temperature is raised, and the Silex of the ore forms with the Calcium oxide a fusible Calcium ortho-silicate, through which the Tin falls and collects at the bottom.

The metal is cast into ingots and submitted to a process of liquation. When brought to incipient fusion, the purer Tin melts first, and is collected apart from the less fusible and more impure Block Tin.

652. Tin forms THREE oxides. 1. STANNOUS OXIDE $SnO = 134$. Is black; its HYDRATE $2SnO,H_2O$, white. This latter is soluble in Potassium hydrate, so that it is better to prepare it by adding Sodium carbonate to a solution of Stannous chloride. By boiling with Potassium hydrate, Stannous hydrate loses water, and becomes black. When heated to redness, it absorbs Oxygen, and burns like tinder to Stannic oxide. It is a basic oxide. 2. STANNIC OXIDE $SnO_2 = 150$. Is pale buff-coloured, insoluble in acids. It constitutes the Tinstone of Mineralogists. Stannic oxide is used for producing enamels, and is the so-called *Putty-powder* of Plate-polishers. With water, it forms two acids. *a.* STANNIC ACID H_2SnO_3. A white, gelatinous precipitate, obtained on addition of Ammonia to Stannic chloride. It is very soluble in the alkalies, and forms Stannates. POTASSIUM STANNATE $K_2SnO_3,3H_2O$, crystallizes in oblique, rhombic prisms. SODIUM STANNATE $Na_2SnO_3,3H_2O$, in 6-sided tables. A solution of this salt is much used for "tinning" copper. When Stannic hydrate is heated to 140° C., it is converted into metastannic acid. *b.* METASTANNIC ACID $H_2Sn_5O_{11},4H_2O$, when dried at 100° C. The air-dried contains $9H_2O$. It is formed by the action of Nitric acid upon tin. When moistened with Stannous chloride, it becomes golden-yellow, owing to the formation of STANNOUS METASTANNATE $SnSn_5O_{11},4H_2O$. Metastannates do not crystallize. 3. STANNOUS STANNIC OXIDE SnO,SnO_2, as a hydrate, is white and soluble in ammonia. It is obtained by boiling Ferric hydrate with Stannous chloride. $Fe_2H_6O_6 + 2SnCl_2 = 2FeCl_2 + 2H_2O + SnO,SnO_2,H_2O$. Stannous Stannic chloride is a test for Gold, owing to the production of *Purple of Cassius* $SnAu_2Sn_2O_6,4H_2O$.

653. There are THREE Sulphides. 1. STANNOUS SULPHIDE

$SnS=150$, by fusing Tin with Sulphur. As a brown hydrate, by precipitating Stannous chloride with Hydrogen sulphide. 2. STANNIC SULPHIDE SnS_2, is yellow, and well known as *Mosaic Gold.* It is used to imitate Bronze. As a hydrate of a dirty-yellow colour, Stannic sulphide is precipitated from solutions of Stannic chloride by Hydrogen sulphide. It is soluble in Ammonium sulphide, and forms Sulphur salts with all the sulphides of the alkaline metals. 3. STANNOUS-STANNIC SULPHIDE SnS,SnS_2.

654. SALTS OF TIN. The only salts of importance are the respective chlorides. STANNOUS CHLORIDE $SnCl_2,2H_2O$, in prisms. Its solution is strongly acid and decomposed by water into STANNOUS-OXY-CHLORIDE $SnO,SnCl_2,2H_2O$. Thus:—$2(SnCl_2,2H_2O)+H_2O=2HCl+SnO,SnCl_2,2H_2O+2H_2O$. It is a great reducing agent. It reduces Mercuric chloride to Mercurous chloride: $SnCl_2+2HgCl_2=SnCl_4+Hg_2Cl_2$. And Mercurous chloride to Mercury: $SnCl_2+Hg_2Cl_2=SnCl_4+2Hg$. *Butter of Tin* $SnCl_2$, is obtained by distilling tin filings with Mercuric chloride. STANNIC CHLORIDE $SnCl_4,5H_2O$. In Rhomboids, soluble in Hydrochloric acid. Its solution is easily obtained by dissolving Tin, or Stannous chloride, in weak Aqua regia. It is decomposed by much water into Hydrochloric and Stannic acids.

655. TESTS FOR TIN. In the blow-pipe flame, Tin salts are reduced to the metallic state. 1. STANNOUS. Colourless, with strong acid reaction. They become milky on exposure to air, owing to the formation of Oxy-chloride, and they are decomposed by much water. In acid solutions, Hydrogen sulphide precipitates brown Stannous sulphide, soluble as Stannic sulphide in Ammonium di-sulphide, or in Ammonium hydrogen sulphide to which Flowers of Sulphur have been added. Sodium carbonate and Ammonic hydrate, a white precipitate of Stannous hydrate. Potassium hydrate, a white precipitate, very soluble in excess. Mercuric chloride is reduced first to Mercurous chloride, and then into minute grey globules of Mercury. Auric chloride, to a solution of Stannous Stannic chloride, gives the Purple of Cassius. 2. STANNIC. The solutions are colourless and strongly acid. In acid solutions, Hydrogen sulphide gives a yellow precipitate soluble in Ammonium sulphide. Sodium

carbonate, a white precipitate of hydrated stannic acid. Potassium hydrate, a white precipitate soluble in excess. Ammonia, likewise a white precipitate.

Tin is always weighed as STANNIC OXIDE SnO_2, 100 parts of which contain 78·66 of the metal.

LVI. Antimony or Stibium Sb=122. Sb_4=488.

656. 56.—ANTIMONY Sb=122. Sb_4 =488=2 vols. This triad element is but rarely found native. It is a very brittle, lustrous, bluish-white metal, crystalline, and of sp. gr, 6·71. It melts at 621° C. In air it undergoes no change, but when strongly heated, it burns into ANTIMONOUS OXIDE Sb_2O_3. In the state of powder, it takes fire in Chlorine, and burns to ANTIMONIC CHLORIDE $SbCl_5$; Hydrochloric acid also slowly dissolves it as ANTIMONOUS CHLORIDE $SbCl_3$, with liberation of Hydrogen. When heated with Sulphuric acid, Sulphurous anhydride is evolved, and ANTIMONOUS SULPHATE Sb_23SO_4, is formed. $2Sb + 6H_2SO_4 = 3SO_2 + 6H_2O + Sb_23SO_4$. Nitric acid does not dissolve it, but oxydizes and converts it into MET-ANTIMONIC ACID $HSbO_3$.

657. Antimony expands at the moment of its congelation, hence are some of its ALLOYS of great value. *Type-metal* consists of 1 part of Antimony with 1 part of Tin and 2 parts of Lead. *Britannia metal* consists of equal parts of Antimony, Tin, Lead, Bismuth, and Brass. Some kinds of *Pewter* contain Antimony.

658. Antimony in various forms is used in medicine. TARTAR EMETIC $[2(KSbC_4H_4O_7)H_2O]$, is the most important salt. It crystallizes in square prisms, soluble in 15 parts of water.

659. The chief ORE of Antimony is *Grey Antimony ore*, a SULPH-ANTIMONOUS ANHYDRIDE Sb_2S_3, found in Granite and Slate. It has a strong metallic lustre, and crystallizes in four-sided prisms. It fuses at a low red-heat, and can thus be easily melted out of the Gangue with which it is mixed.

Sulph-antimonous anhydride is readily soluble in alkaline Sulphides. When roasted, the greater portion oxydizes as Antimonou oxide Sb_2O_3, and forms a fusible mixture with the remainder, known as *Glass of Antimony*.

If, before this fusion takes place, the roasted mass be mixed with Charcoal, made into a paste with Sodium carbonate and heated to redness in crucibles, the metal ANTIMONY is obtained. The object of the Sodium carbonate is to convert sulph-antimonous anhydride into oxide, which latter is reduced by Carbon. *a.* $Sb_2S_3 + 3Na_2CO_3 = 3Na_2S + 3CO_2 + Sb_2O_3$. *b.* $Sb_2O_3 + 3C = 3CO + 2Sb$.

660. Antimony unites with Hydrogen, to form ANTIMONIETTED HYDROGEN $SbH_3 = 125$. It is a gas, burning with greenish flame to Water and ANTIMONOUS OXIDE Sb_2O_3. If heated in absence of air, it is reduced into its elements, a crust of metallic antimony being deposited. This crust is easily soluble in Ammonium sulphide. If the gas be passed through a solution of Argentum nitrate, Silver antimonide is deposited: $SbH_3 + 3AgNO_3 = 3HNO_3 + Ag_3Sb$. Antimonietted hydrogen is always formed when a solution of any salt of antimony is brought together with nascent Hydrogen (Marsh's test, p. 154, par. 267).

661. Antimony has THREE oxides, but only one basic oxide. *a.* ANTIMONOUS OXIDE $Sb_2O_3 = 292$. A greyish white powder, assuming a yellow colour when heated. It may be sublimed at high temperatures, in absence of air, in the form of brilliant needles. It is known to Mineralogists as *White-antimony ore*, crystallized in prisms. It is most easily prepared by treating Antimonous sulphate with Sodium carbonate, and washing with water. As METANTIMONOUS ACID $HSbO_2$, it is obtained by pouring a solution of Antimonous chloride into Sodium carbonate: it is then soluble in Potassium hydrate. Hydrogen chloride and tartrate dissolve it easily. It forms salts. When strongly heated in air, it acquires Oxygen and becomes ANTIMONOUS-ANTIMONIC OXIDE Sb_2O_4. *b.* ANTIMONIC ANHYDRIDE $Sb_2O_5 = 324$. A pale-yellow powder, insoluble in water. Made from METANTIMONIC ACID $HSbO_3$ at a low red-heat. $2HSbO_3 = H_2O + Sb_2O_5$. When more strongly heated, it loses Oxygen and is changed into: *c.* ANTIMONOUS-ANTIMONIC OXIDE Sb_2O_4. Neither basic nor acid. Hydrogen tartrate dissolves Antimonic oxide and leaves Antimonic anhydride. Antimonic anhydride, like

Phosphoric and Arsenic anhydrides, forms three acids. ORTH-ANTIMONIC ACID H_3SbO_4, is but little known. It is probably formed by the action of water upon Antimonic chloride: $4H_2O+SbCl_5=5HCl+H_3SbO_4$. METANTIMONIC ACID $HSbO_3$. Is obtained by the action of nitric acid upon metallic Antimony. It is Frémy's antimonic acid. PYR-ANTIMONIC $H_4Sb_2O_7$. Is the metantimonic acid of Frémy. POTASSIUM DI-HYDROGEN PYR-ANTIMONATE $K_2H_2Sb_2O_7,6H_2O$, is interesting as a test for Sodium, which it precipitates as SODIUM DI-HYDROGEN PYR-ANTIMONATE $Na_2H_2Sb_2O_7,6H_2O$.

662. ANTIMONY and SULPHUR. *a.* SULPH-ANTIMONOUS ANHYDRIDE $Sb_2S_3=340$. Is found native, and is the ore of antimony. It is precipitated as a deep orange-red, bulky solid, on the addition of Hydrogen sulphide to a solution of Antimonous chloride; $2SbCl_3+3H_2S=6HCl+Sb_2S_3$. It is soluble in Hydrogen chloride, in Potassium hydrate and in Ammonium sulphide. The mineral known as *Red antimony-ore*, is an OXY-DI-SULPHIDE Sb_2OS_2. When sulph-antimonous anhydride is boiled with Sodium carbonate, "Mineral kermes" is deposited on cooling: $6K_2CO_3+3Sb_2S_3+3H_2O=3K_2S+6KHCO_3+2Sb_2S_3,Sb_2O_3$. Sulph-antimonous anhydride is also used for making Hydrogen sulphide (p. 91, par 109). $Sb_2S_3+6HCl=3H_2S+2SbCl_3$. Salts of Sulph-antimonous anhydride are called SULPH-ANTIMONITES. *b.* SULPH-ANTIMONIC ANHYDRIDE Sb_2S_5, is orange-coloured. It forms SULPH-ANTIMONATES. TRI-SODIUM SULPH-ANTIMONATE $Na_3SbS_4,9H_2O$, crystallizes in brilliant tetrahedra: it is known as Schlippe's salt.

663. ANTIMONY and CHLORINE. *a.* ANTIMONOUS CHLORIDE $SbCl_3=228{\cdot}5$. A white, fusible, volatile compound, known as Butter of antimony. It is prepared either by evaporating the solution of Antimonous chloride until it can be distilled, or by distilling a mixture of 3 parts of pounded antimony with 8 parts of Mercuric chloride: $3Sb+3HgCl_2=Hg_2Cl_2+SbHg+2SbCl_3$. Antimonous chloride is soluble in Hydrogen chloride and in little water; but by much water it is decomposed into ANTIMONOUS OXY-CHLORIDE $SbOCl$ and Hydrogen chloride. $SbCl_3+H_2O=2HCl+SbOCl$. *b.* ANTIMONIC CHLORIDE $SbCl_5=299{\cdot}5$. Is a volatile, colourless liquid, fuming in air. It is readily obtained by burning Antimony in an excess of Chlorine. It

is changed by a little water into ANTIMONIC OXY-CHLORIDE $SbOCl_3$; by much water into ORTH-ANTIMONIC ACID H_3SbO_4.

664. TESTS. In solutions of Antimonous chloride, which are always strongly acid, water occasions a white precipitate of Oxychloride, soluble in Tartaric acid, and precipitated as orange-red Sulph-antimonous anhydride, soluble in Ammonium sulphide and in Hydrochloric acid. Potassium hydrate throws down and redissolves antimonous acid. Ammonia, a white precipitate.

In solutions of Tartar emetic, Hydrochloric acid first gives a white precipitate easily soluble in excess. Marsh's test and Reinsch's test should be employed in cases of suspected poisoning (p. 157, par. 274). The Copper-slip containing the metallic deposit, should be warmed with a solution of Potassium hydrate, taking care to expose it as much as possible to the air: the solution should be diluted, poured off from the Copper, a stream of Hydrogen sulphide passed through, filtered from Cupric sulphide and decomposed by Hydrochloric acid. If antimony be present, there will be an orange red precipitate of Antimonous sulphide.

The separation of Antimonous sulphide from Arsenious sulphide is readily accomplished by means of Ammonium hydro-carbonate, which dissolves Arsenious sulphide.

665. Antimony may be weighed as Antimonous-antimonic oxide, 100 parts of which contain 79·22 per cent. of metal. Antimony is precipitated from its chlorides by a bar of metallic tin.

CHAPTER XXX.

THE CHEMISTRY OF MOLYBDENUM, AND TUNGSTEN, WITH NOTES ON VANADIUM, COLUMBIUM, TANTALUM, ZIRCONIUM AND THORINUM.

LVII.—**Molybdenum** Mo=96. [666. Properties. Sp. gr. 8·63. 667. Never native. Chief mineral Molybdenite, a SULPHIDE MoS_2. 668. Three oxides. MOLYBDOUS OXIDE Mo_2O_3. MOLYBDIC OXIDE MoO_2. MOLYBDIC ANHYDRIDE MoO_3. LEAD MOLYBDATE $PbMoO_4$. 669. MOLYBDIC SULPHIDE MoS_2. SULPHO-MOLYBDIC ANHYDRIDE MoS_3. 670. MOLYBDIC CHLORIDE $MoCl_4$. 671. Tests.] LVIII. **Tungsten** or **Wolfram** W=184. [672. Properties of Tungsten, Sp. gr. 17·6. 673. Chief minerals, Scheelite and Wolfram or Tungsten. Scheelite is CALCIUM TUNGSTATE $CaWO_4$. Wolfram is FERROUS MANGANOUS TUNGSTATE $3FeWO_4,MnWO_4$. 674. Two oxides. TUNGSTEN OXIDE WO_2. TUNGSTIC ANHYDRIDE WO_3. TUNGSTIC ACID H_2WO_4. TUNGSTOUS TUNGSTATE WO_2,WO_3. 675. Two sulphides. TUNGSTEN SULPHIDE WS_2. SULPHO-TUNGSTIC ANHYDRIDE WS_3. 676. Chlorides WCl_4 and WCl_6. 677. Tests.] [LIX. **Vanadium** V=51·3. 678. Properties of Vanadium. Allied to Phosphorus. 679. Four oxides. VANADIOUS OXIDE V_2O_2. VANADIC OXIDE V_2O_3. VANADIC ANHYDRIDE V_2O_5. METAVANADIC ACID HVO_3. 680. Oxy-chlorides resemble those of Phosphorus. 681. Tests.] [LX. **Columbium** or **Niobium** Nb=94. 682. In Columbite. COLUMBIC ANHYDRIDE Nb_2O_5. COLUMBIC ACID $HNbO_3$. COLUMBIC CHLORIDE $NbCl_5$. OXY-CHLORIDE $NbOCl_3$. COLUMBIC FLUORIDE NbF_5. OXY-FLUORIDE $NbOF_3$.] [LXI. **Tantalum** Ta=182. 683. Discovered by Ekeberg in 1802. TANTALIC ANHYDRIDE Ta_2O_5.] [LXII. **Zirconium** Zr=89·5. 684. The metal resembles Titanium. Sp. gr. 4·15. 685. Preparation from POTASSIUM FLUOZIRCONIDE $2KF,ZrF_4+2K_2=6KF+Zr$. 686. ZIRCONIUM OXIDE ZrO_2. 687. ZIRCONIUM SULPHIDE ZrS_2. 688. SALTS of Zirconium. ZIRCONIUM CHLORIDE $ZrCl_4$. FLUORIDE ZrF_4. SILICATE $ZrSiO_4$. Found as Zircons and Hyacinths.] [LXIII. **Thorinum** Th=238. 689. Somewhat resembles Zirconium. 690. THORINUM OXIDE ThO_2. CHLORIDE $ThCl_4$. THORINUM SULPHATE $Th2SO_4$.]

[LVII. Molybdenum Mo = 96.

666. 57.—MOLYBDENUM Mo=96. It is a white metal, brittle, and with difficulty fusible. Sp. gr. 8·63. Strongly heated in the air, it oxydizes to MOLYBDIC ANHYDRIDE MoO_3. Nitric acid in excess converts the metal also into the anhydride; otherwise into nitrate of Molybdic oxide.

667. Molybdenum is never found native. Its chief ore is Molybdenite, a MOLYBDIC SULPHIDE MoS_2. It resembles Plumbago, whence the name of the metal μολύβδαινα, *molybdaina*, a mass of lead. When roasted in the air, Sulphurous anhydride is evolved, and Molybdic anhydride remains. Heated with Charcoal in a crucible in a Smith's forge, Carbonic oxide is evolved and Molybdenum remains.

668. Molybdenum has THREE oxides, two of which are basic. *a*. MOLYBDOUS OXIDE Mo_2O_3=240. It is precipitated in the form of a black hydrate, soluble in Ammonium sesqui-carbonate. It may be obtained as Mo_2O_3, by digesting Molybdic anhydride with Zinc and Hydrochloric acid. *b*. MOLYBDIC OXIDE MoO_2=128. Is dark-brown and nearly insoluble in acids. Its hydrate is red-brown, and its salts have a dark brown colour. *c*. MOLYBDIC ANHYDRIDE MoO_3=144. A pale yellow powder, but little soluble in water, with yellow colour. Its solution reddens litmus. LEAD MOLYBDATE $PbMoO_4$, is found native in bright-yellow scales, as Wulfenite. AMMONIUM MOLYBDATE $(NH_4)_2MoO_4$, crystallizes in colourless prisms. Its solution is employed for detecting and estimating Phosphoric acid. The ortho-phosphate is acidulated with Nitric acid, and the Molybdate added. The liquid turns yellow, and deposits on boiling a yellow precipitate of Molybdic and Phosphoric acids, combined with Ammonia: it contains 3 per cent. of Phosphoric anhydride. Zinc, in presence of Hydrochloric acid. reduces a molybdate to the state of a molybdous salt, with the production of a blue, green, and black colour; and from which Ammonia precipitates the black hydrate. When two parts of Sodium molybdate are heated in a crucible with one part of

Ammonium chloride, we obtain MOLYBDIC OXIDE MoO_2, on digesting the mass with Potassium hydrate.

669. MOLYBDIC SULPHIDE MoS_2, is found native. SULPHO-MOLYBDIC ANHYDRIDE MoS_3, is dark-brown, and forms Sulphur-salts. It is precipitated in acid solutions, by Hydrogen sulphide, and is soluble in Ammonium sulphide.

670. MOLYBDIC CHLORIDE $MoCl_4$, resembles Iodine in appearance. It may be prepared by heating the metal in Chlorine, or dissolving molybdic oxide in Hydrochloric acid.

671. TESTS. Besides the action of Zinc upon a molybdate, Stannous chloride reduces a solution of a molybdate to Molybdic molybdate $MoO_2,4MoO_3$, of a blue colour.]

[LVIII. Tungsten or Wolfram W = 184.

672. 58.—TUNGSTEN OR WOLFRAM W=184. An iron-grey metal of sp. gr. 17·6; very hard, and difficult to melt. As powder, it is easily burnt to TUNGSTIC ANHYDRIDE WO_3. Nitric acid oxydizes Tungsten to TUNGSTIC ACID H_2WO_4.

673. The metal is found both as *Scheelite* and *Wolfram*, or Tungsten, signifying, in Swedish, a heavy stone. Scheelite is CALCIUM TUNGSTATE, $CaWO_4$; Wolfram a FERROUS-MANGANOUS TUNGSTATE ($3FeWO_4,MnWO_4$). From Tungstic anhydride, Tungsten may be obtained, by heating it with Charcoal in the intense heat of a smith's forge.

674. There are TWO oxides, and probably an intermediate one. TUNGSTEN OXIDE WO_2 is brown or black, according to the mode of preparation. It is obtained by reducing Tungstic anhydride. It is not basic. TUNGSTIC ANHYDRIDE WO_3=232. It forms a yellow powder, insoluble in water and in acids, but soluble in alkaline hydrates. It is prepared from Calcium tungstate by Hydrochloric acid; the latter dissolves the Calcium and leaves Tungstic anhydride. When POTASSIUM TUNGSTATE $K_2WO_4,2H_2O$, is decomposed by Hydrochloric acid, we obtain TUNGSTIC ACID H_2WO_4. Besides tungstic acid, we have META-TUNGSTIC ACID $H_2W_4O_{13}$. TUNGSTOUS TUNGSTATE

WO_2,WO_3, obtained by digesting Tungstic acid with Zinc and Sulphuric acid, is blue.

675. Tungsten has two sulphides. TUNGSTEN SULPHIDE WS_2, and SULPHO-TUNGSTIC ANHYDRIDE WS_3. The latter unites with the former to TUNGSTEN TUNGSTIC SULPHIDE WS_2,WS_3, and forms Sulphur salts with alkaline sulphides.

676. Both CHLORIDES, WCl_4 and WCl_6 are volatile: they are decomposed by water into Hydrochloric acid, and their respective oxides.

677. TESTS. The tungstates are not precipitated by Hydrogen sulphide, or by Ammonium sulphide. Besides tests incidentally mentioned, Tin, in presence of free acid, reduces a tungstate to the condition of a tungsten tungstate of a violet colour. The acids precipitate Tungstic acid of a yellow colour.]

[LIX. Vanadium V = 51·3.

678. 59.—VANADIUM V=51·3. Was discovered by Sefström, in 1830. It is, according to Roscoe, most nearly related to Phosphorus. The properties of the metal are unknown.

679. Four oxides are known. VANADIOUS OXIDE V_2O_2, forms lavender-coloured salts. VANADIC OXIDE V_2O_3, is black. It is made by heating Vanadic anhydride in a stream of Hydrogen. VANADIC ANHYDRIDE V_2O_5, is brownish-red, sparingly soluble in water, to which it communicates a yellowish tint. The most abundant ore of Vanadium is *Vanadinite*, $3Pb_32VO_4,PbCl_2$. METAVANADIC ACID HVO_3.

680. The CHLORIDES are not well known. The OXY-CHLORIDES resemble those of Phosphorus.

681. TESTS. When Vanadinates, which are red or yellow, are acted upon by Hydrogen sulphide, a purple solution of DI-VANADIUM TETROXIDE V_2O_4, results.]

[LX. Columbium or Niobium Nb = 94.

682. 60.—COLUMBIUM or NIOBIUM Nb. Was discovered in 1801

by Hatchett, in Columbite. COLUMBIC ANHYDRIDE Nb_2O_5 and COLUMBIC ACID $HNbO_3$. COLUMBIC CHLORIDE $NbCl_5$, melts at 194° C. OXY-CHLORIDE $NbOCl_3$. COLUMBIC FLUORIDE NbF_5. OXY-FLUORIDE $NbOF_3$. All these compounds shew the close relationship of Columbium to Phosphorus.]

[LXI. Tantalum Ta = 182.

683. 61.—[TANTALUM Ta=182. Was discovered by Ekeberg, in 1802, in Tantalite and Yttro-tantalite. TANTALIC ANHYDRIDE Ta_2O_5. Also related to Phosphorus.]

[LXII. Zirconium Zr = 89·5.

684. 62.—ZIRCONIUM Zr=89·5. The metal closely resembles Silicon and Titanium, although it has not been studied with the same carefulness. In the form of a black amorphous powder, it is yet made to assume metallic lustre under the burnisher. Its great solvent is Hydro-fluoric acid. Troost has obtained Zirconium in the form of crystalline plates, of sp. gr. 4·15.

685. The metal is obtained by heating POTASSIUM FLUO-ZIRCONIDE $2KF,ZrF_4$, with Potassium. $2KF,ZrF_4+2K_2=6KF+Zr$.

686. When amorphous Zirconium is heated in air, it burns with brilliancy to ZIRCONIUM OXIDE ZrO_2=121·5. This is the only basic oxide. It is a dense white substance, insoluble in all acids, except Sulphuric acid. Its hydrate is white, and readily soluble. It is insoluble in alkaline hydrates. When heated with Potassium carbonate, Zirconium oxide expels Carbonic anhydride, and forms a compound with the Potassium oxide.

687. ZIRCONIUM SULPHIDE ZrS_2; is brown. It may be obtained by burning Zirconium in Sulphur vapour.

688. The SALTS of Zirconium worthy of mention are: ZIRCONIUM CHLORIDE $ZrCl_4$=231·5, in needles, with water of crystallization. ZIRCONIUM FLUORIDE ZrF_4; with water. FLUO-ZIRCONIC ACID 2HF, ZrF_4. FLUO-ZIRCONIDES. ZIRCONIUM SILICATE $ZrSiO_4$, is found

in the form of "Zircon" and "Hyacinth." Fused with Potassium hydrate, and decomposed by Hydrochloric acid, on evaporation to dryness and re-solution, Zirconium chloride is obtained, while Silicic anhydride remains undissolved. Excess of Ammonic hydrate precipitates Zirconium hydrate.]

[LXIII. Thorinum Th = 238.

689.—THORINUM Th=238. Somewhat resembles Zirconium. It was discovered in 1829, by Berzelius, in the mineral "Thorite."

690. When ignited, it burns readily to THORINUM OXIDE ThO_2, of sp. gr. 9·4. THORINUM CHLORIDE $ThCl_4$, is volatile. By passing its vapour over heated Sodium, we obtain Thorinum. THORINUM SULPHATE $Th2SO_4$, is precipitated on boiling, but re-dissolves on cooling.]

INDEX.

END OF PART I.